T5-AFS-713

CORROSION OF REINFORCEMENT IN CONCRETE CONSTRUCTION

Papers in this volume were presented at a meeting on 'Corrosion of reinforcement in concrete construction' held in London, England on 13–15 June 1983 and organised by the Society of Chemical Industry Materials Preservation Group and Road and Building Materials Group in association with The Concrete Society, European Federation of Corrosion, Institution of Civil Engineers, and National Association of Corrosion Engineers (USA).

CORROSION OF REINFORCEMENT IN CONCRETE CONSTRUCTION

Edited by
ALAN P. CRANE
Immediate Past Chairman
Materials Preservation Group
Society of Chemical Industry

Published for the
SOCIETY OF CHEMICAL INDUSTRY
London, by

ELLIS HORWOOD LIMITED
Publishers · Chichester

First published in 1983 by

ELLIS HORWOOD LIMITED
Market Cross House, Cooper Street, Chichester, West Sussex, PO19 1EB, England

The publisher's colophon is reproduced from James Gillison's drawing of the ancient Market Cross, Chichester.

Distributors:

Australia, New Zealand, South-east Asia:
Jacaranda-Wiley Ltd., Jacaranda Press,
JOHN WILEY & SONS INC.,
G.P.O. Box 859, Brisbane, Queensland 40001, Australia

Canada:
JOHN WILEY & SONS CANADA LIMITED
22 Worcester Road, Rexdale, Ontario, Canada.

Europe, Africa:
JOHN WILEY & SONS LIMITED
Baffins Lane, Chichester, West Sussex, England.

North and South America and the rest of the world:
Halsted Press: a division of
JOHN WILEY & SONS
605 Third Avenue, New York, N.Y. 10016, U.S.A.

British Library Cataloguing in Publication Data
Crane, A.
Corrosion of reinforcement in concrete construction.
1. Reinforced concrete
I. Title
624.1'8341 TA444

Library of Congress Card No. 83-10796

ISBN 0-85312-600-3 (Ellis Horwood Limited, Publishers)
ISBN 0-470-27477-8 (Halsted Press)

Typeset in Press Roman by Ellis Horwood Ltd.
Printed in Great Britain by R. J. Acford, Chichester.

Table of Contents

Foreword 9

PART 1 – EXTENT AND NATURE OF THE PROBLEM

Chapter 1 **The implications of reinforcement corrosion for safety and serviceability of structures** 11
R. J. CURRIE, Building Research Establishment, Watford WD2 7JR, UK

Chapter 2 **Durability of reinforced concrete in sea water** 19
SANDOR POPOVICS, Drexel University, Philadelphia, PA 19104, USA, Y. SIMEONOV, Bulgarian Academy of Science, G. BOZHINOV, N. BAROVSKY, Central Laboratory of Physical and Chemical Mechanics, Sofia, Bulgaria.

Chapter 3 **Exposure tests on concrete for offshore structures** 39
K. M. BROOK and J. A. STILLWELL, Wimpey Laboratories Ltd, Hayes, Middlesex UB4 0LS, UK.

Chapter 4 **Corrosion of steel reinforcement in offshore concrete – Experience from the Concrete-in-the-Oceans Programme** 59
M. B. LEEMING, CIRIA/UEG, Westminster, London SW1P 3AU UK.

Chapter 5 **Failure analysis of the collapse of the Berlin Congress Hall** 79
B. ISECKE, Bundesanstalt für Materialprüfung, 1000 Berlin 45, Federal Republic of Germany.

Chapter 6 **Deterioration of marine concrete structures with special emphasis on corrosion of steel and its remedies** 91
M. A. AZIZ, M. A. MANSUR, National University of Singapore, Kent Ridge, Singapore

Chapter 7 **The influence of concrete quality on carbonation in Middle Eastern conditions – A preliminary study** 101
K. W. J. TREADAWAY, Building Research Establishment, Watford WD2 7JR, UK, G. MACMILLAN, P. HAWKINS, Ministry of Works, Power and Water, Bahrain, C. FONTENAY, Haji Hassan Ready Mix C., Bahrain.

PART 2 – MECHANISMS

Chapter 8 **The corrosion of steel reinforcements in concrete immersed in seawater** 119
N. J. M. WILKINS and P. F. LAWRENCE, AERE Harwell, Oxon OX11 0RA, UK.

Chapter 9 **The influence of chlorides and sulphates on durability** 143
W. R. HOLDEN, C. L. PAGE and N. R. SHORT, University of Aston, Birmingham, B4 7ET, UK.

Chapter 10 **The mechanisms of the protection of steel by concrete** 151
H. ARUP, Korrosionscentralen, DK2600 Glostrup, Denmark

Chapter 11 **Corrosion rate of reinforcements during accelerated carbonation of mortar made with different types of cement** 159
J. A. GONZALEZ, C. ALONSO and C. ANDRADE, Instituto Eduardo Torroja de la Construccion y del Cemento, Madrid, Spain

PART 3 – DIAGNOSIS, MONITORING AND REPAIR

Chapter 12 **Corrosion monitoring of steel in concrete** 175
J. L. DAWSON, UMIST, Manchester M60 1QD, UK.

Chapter 13 **Analysis of structural condition from durability results** 193
R. D. BROWNE, M. P. GEOGHEGAN and A. F. BAKER, Taywood Engineering Ltd, Southall, Middlesex UB1 2QX, UK.

Chapter 14 **The investigation and repair of damaged reinforced concrete structures** 223
R. T. L. ALLEN and J. A. FORRESTER, Cement & Concrete Association. Cardiff CF1 3AS, UK.

Chapter 15 **Mortar repair systems – Corrosion protection for damaged reinforced concrete** 235
L. H. McCURRICH, C. KEELEY, L. W. CHERITON, K. F. TURNER, Fosroc Technology Ltd., Leighton Buzzard, Beds LU7 7ER, UK.

Chapter 16 **Corrosion monitoring in the industrial-related research area** . . . 255
GUSTAV BRACHER, Sika AG, CH-8048 Zurich, Switzerland

Chapter 17 **The repair of concrete – A laboratory and exposure site investigation** . . . 263
D. G. JOHN, UMIST, Manchester M60 1QD, UK, A. T. COOTE, K. W. J. TREADAWAY, Building Research Establishment, Watford WD2 7JR, UK, J. L. DAWSON, UMIST, Manchester M60 1QD, UK.

PART 4 – AVOIDANCE AND DESIGN RECOMMENDATIONS

Chapter 18 **Criteria for the cathodic protection of bridge decks** . . . 287
RICHARD F. STRATFULL, Corrosion Engineering Inc., CA95691, USA

Chapter 19 **Densified silica-cement coating as an effective corrosion protection** . . . 333
THEODOR A. BÜRGE, Sika AG, CH-8048 Zurich, Switzerland.

Chapter 20 **Relation between the alkali content of cements and the corrosion rates of the galvanized reinforcements** . . . 343
C. ANDRADE, A. MOLINA, F. HUETE, J. A. GONZALEZ, Instituto Eduardo Torroja de la Construcccion y del Cemento, Madrid, Spain

Chapter 21 **Long-term corrosion resistance of epoxy-coated reinforcing bars** . . . 357
JIRO SATAKE, MASAHI KAMAKURA, KIYOSHI SHIRAKAWA, NAOTO MIKAMI, Sumitomo Metal Industries Ltd., Japan, NARYAAN SWAMY, University of Sheffield, Sheffield S1 3JD, UK.

Chapter 22 **The influence of constructional and manufacturing conditions on the corrosion behaviour of prestressed wires before grout injection** . . . 379
B. ISECKE, Bundesanstalt für Materialprüfung, 1000 Berlin 45, Federal Republic of Germany.

Chapter 23 **The influence of cement type on the electrochemical behaviour of steel in concrete** . . . 393
C. M. PREECE, F. O. GRØNVOLD, T FRØLUND, Danish Corrosion Centre, DK 2600 Glostrup, Denmark.

Chapter 24 **Sea water corrosion attack on the concrete blocks embedding zinc galvanized steel rebars** 407
HARUO SHIMADA, Nippon Steel Corporation, Kawasaki 211, Japan, SEIYA NISHI, Onoda Cement Co., Japan.

Chapter 25 **The promising concrete rebars for construction in offshore, seaside and desert environments** 419
KŌICHI KISHITANI, Tokyo University, Japan, ISAO FUKUSHI, Housing and Urban Development Corporation, Japan, HARUO SHIMADA, Nippon Steel Corporation, Kawasaki 211, Japan.

INDEX 435

Foreword

The corrosion of steel reinforcement in concrete construction is a world-wide problem which can, on the small scale, cause disfigurement and, on the large scale, lead to structural catastrophe. Its implications were of such importance that until quite recently, open discussion of the problem was rarely feasible. However, in 1978 the Materials Preservation and Road and Building Materials Groups of the Society of Chemical Industry successfully organised their first meeting on the subject [1]. Interest in the meeting was such that the number of potential delegates far exceeded the available accommodation. That interest and the degree of concern which surrounds the corrosion problem have not waned in the intervening period, but fortunately healthy discussion has become more public. These points are proved by the willing co-operation of the Institute of Civil Engineers, the Concrete Society and the National Association of Corrosion Engineers of the USA in coming together to co-sponsor with the Society of Chemical Industry the 1983 Conference. Offers of papers were received from Continental Europe, the Far East, the Middle East, North America, Scandinavia and the United Kingdom and many have been collected together in this book. The combined papers provide a breadth of scientific knowledge and practical experience of corrosion of reinforcement in concrete construction which is believed to be unequalled elsewhere.

The vital topics which are covered include:

(i) evaluations of past failures
(ii) the role of cement composition
(iii) the control of concrete permeability
(iv) design criteria
(v) the condition and nature of the reinforcing bar material
(vi) mechanistic studies
(vii) monitoring techniques
(viii) methods of repair.

It is hoped that this further dissemination of information and its accompanying multi-national discussions will prove to be of value in the quest to understand and provide practical, economic solutions to the problem.

On a personal note, I must thank my fellow members of the Conference Organising Committee together with Miss Jane Bovier and the rest of her staff at the Society of Chemical Industry. Without their counselling and endeavour neither the Conference nor this book would have been possible.

Alan P. Crane
Chairman, Organising Committee

[1] *Corrosion of Steel Reinforcements in Concrete Construction.* Society of Chemical Industry, 1979.

Part 1

EXTENT AND NATURE OF THE PROBLEM

Chapter 1

The implication of reinforcement corrosion for safety and serviceability of structures

R. J. CURRIE, Building Research Establishment, Watford, UK

A LINK BETWEEN SAFETY AND SERVICEABILITY

In the design process for structures a distinction is made between the circumstances which affect the building's safety and those which do not constitute an immediate hazard but may make the building unusable. This later condition is referred to as the structure's limit state of serviceability. Serviceability faults may also make the structure expensive to maintain.

Traditionally the structural and non-structural factors affecting the serviceability of buildings have been treated as separate issues by designers. For example, deflection of roof beams has been considered solely as a structural engineer's concern whereas the specification of roof coverings and roof construction has been specified by the architect.

As a result flat roofs have been laid to very shallow falls supported on very slender beams. Whilst in theory both the independent objectives of adequate drainage and satisfactory deflection should be achievable, in practice these independent requirements have combined to give a widespread source of building deterioration and loss of serviceability.

It has been argued that problems of this type are not serious and that the benefits of convenient construction outweigh the higher maintenance and repair consequences. An alternative and possibly more realistic philosophy is one which asserts that, 'in the long-term the cheapest building is the one you do not have to keep repairing'.

Setting these opposing views apart, there is a far more serious consequence of adopting the purely short-term economic criteria as a basis for building design. This is the possibility of neglected serviceability faults becoming structural safety hazards. In the case of the flat roof, many instances have been investigated (for example, see Ref. [1]) where ponding of water on the roofs has led to water penetration and as a result the structural supporting system has suffered deterioration and corrosion of its reinforcement. Again, in theory, it is possible to design concrete so that embedded steel should not corrode; however, in practice, joints in precast members can form direct paths to the steel for moisture and oxygen, concrete not strictly made to specification nor compacted well will carbonate to the steel reinforcement after a relatively few years and the reinforcement will not always be embedded fully or to the correct depth. The presence of chlorides either by chance or by deliberate additions may also aggravate the problems.

EXPERIENCE FROM DEFECTS

What factors become apparent from the variety of defects now occurring, many of which involve corrosion of reinforcement in concrete construction?

The first is that if there is any chance of something being done incorrectly or inadequately, sooner or later it will be done incorrectly or inadequately.

The second is that it is important that designers come to terms with the reality that all structures including reinforced concrete ones do not necessarily last for ever.

The third is that maintenance/repair and periodic structural inspections appear to be the most practicable means available to prevent local serviceability deterioration leading to major structural failure. Even if periodic inspections are carried out, they will not necessarily identify corrosion problems in some types of sensitive structures, i.e. structures in which collapse can occur without warning following local corrosion. Should such structures be built?

These factors lead to the view that there are implications from in-service corrosion and deterioration for both the general approach used by engineers and architects to the design of buildings and for the detailed appraisal of existing structures suffering corrosion of reinforcement.

IMPLICATIONS FOR DESIGN OF NEW BUILDINGS

The more general implications are that the existing balance between strength and durability requirements in our codes of practice may be too heavily biased towards strength criteria. Indeed for concrete construction some aspects of durability are considered to be automatically met provided that the cube strength is adequate while the connection between the quality of the concrete in the structure and strengths of sample cubes is far from direct.

Such attributes lead to situations where questions such as 'How long will it last?' or 'Is the component suitable for an 80-year life?' are not even asked, let alone answered. The more general durability clauses included in codes of practice do not constitute sufficiently strict criteria to guarantee an adequate building life and their presence may actually inhibit engineers from thinking positively about their specific requirements on the ground that 'durability is covered in the code'.

An extension of this line of thought is often carried over into the detailing of building construction itself. The assumption is often made that the structure will never need to be repaired or even inspected. The result is that when it eventually becomes necessary to inspect or repair the structure, the process is made difficult and expensive because no provision has been made for access or for removal of fixings or finishes. Some corrosion problems with reinforced concrete cladding panels have already led to demolition of the whole building being the most attractive solution, owing to the high cost and difficulty in removing the affected panels.

MEASURES FOR IMPROVEMENT IN DESIGN

In order to redress the balance between strength and durability, chemists and material scientists have an important part to play in pressing for the education of engineers and architects to include thorough understanding of the mechanisms involved in corrosion processes and of protection systems in general. This is especially true in the context of reinforced concrete construction.

In the short-term, codes of practice could draw the attention of practising engineers and architects to the need to consult specialist literature on corrosion and durability aspects of construction.

THE APPRAISAL OF EXISTING BUILDINGS SUBJECT TO REINFORCEMENT CORROSION

1. Performance of reinforcement

The implications for safety and serviceability in concrete structures undergoing appraisal as a result of reinforcement corrosion can best be assessed by considering the effects of the attack at three levels within the structural system. Firstly consider the effects of corrosion on the performance of the reinforcement itself. General corrosion which appears uniform along a length of reinforcement will have two effects: it will reduce the cross-sectional area of the steel and it will create local discontinuities in the steel surface. This type of corrosion reduces the tensile capacity of the steel in proportion to the loss of its cross-sectional area and deduces the steel's resistance to fatigue damage.

(a) *Uniform corrosion*

Where concrete has been carbonated to the depth of the steel reinforcement and a small but uniform amount of moisture is present, the steel is likely to corrode uniformly. Such deterioration is often indicated by fine hair cracking parallel to the direction of the reinforcement throughout the length of the structural component. Fortunately, because the corrosion is fairly uniform, cracking of the concrete cover in normally reinforced or pretensioned prestressed solid components usually occurs prior to any particular structural cross-section becoming excessively weak, thus giving visual warning of the deterioration. Provided that it is possible to inspect the structural frame and inspections are carried out, this deterioration can be identified and action taken before a structural hazard occurs.

(b) *Localised corrosion*

Alternatively, corrosion may occur at local points within the bar lengths and form severe notching or thinning whilst the remainder of the bar is left uncorroded. This preferential corrosion is often found in association with high levels of chloride in the concrete which may be present as a direct result of the deliberate addition of calcium chloride to the concrete mix, or exceptionally as a result of sea-dredged aggregate being used without adequate washing. Structural cracking, or honeycombing, can also create conditions favourable to preferential corrosion by allowing the ingress of the atmosphere causing local carbonation or by allowing chlorides (in the form of sodium chloride from external de-icing salts) access to the steel. Expansion and construction joints also provide a series of potential paths for the ingress and retention of chlorides in structures. This preferential corrosion has the effect of reducing considerably the cross-sectional area of the bars locally and hence reducing their load-carrying capacity. It also alters the overall strain characteristics of the steel.

When a notched or locally thinned bar is subjected to a tensile force, the strain is concentrated at the notch and the overall strain of the bar will be less at failure than with a perfect bar. Hence, as notching becomes more severe, the reinforcement effectively becomes more brittle and large deformations which might normally give a warning of failure may not occur.

Cracking of the concrete cover will be more severe with this type of corrosion, sometimes rust staining will be evident at the cracks. They will be wider than hair cracking and will usually taper in thickness away from the point of corrosion. The surface of the concrete may also bulge locally. Under normal circumstances as far as the structural engineer is concerned, the effects of chlorides, carbonation and moisture content have, for all practical purposes, no influence on the physical properties of the cured concrete material although carbonation may improve the compressive strength of concrete slightly.

2. Performance of the component

The second level of influence is the effect of corrosion on the performance of the component. It is possible to classify component failure into two general classes. The first occurs when either the steel or the concrete is degraded by corrosion or cracked to a point where the materials can no longer support the stresses imposed on them. The second is where disruption due to cracking alters the geometry of the section significantly, e.g. a longitudinal crack between the flange and web of an I-beam.

Provided that the fault is identified the first class of failure is usually a serviceability rather than safety matter since visual signs of corrosion can be expected in normally reinforced and prestressed units before the material strengths become critical. The larger the diameter of the reinforcing bar the earlier cracking will result from corrosion. As cracking proceeds the components will become less stiff, leading in turn to continued deflection in beams or bowing in slender columns.

The second class of failure is not easy to assess because relatively little corrosion cracking may be required to alter a component's behaviour radically. Hollow beam units are particularly susceptible to this form of deterioration since cracking may occur only on the inner face of the unit and eventually may lead to detachment of the soffit containing the reinforcement in its entirety without warning (Fig. 1).

Fig. 1 – Cracking in hollow units.

Similarly, longitudinal cracking at the ends of prestressed beams may be important because in some designs no shear steel is provided, the shear resistance being provided by the plain concrete section and the prestressing force. In these circumstances, cracking at the ends of the units may reduce the shear capacity of the beams considerably and failure may occur without warning if the beam has been used as an isolated component.

In post-tensioned components behaviour is similar in many respects because corrosion can occur in poorly grouted ducts giving no external sign of the deterioration. The consequences of hidden corrosion are usually more severe in post-tensioned components because tendons may be grouped together and can be unbonded. In these circumstances local corrosion will affect more tendons at any one cross-section and when the first tendon fails the additional load may be too big to be sustained by the remaining ones. The combination of these two factors provides the conditions in which progressive failure of all the tendons in one duct can occur without necessarily giving external signs of deterioration. However, there have been cases where ducts have been completely ungrouted and warning of beam failure has been given through the corroded tendons being projected out of their anchorage as the prestressing force is released.

Post-tensioned units are sometimes precast in sections to facilitate transportation. Failure of prestressing tendons in isolated components of this type leads to a sudden breakdown of the beam into its precast sections resulting in a total loss of performance, and hence a relatively small amount of localised corrosion can bring about a serious safety hazard without warning.

3. Overall structural performance

Lastly and most significantly is the effect of corrosion on the overall structural performance. Having considered the corrosion effects on the reinforcement and how these in turn affect the performance of the components, the next step is to form a view on how the components are incorporated into the structure and how the deterioration may affect the performance of the structure. This becomes a study of the stability of structures and their susceptibility to progressive collapse which is too complex to deal with here. However, there are a few general points which can be made. Redundancy in structures is widely accepted as beneficial for robustness and stability. Redundancy is usually achieved by introducing additional connections between components either explicitly or by virtue of the construction type. For example, structural concrete floor screeds in effect connect precast floor beams provided that the standard of workmanship is good. The load distribution and the strength benefits of redundancy can be obtained provided that most of the floor or surrounding structure is in good condition. This will often be the case since corrosion caused by adding chlorides may only be present in one type of precast component or one area of *in situ* construction. Also because of the varying conditions which exist within structures, corrosion proceeds at different rates at different locations. Hence it is usually possible to identify the problem in advance before widespread damage occurs. However, it must also be recognised that if general deterioration has occurred throughout the floor or roof, connections between components may serve to spread partial collapse and not prevent it.

As discussed earlier, corrosion deterioration may be found almost anywhere in a structure affecting bending and shear capacities of components. It is therefore important to examine the overall structure for the consequences of both types of failure mode and to take appropriate action giving priority to the consequences of failure rather than the severity of the corrosion. Particularly sensitive are structures whose integrity and stability are wholly or partially dependent on high-strength tendons. Such situations where a large mass of construction is held in equilibrium by a small area of steel are suspended shell roofs, long-spanned post-tensioned construction and tendon-stayed cantilevered structures.

CONCLUSIONS

Corrosion of reinforcement in structures is now a feature which figures heavily in the maintenance of existing buildings and has contributed to a number of structural collapses [2].

There is a need for structural and architectural designers to acquire a greater understanding of factors influencing reinforcement corrosion and for codes of practice to place more emphasis on the need to design for durability.

Safety implications of reinforcement corrosion depend primarily on the structural form or system of the construction, secondly, on the way in which geometry of the structural components may be affected, and to a lesser extent on the total amount of corrosion of the bars.

Serviceability implications are reflected mainly in the economic losses which may be incurred through temporary loss of use, cost of repairs or the need for premature demolition.

These issues together form a valid argument for greater awareness of the need for designers to consider explicitly the implication of their requirements for durability.

REFERENCES

[1] *The Structural Condition of Intergrid Buildings of Prestressed Concrete,* HMSO, 1978.

[2] Building collapses: investigation and prevention, *BRE News,* Winter 1979.

Chapter 2

Durability of reinforced concrete in sea water

SANDOR POPOVICS, Drexel University, Philadelphia, USA, **Y. SIMEONOV,** Bulgarian Academy of Science, **G. BOZHINOV, N. BAROVSKY,** Central Laboratory of Physical and Chemical Mechanics, Sofia, Bulgaria

ABSTRACT

This paper presents the state of the art on the durability of reinforced concrete in marine environment. Specifically the following two topics are discussed:

(1) The corrosion mechanism of steel embedded in concrete in marine environment.
(2) Durability of reinforced concrete structures in sea water.

The role of concrete in the protection of reinforcement is emphasized in this report and the importance of improvement of the properties of this concrete is demonstrated. A sizable list of references completes the paper.

HISTORY

The problem of concrete deterioration by sea water was discussed as early as 1840 by J. Smeaton and L. J. Vicat.

A series of papers under the general heading of 'What is the Trouble with Concrete in Sea Water', prepared by R. J. Wig of the US Bureau of Standards and L. R. Ferguson of the Portland Cement Association, was published in *Engineering News-Record*, **79**, 1917. These papers report on an examination, over a two-year period, of nearly every concrete structure in sea water in the continental limits of the United States as well as a great number in Canada, Cuba and Panama. A summary of the conclusions of these papers were re-published recently by Tibbetts [1].

Since these early times, the volume of literature on the corrosion of concrete and reinforced concrete in marine environment has increased tremendously indicating the ever-growing importance of this topic [2].

CORROSION OF STEEL IN CONCRETE

Instances of distress due to steel corrosion can be found in most applications of reinforced or prestressed concrete: buildings, slabs, beams, bridge decks, piles tanks, pipes, etc. [3, 4]. Sometimes the first evidence of distress is brown staining of the concrete around the embedded steel. This brown staining, resulting from corrosion (rusting) of the steel, may permeate to the concrete surface without cracking of the concrete but usually it accompanies cracking, or cracking of the concrete occurs shortly thereafter. Concrete cracking occurs because the corrosion product of steel, an iron oxide or 'rust', has a volume much more than the metallic iron from which it is formed. The forces generated by this expansive process can far exceed the tensile strength of the concrete with resulting cracking. Steel corrosion not only causes distress because of staining, cracking, and ravelling of the concrete but may also cause structural failure resulting from the reduced cross-section and hence reduced tensile capacity of the steel, this normally being more critical with thin prestressing steel tendons than with larger reinforcing bars [5, 6].

Steel or iron corrosion may be produced by several mechanisms. Although 'corrosion' of iron by direct oxidation (burning) or by acid attack can occur, such corrosion is of little concern in concrete. Indirect oxidation (electrochemical corrosion) can result from dissimilar or non-uniform metals or dissimilar environments. Corrosion of steel in concrete due to 'stress corrosion', hydrogen embrittlement, or electrolysis due to 'stray' electrical currents have rarely been reported as possible cause of distress. Therefore, particular emphasis will be placed in the following discussion on electrochemical corrosion, believed to be the cause for essentially all of the corrosion distress that occurs.

The preponderance of instances of electrochemical corrosion in concrete results because of the existence of differences in metals or non-uniformities of the steel (different steels, welds, active sites on the steel surface) or non-uniformities in the chemical or physical environment afforded by the surrounding concrete. These non-uniformities under certain specific conditions can produce significant electrical potential differences and resultant corrosion [7].

Even though the potentiality for electrochemical corrosion might exist because of non-uniformity of the steel in the concrete (active sites), the traditional view is that the corrosion is normally prevented by a 'passivating' iron oxide (gamma Fe_2O_3) film which rapidly forms on the steel surface in the presence of moisture, oxygen, and the water-soluble alkaline products during the hydration of the cement [8, 9]. The principal soluble product is calcium hydroxide, $Ca(OH)_2$, and the initial alkalinity of the concrete is at least that of saturated water (pH of about 12.4 depending upon the temperature). In addition, the seemingly relatively small amounts of sodium and potassium oxides in cement further increase the initial alkalinity of concrete or paste extracts and pH values of 13.2 and higher have been reported [10, 11].

It is worthwhile to mention that there is a different, more recent view concerning the high corrosion resistance of steel in concrete. This assumes no protective oxide film, thus the corrosion may be initiated by formation of a chloride-ion film immediately. Since, however, the walls of pores in a cement paste adsorb great amounts of Cl^-, the chloride film can form on the steel surface only if the Cl^- concentration exceeds a certain threshold value [12].

In either case, there are two general mechanisms by which the corrosion of steel in concrete can be initiated and maintained:

(1) Reduction of alkalinity by leaching of alkaline substances with water or their reactions with a pozzolanic material, or partial neutralization by reaction with carbon dioxide [13, 14] or other 'acidic' materials; and
(2) by electrochemical action involving chloride ions in the presence of oxygen [15, 16].

The simplest expression of the basic electrochemical cycle considers the reactions of the anodic and cathodic zones of corroding steel, and the electrical flow between these two zones [17–20].

A flow of current in the steel from an anodic to a cathodic area, in the presence of oxygen and water, results in the production of hydroxyl ions at the cathode. As these migrate to the anode, they react with ferrous iron and form hydrous iron oxides (Fig. 1).

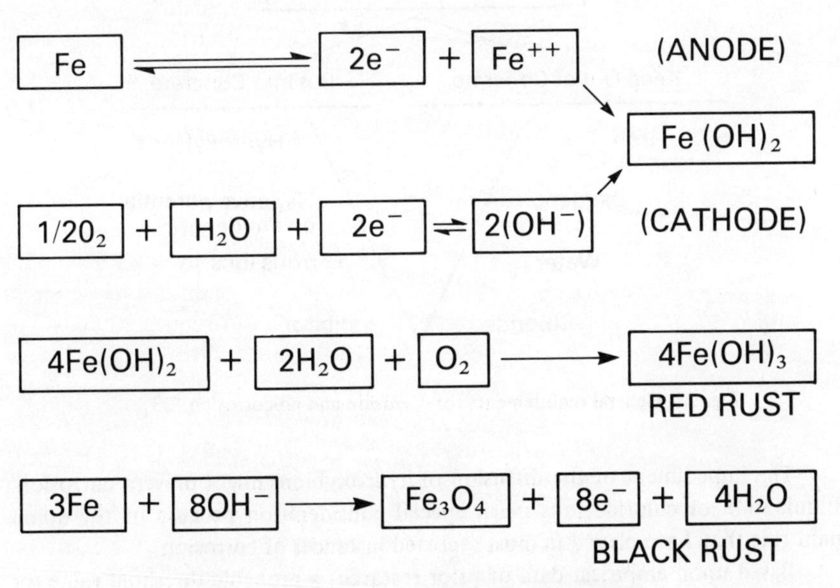

Fig. 1 – Electrochemical reactions producing rust.

The protective methods are techniques that force in principle the equations in Fig. 1 from right to left [21, 22]. The corrosion reactions can also be disrupted by eliminating any one of the reactants on the left side of either of the two equations. For example, the use of non-corrosive reinforcement is an obvious means of eliminating corrosion. A supply of oxygen and water is obviously necessary for corrosion to occur. The conditions producing corrosion are shown schematically in Fig. 2. It is possible to eliminate oxygen from the system by effective coatings on the concrete or steel.

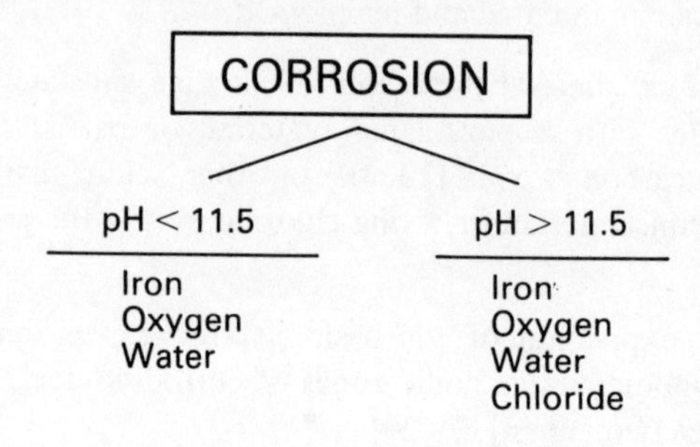

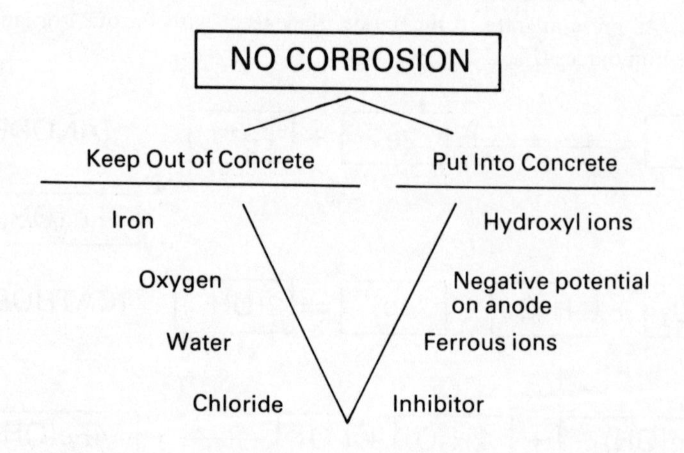

Fig. 2 – General requirements for corrosion and no-corrosion [23].

The impediment of the diffusion of hydroxyl ions might prevent corrosion. Elimination of chloride ions merit special consideration because of the dominant role thay have played in most reported instances of corrosion.

Based upon empirical data of prior research, a probable threshold value for destroying the passivity in the iron–cement paste system is at about a chloride: hydroxyl ion molar activity ratio of 0.6 in solution at the iron–paste interface [7]. The amount of chloride required for initiating corrosion is thus, in part, dependent upon the pH of the liquid in paste. At a pH less than 11.5, corrosion may occur without chlorides; at a concrete pH greater than 11.5, a measurable amount of chloride is required, and that amount increases as the pH at the iron–liquid interface increases (Fig. 3). Calcium chloride may slightly lower the

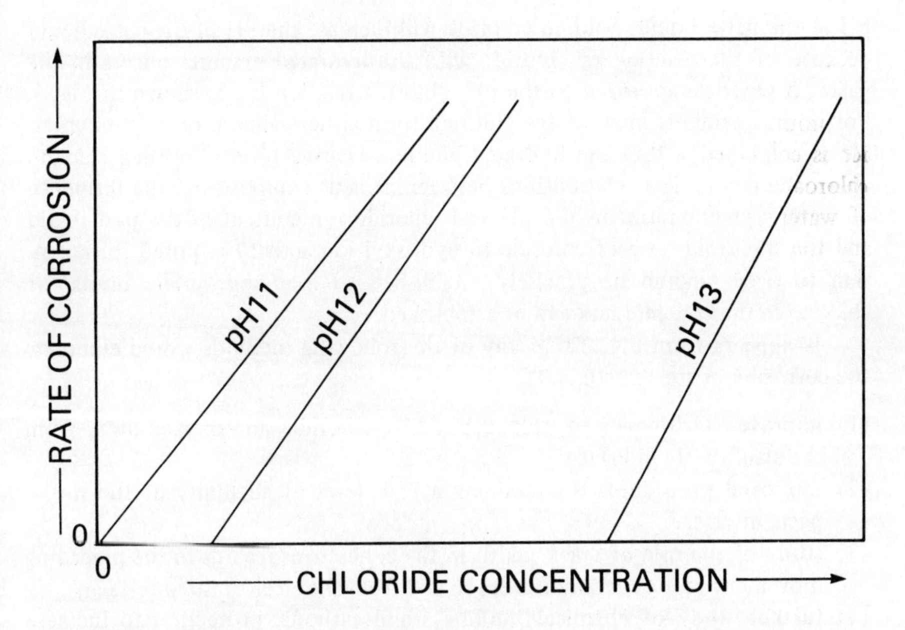

Fig. 3 – Threshold chloride in concentration for corrosion to occur increases as the alkalinity, pH, of the cement paste liquid increases [7].

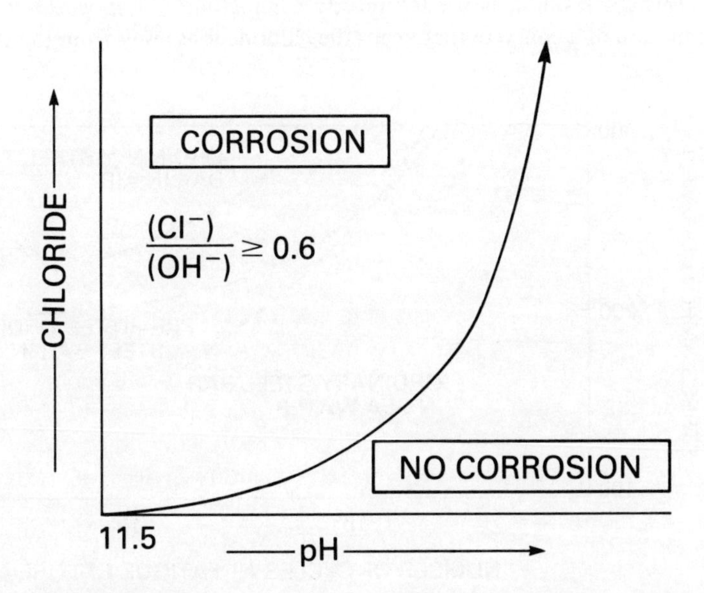

Fig. 4 – Threshold chloride ion concentration for corrosion to occur as a function of the alkalinity [23].

pH in the paste liquid. Sodium chloride will increase the pH in the paste liquid because of the reaction of chloride with the hydrated alumina phases in the paste. A practical approach to the pH–chloride relationship is shown in Fig. 4. For normal cements most of the chloride from either calcium or sodium chloride is combined within the hydrated alumina bearing phases forming calcium chloroaluminate. The relationships between cement composition, the influence of water–cement ratio on the pH and chloride ion content of the past liquid and the threshold ratio of chloride to hydroxyl ion activity required for corrosion to occur should be precisely established so that appropriate limits for chloride in the concrete mix can be established.

It appears from Fig. 2 that any of the following methods would eliminate the corrosion of steel in concrete:

(1) improved techniques to keep destructive chlorides and oxygen away from the steel–paste interface;
(2) improved procedures for retaining a high level of alkalinity at the steel–paste interface;
(3) study of the role of crack width in the concrete as regards to the preceding movement of ions and molecules to and from the steel–paste interface;
(4) further study of chemical inhibitors and cathodic protection to increase the degree of assurance in these methods under a variety of conditions of use [23].

Perhaps research is needed mostly for method (1) above, that is, for the production of a concrete that keeps the chloride ions away from the steel.

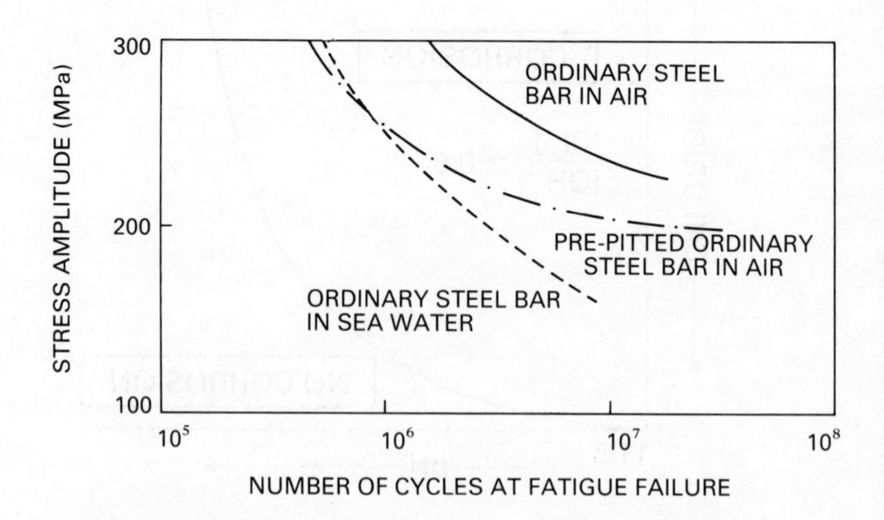

Fig. 5 – Fatigue of reinforcement as influenced by the ambience [82].

A frequently used method for the reduction of corrosion of reinforcement is the coating of steel bar with a suitable, non-corrosive material. Zinc and other metal coatings have been used with certain success for a long time [24–29]. Recently non-metallic (epoxy) coatings have been introduced for the same purpose that have shown good results [30–32], although even these cannot assure perfect protection.

Note that the fatigue strength of a steel bar is considerably lower in sea water than in air [33]. It appears from Fig. 5 that the sea water creates small defects on the surface of reinforcement which then grow under the repeated load and thus reduce the fatigue strength [34]. Corrosion of polymer-coated steel in concrete beams subjected to fatigue testing in sea water [35] also seems to support this explanation.

REINFORCED CONCRETE IN SEA WATER

The diffusion of dissolved oxygen through the pores of the concrete cover is an important factor for the durability of reinforced structures in sea water since the reinforcement would not rust without oxygen [36]. Fig. 6 illustrates this process in terms of cover thickness and water–cement ratio. The effect of cover thickness on the oxygen supply is clearly demonstrated. Results by Szilard and Wallevik show similar trends [37]. Note, however, that this diffusion depends also on the oxygen saturation of the concrete.

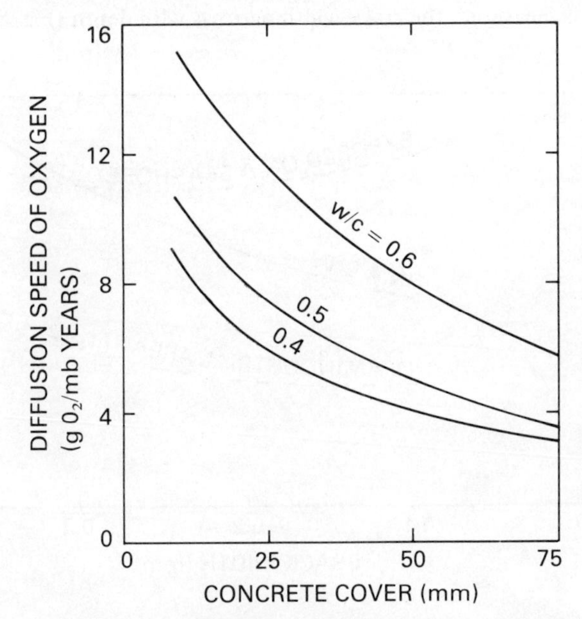

Fig. 6 – Effect of the thickness of concrete cover on the diffusion speed of oxygen in concretes of various water–cement ratios.

As was mentioned earlier, all iron oxide has a volume much more than the metallic iron from which it is formed. This is especially true for the 'red rust' which has the volume of four times as large as that of the iron, while the 'black rust' volume is twice as large as the iron's. The extra volume presses the concrete from inside and causes tensile stresses in the concrete which ultimately result in cracking. This starts a chain reaction in that oxygen, chlorides and water enter faster into the concrete through these cracks, thereby accelerating the corrosion of reinforcement and causing further cracking. The same acceleration is obtained when cracking is induced by other factors such as applied load (waves for instance), thermal stresses, shrinkage, etc., or when the concrete has excessive porosity due to incomplete compaction [38].

Cracking is particularly important because it has been noticed that cracks under repeated loading act as pumps supplying new sea water inside the concrete and to the reinforcement. It seems also logical to expect that the wider the crack, the more corrosion it will cause. Indeed, there have been specifications where the permissible upper limit of crack width was specified to prevent excessive steel corrosion [39]. However, according to a German report which presented the results of a ten-year exposure test [40], the extent of corrosion of the reinforcement in concrete exposed on a seashore was not related to the width of the crack (Fig. 7). A possible explanation of this phenomenon is that the meaningful factor is the width of the crack where it reaches the steel bar and not the width on the outside surface of the concrete. Although it is usually the latter that is measured, the crack width narrows with depth (Fig. 8).

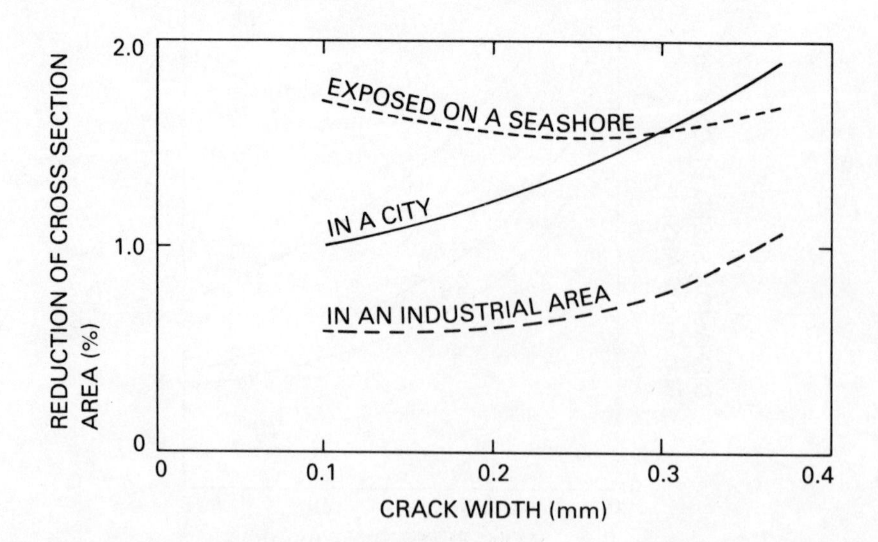

Fig. 7 – Relationship between crack width and corrosion in different environments [82].

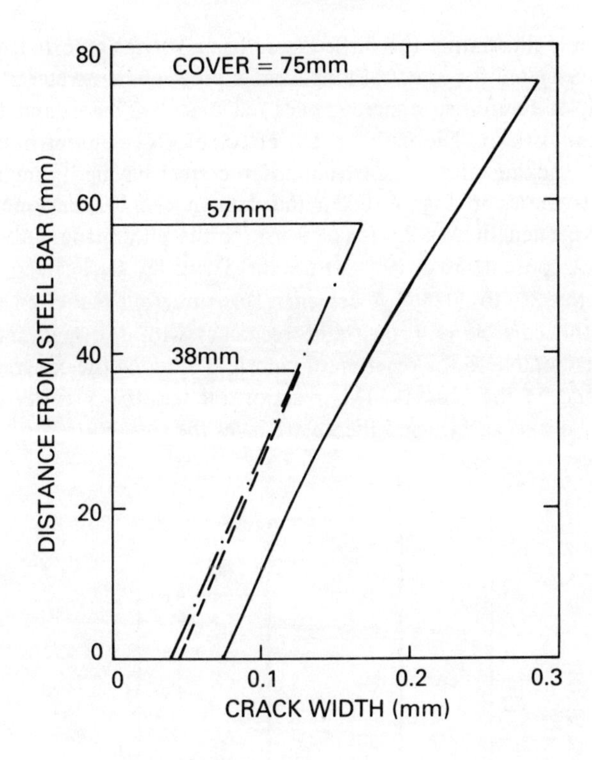

Fig. 8 – Crack width as a function of depth [82].

The development of cracks under repeated loadings (wave action, etc.) is the main reason for the observation that traditional concretes alone cannot prevent corrosion of the reinforcement in sea water [41].

The rate and extent of cracking caused by the corrosion of reinforcement depend partly on the properties of hardened concrete (tensile strength, modulus of elasticity, thickness of cover), and partly on the rust (black or red, quantity). This latter is influenced again by the concrete properties (type of cement, thickness of cover, permeability, diffusivity, presence of excessive porosity and/or cracks) as well as on the availability of certain ions, primarily oxygen and chloride ions, including the characteristics of the environment [42]. At present it is not possible to evaluate the effects of each of these factors separately on the corrosion of reinforcement and the resulting cracking of the concrete. The reason for this is that the cracking process is quite complex and the basic mechanisms of the individual factors are not understood completely in many cases. Another consequence is that there are many contradictory experimental results and opinions reported concerning the behaviour of reinforced concrete structures in sea water.

An example illustrating the difficulties is a test series reported by Verbeck [43]. He investigated the effect of long-term exposure to sea water at St. Augustine, Florida, of reinforced concrete piles made with 22 portland cements of different compositions. The data on the effect of C_3A content of the cement on concrete cracking due to corrosion of the steel having $1\frac{1}{2}$ inch-(38 mm-) thick cover is shown in Fig. 9. From the data, it can be concluded that the average cracking length was 36 ft (11.0 m) for the piles made with cement of 2% to 5% C_3A content, 30 ft (9.1 m) for the 5% to 8% C_3A cement, and 13 ft (4.0 m) for the 8% to 11% C_3A cement. This threefold reduction in cracking (caused by the corrosion of the reinforcement) with the high C_3A content cement is attributable to the reaction of chlorides entering the concrete with the C_3A compound of the cement. This reaction reduces the quantity of chloride ions around the steel surface and thereby reduces the steel corrosion.

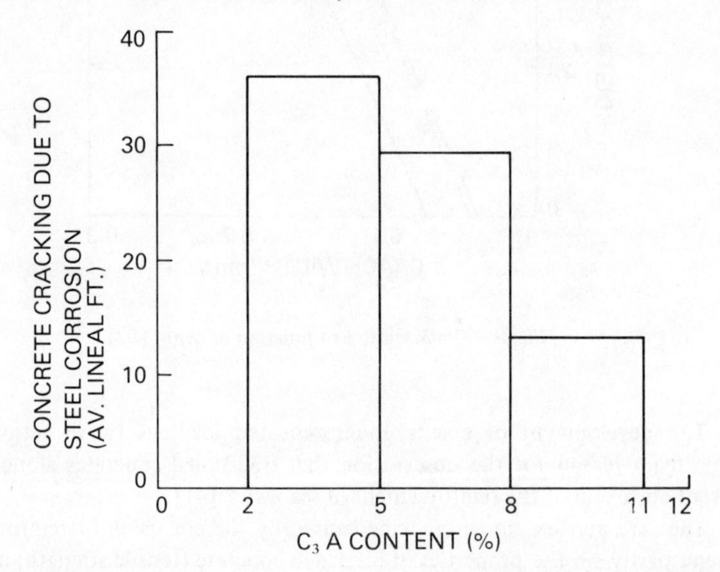

Fig. 9 – Effect of C_3A content on concrete cracking due to steel corrosion in marine environment [86].

Bazant published a numerical method which can be used for the prediction of the approximate corrosion rates of the reinforcement, and the times of the resulting cracking of the concrete cover, from the properties of the concrete and the environment [12, 44]. He formulates the transport of oxygen and chloride ion through the concrete cover, the mass sinks and sources of oxygen, ferrous hydroxide and hydrated red rust due to chemical reactions, the depassivation of steel due to critical chloride ion concentration, the cathodic and anodic potential with the concentration polymerization of electrodes, and the flow of electric currents through the electrolyte in the concrete pores.

Under certain circumstances the sea water, primarily the magnesium salts solved in sea water, can deteriorate the concrete without the corrosion of the reinforcement. This is, however, rare when the structure is properly designed [45–47], the composition is well selected (cement with not too high C_3A content, low water–cement ratio, proper air content [48, 49], adequate cement content [50], good aggregate [51, 52], and the construction is properly performed (careful placing and compaction, adequate curing) [53–67].

A report of the Portland Cement Association provides the following conclusions concerning the durability of reinforced concrete piles made with many different portland cements after exposure for 20 years to sea water in four locations in the United States [68]:

(1) $1\frac{1}{2}$ inches (38 mm) of concrete cover over reinforcing steel is inadequate for any sea water exposures.
(2) Rusting and expansion of reinforcing steel may obscure some of the results of other forms of deterioration in severe exposures.
(3) Low water–cement ratio concrete mixtures (0.4 to 0.45 by weight) of low slump and high cement content are necessary for durable concrete. A mix of 660 lb/cu yd (= 400 kg/m^3) represents the minimum cement that can be recommended for severe exposures.
(4) Air entrainment is moderately beneficial in severe exposures to sea water and at least equally so in the mild exposures. However, the improvement effected in the piles in air entrainment under freezing conditions is less than it is in other cases of freezing and thawing.
(5) Sea water attack has taken place in the mild exposure with only one cement and this was of high sulphate susceptibility having 13.2% C_3A content.
(6) Rusting and expansion of reinforcing steel is more pronounced in mild climate than in severe climate.

A similar set of conclusions of a pertinent RILEM Symposium [69] states the following concerning the durability of concrete structures, with or without reinforcement, in sea water:

(1) The most important environmental factor is the position of the concrete in relation to sea water level. Concrete between tide levels is subject to frost action, to wetting and drying, to the capillary rise of sea water in the pores of the concrete, and to attrition from wave action.
(2) In many instances, damages to concrete are caused by sea spray. Corrosion of the steel with attendant spalling is attributed to differential salt concentration in the concrete leading to electrolytic corrosion.
(3) Slow sulphate action of sea water on dense concretes relates to high concentrations of chlorides present. Higher salt concentrations exist in the tidal zone because of evaporation.

(4) Frost action is the main factor in cold climates and here air entrainment is of prime importance.
(5) For reinforced concrete, the severity of exposure immediately above high water level or where sea spray can reach the concrete may be even greater than between tide levels. It is in the upper portions of reinforced concrete piles and on the underside of decks, or beams supporting them, that corrosion is most common.
(6) British trials on reinforced concrete piles showed the prime cause of failure to be corrosion leading to cracking of the concrete.
(7) A minimum protective cover of 5 cm (approximately 2 in) of dense concrete is required over reinforcing steel.
(8) Special emphasis should be placed upon producing a concrete of high uniformity and density. It should be relatively stiff and vibrated.
(9) A fundamental factor in the resistance of concrete to sea water is to let it be penetrated as little as possible.
(10) Composition of concretes for exposure to the sea must allow a great compactness after placing. Placing methods must approach perfection.
(11) Concreting under water or within reach of the tide should be avoided; prefabrication is much preferred.
(12) Sharp edges should be avoided and in the case of piles, cylindrical shapes are preferred.

These two sets of conclusions clearly show the importance of the quality of concrete cover in the protection of reinforcement in sea water. Low micro- and macro-porosity, low ion and water permeability, high tensile strength, low modulus of elasticity, good resistance to sulfate action as well as freezing and thawing are essential for an adequate protection of reinforcement against corrosion in sea water (Table 1), even when the reinforcement is coated with a material that is not corrosive in marine environment. Adequate thickness of the concrete cover (Table 2), proper selection of the quality and quantity of ingredients of the concrete, careful placing, compaction and curing contribute to the required quality of the concrete. It is also clear, however, that further improvement is needed in the properties of concrete, even when coated reinforcement is used. A promising way to achieve this improvement is the addition of a suitable epoxy to the concrete in the mixer [70–81].

Concrete and reinforcing steel usually do not deteriorate at the same rate in sea water. The reason for this is that conditions leading to the deterioration of the concrete do not necessarily corrode the reinforcement, or at least not to the same extent, and vice versa. It is true, however, that the deterioration of the concrete usually speeds up the corrosion of the embedded steel and, again, vice versa.

Table 1 – Permissible minimum cement content and maximum water–cement ratio for concrete in sea water.

	ACI–357		FIP		
Minimum cement content	356 (kg/m^3)		Tide zone		400 (kg/m^3)
			Others maximum aggregate size D	D = 40 mm	320
				D = 20 mm	360
Maximum water–cement ratio by weight	in sewater	0.45	0.45 (less than 0.4 is preferred)		
	in air or tide zone	0.4			

Table 2 – Permissible minimum thickness of concrete cover for structures in sea water.

Location	ACI–357		FIP	
	Steel bar	Duct for post-tension	Steel bar	PC bar
In air	50 mm	75 mm	–	–
In air and tide zone	65 mm	90 mm	75 mm	100 mm
In sea water	50 mm	75 mm	60 mm	75 mm
Cover of stirrups	not less than above values minus 13 mm			

CONCLUSIONS

The mechanism of the well documented corrosion of steel reinforcement in concrete in sea water is understood mostly in qualitative terms. There are many details of the process where further research is needed for clarification. This includes the exact role of concrete in the protection, and how to achieve the best protection.

ACKNOWLEDGEMENT

This paper is a part of a cooperative research project sponsored partially by the National Science Foundation, NSF Grant Number: INT-8008869.

REFERENCES

[1] Tibbetts, D. C., 'Performance of concrete in sea water: some examples from Halifax, N.S.', *Performance of Concrete,* E. G. Swenson, Technical Editor, Canadian Building Series No. 2, University of Toronto Press, 1968, pp. 159–180.

[2] Escalante, E. and Ito, S., 'A bibliography on the corrosion and protection of steel in concrete', *NBS Special Publication 550,* National Bureau of Standards, Washington, DC, 1979.

[3] Idorn, G. M., 'Viewpoints on research of the durability of concrete exposed to sea water', *RILEM Bulletin,* No. 16, Paris, September 1962, pp. 59–64.

[4] ACI, *Performance of Concrete in Marine Environment, Publication SP65,* American Concrete Institute, Detroit, 1980.

[5] Idorn, G. M., 'Research activities concerning the durability of concrete in sea water', *RILEM Durability of Concrete,* Final Report of the International Symposium, Praha, 1962, pp. 351–355.

[6] Calleja, J., 'Durability', *7th International Congress on the Chemistry of Cement,* Vol. I, Paris, 1980, pp. VII. 2/1–VII. 2/48.

[7] Verbeck, G. J., 'Mechanism of corrosion of steel in concrete', *Corrosion of Metals in Concrete, Publication SP49,* American Concrete Institute, Detroit, 1975, pp. 21–38.

[8] Steinour, H. H., 'Influence of the cement on the corrosion behavior of steel in concrete', *Research Department Bulletin 168,* Research and Development Laboratories of the Portland Cement Association, May 1964, Skokie.

[9] Haussmann, D. A., 'Electrochemical behvior of steel in concrete', *ACI Journal,* Proc. **61**, (2), February 1964, pp. 161–188.

[10] Shalon, R. and Raphael, M., 'Influence of sea water on corrosion of reinforcement', *ACI Journal,* Proc. **55**, (6), June 1959, pp. 1251–1268.

[11] Shalon, R. and Raphael, M., 'Corrosion of reinforcing steel in hot countries', *RILEM Bulletin,* No. 24, Paris, September 1964, pp. 29–45.

[12] Bazant, Z. P., 'Physical model for steel corrosion in concrete sea structures – theory', *Journal of the Structural Division, ASCE,* Proc. **105** (ST6), June 1979, pp. 1137–1153.

[13] Hamada, M., 'Neutralization (carbonation) of concrete and corrosion of reinforcing steel', *Proceedings of the Fifth International Symposium on the Chemistry of Cement, Part III, Properties of Cement Paste and Concrete,* Tokyo, December, 1969, pp. 343–383.

[14] Volkwein, A. and Springenschmid, R., 'Corrosion of reinforcement in concrete bridges at different ages due to carbonation and chloride penetration', *Proceedings, Second International Conference on the Durability of Building Materials and Components, September 14–16, 1981,* National Bureau of Standards, Gaithersburg, MD, pp. 199–209.

[15] Gunther, F. and Ruprich, G., 'Einfluss des Cl-Gehaltes des Betons auf das Korrosionverhalten von Spannstahl' (Influence of Cl content of concrete on the corrosion of prestressed tendon), *Baustoffindustrie* **8**, (9), August 1966, pp. 226–229.

[16] Aschan, N. and Palm, S., 'Sahkokeniallinen tutkimus lisaaineiden vaikutuksen silvitta miseksi betoniterasted korroosiossa' (Electrochemical testing of the effects of admixtures on corrosion of reinforcement), *Tiedotue,* Sarja III – Rakennus 124, The State Institute for Technical Research, Finland, Helsinki, 1968.

[17] Figg, J., 'Rusting reinforcement – the no. 1 problem of durability', *Concrete,* **14**, (5), London, May 1980, pp. 34–36.

[18] Spellman, Donald L. and Stratfull, Richard F., 'Chlorides and bridge deck deterioration', *Highway Research Record, No. 328, Concrete Durability, Cement Paste, Aggregates, and Sealing Compounds,* Highway Research Board, Washington, DC, 1970, pp. 38–49.

[19] Tremper, B., Beaton, J. L. and Strafull, R. F., 'Causes and repair of deterioration to a California bridge due to corrosion of reinforcing steel in a marine environment', Part II: Fundamental Factors Causing Corrosion, *Highway Research Board Bulletin 182,* Washington, DC, 1958, pp. 18–41.

[20] Locke, C. E. and Siman, A., 'Electrochemistry of reinforcing steel in salt-contaminated concrete', *Corrosion of Reinforcing Steel in Concrete,* D. E. Tonini and J. M. Gaidis, Editors, ASTM STP713, Philadelphia, PA, 1980, pp. 3–16.

[21] Browne, R. D. and Baker, A. F., 'The performance of structural concrete in a marine environment', *Development in Concrete Technology* 1, F. D. Lydon, Editor, Applied Science Publishers Ltd, London, 1979, pp. 111–149.

[22] Browne, R. D., 'Corrosion of steel in reinforced concrete in marine and other chloride environments', *Proceedings, Second International Conference on the Durability of Building Materials and Components,* National Bureau of Standards, Gaithersburg, MD, pp. 210–224.

[23] Erlin, B. and Verbeck, G. J., 'Corrosion of metals in concrete – needed research', *Corrosion of Metals in Concrete, Publication SP49,* American Concrete Institute, Detroit, 1975, pp. 39–46.

[24] Backstrom, T. E., 'Use of coatings on steel embedded in concrete', *Corrosion of Metals in Concrete, Publication SP49,* American Concrete Institute, Detroit, 1975, pp. 103–114.

[25] Baker, E. A., Money, K. L. and Samborn, C. B., 'Marine corrosion behavior of bare and metallic-coated steel reinforcing rods in concrete', *Chloride Corrosion of Steel in Concrete,* STP629, American Society for Testing and Materials, 1977, pp. 30–50.

[26] Cook, A. R. and Radtke, S. F., 'Recent research on galvanized steel for reinforcement of concrete', *Chloride Corrosion of Steel in Concrete,* STP629, American Society for Testing and Materials, 1977, pp. 51–60.

[27] Zinc Institute, *Galvanized Reinforcement for Concrete,* International Lead Zinc Research Organization, Inc., New York, 1970.

[28] Zinc Institute, *Galvanized Reinforcement for Concrete – II,* International Lead Zinc Research Organization, Inc., New York, 1981.

[29] Treadaway, K. W. J., Brown, B. L. and Cos, R. N., 'Durability of Galvanized steel in concrete', *Corrosion of Reinforcing Steel in Concrete,* D. E. Tonini and J. M. Gaidis, Editors, ASTM STP713, Philadelphia, PA, 1980, pp. 102–131.

[30] Clifton, J. R., Beeghly, H. F. and Mathey, R. G., 'Nonmetallic coatings for concrete reinforcing bars. Coating materials', *National Bureau of Standards Technical Note 768,* Washington, DC, April 1973.

[31] Clifton, J. R., Beeghly, H. F. and Mathey, R. G., 'Protecting reinforcing bars from corrosion with epoxy coatings', *Corrosion of Metals in Concrete,* ACI Publication, SP49, Detroit, 1975, pp. 115–132.

[32] Kilareski, W. P., 'Epoxy coatings for corrosion protection of reinforcement steel', *Chloride Corrosion of Steel in Concrete,* STP629, American Society for Testing and Materials, Philadelphia, PA, 1977, pp. 82–88.

[33] Paterson, W. S., 'Fatigue of reinforced concrete in sea water', *Performance of Concrete in Marine Environment, ACI Publication SP65,* American Concrete Institute, Detroit, 1980, pp. 419–436.

[34] Sommerville, G. *et al.*, 'Concrete properties', *Proceedings, 8th FIP Congress,* Part 1, May 1978, pp. 8–18.

[35] Roper, H., 'The influence of protection mechanism of coated bar on structural perfomance', Presented at the Annual Convention of the American Concrete Institute, Atlanta, GA, January, 1982.

[36] Gjorv, O. E., Vennesland, O. and El-Busaidy, A. H. S., 'Diffusion of dissolved oxygen through concrete', *NACE Corrosion 76, Paper No. 17,* Houston, TX, 22–26 March 1976, 13 pp.

[37] Szilard, R. and Wallevik, O., 'Effectiveness of concrete cover in corrosion protection of prestressed steel', *Corrosion of Metals in Concrete, Publication SP49,* American Concrete Institute, Detroit, 1975, pp. 47–70.

[38] Thomas, K. and Burney-Nicol, S., 'Survey of the corrosion of steel reinforcement in the tropics', *RILEM Bulletin,* No. 24, Paris, September

1964, pp. 47–50.
[39] Okada, K. and Miyagawa, T., 'Chloride corrosion of reinforcing steel in cracked concrete', *Performance of Concrete in Marine Environment, ACI Publication SP65,* American Concrete Institute, Detroit, 1980, pp. 237–254.
[40] Schiessel, P., 'Zusammenhang swischen Rissbreite und Korrosionsabtragung and der Bewehrung' (Relationship between crack width and corrosion on the reinforcement), *Betonwerk + Fertigteil-Technik,* Heft 12, 1975.
[41] Kari, A., 'Investigation on the improvement of corrosion resistance and strength of ribbed reinforcing steel bars in concrete, especially when subject to fatigue loading', *Publication 17,* Technical Research Centre of Finland, 1980.
[42] Beaton, J. L. and Stratfull, R. F., 'Environmental influence on corrosion of reinforcing in concrete bridge sub-structures', *Highway Research Record,* No. 14, *Concrete Bridge Decks and Pavement Surfaces,* Highway Research Board, Washington, Washington, DC, January, 1963.
[43] Verbeck, G. J., 'Field and laboratory studies of the sulphate resistance of concrete', *Performance of Concrete,* E. G. Swenson, Technical Editor, Canadian Building Series No. 2, University of Toronto Press, 1968, pp. pp. 113–124.
[44] Bazant, Z. P., 'Physical model for steel corrosion in concrete sea structures – application', *Journal of the Structural Division, ASCE,* **Proc. 105** (ST6), June 1979, pp. 1155–1166.
[45] Masabuchi, K., *Materials for Ocean Engineering,* Massachusetts Institute of Technology, 1970.
[46] *ACI Committee,* 'Guide for the design and construction of fixed offshore concrete structures', *ACI Journal,* Proc. **75** (12), December 1978, pp. 684–709.
[47] Schupack, M. 'Design of permanent seawater structures to prevent deterioration', *Concrete International Design and Construction,* **4** (3), March 1982, pp. 19–27.
[48] Lyse, I., 'Durability of concrete in sea water', *ACI Journal,* Proc. **57,** June 1961, pp. 1575–1584.
[49] Lyse, I., 'Durability of concrete in sea water', *RILEM Durability of Concrete,* Preliminary Report of the International Symposium, Praha, 1961, pp. 183–192.
[50] Moksnes, J., 'Condeep platforms from the North Sea – some aspects of concrete technology', *Proceedings, 1975 Offshore Technology Conference, May 5–8, Houston, TX,* Vol. III, pp. 339–350.
[51] Coutinho, A. de Sousa and Rodriges, F. M. Preres, 'Tenue a la mer de betons confectionnes avec des agrégats decomposables influence de la couche protectrice', *RILEM International Symposium on Behaviour of*

Concretes Exposed to Sea Water, May 24–26, 1965, Palermo, Italy, pp. 78-9–77-19.

[52] Coutinho, A. de S., 'L'influence de la nature mineralogique de l'agrégat sur la decomposition des mortiers et betons de ciment Portland a la mer', *RILEM-AIPCN International Symposium on Behaviour of Concretes Exposed to Sea Water, Palermo, Italy, May, 1965,* pp. 78–20.

[53] Whitting D., 'Influence of concrete materials, mix and construction practices on the corrosion of reinforcing steel', *Materials Performance,* **17** (12), December 1978, pp. 9–15.

[54] Mather, B., 'Concrete need not deteriorate', *Concrete International: Design and Construction,* **1**, (9), September 1979, pp. 32–37.

[55] Campus, F., 'Instructions and recommendations for building experimental observations', RILEM, 'The behavior of concretes exposed to sea water', Topic III – General Report, *RILEM Bulletin,* New Series No. 30, March 1966, Paris, pp. 75–86.

[56] Campus, F., Dantinne, R., Dzulynski, M. and Gamski, K., 'Constatation effectuées après trente années d'immersion marine d'éprouvettes de mortiers, de betons et de betons armes dans la mer du Nord à Ostende', *Memoires du CERES,* No. 16, September 1966, pp. 9–34.

[57] Campus, F., 'Rapport d'ensemble relatif aux essais et observations effectues sur des eprouvettes de mortiers et de betons pendant une duree de trente ans (1934–1964), dont un grand nombre ont ete immergées en permanence dans la mer du Nord à Ostende', *Memoires du CERES,* (Nouvelle Serie) No. 24, Université de Liège, February, 1968.

[58] Campus, F., 'Essais de resistance des mortiers et betons a la mer', *RILEM Durability of Concrete,* Preliminary Report of the International Symposium, Praha, 1961, pp. 90–102.

[59] Campus, F., Dantinne, R. and Dzulynski, M., 'Constatations effectuées après trente anées d'immersion marine d'eprouvettes de mortier, de betons et de betons armes dans la mer du Nord', *RILEM-AIPCN International Symposium on Behaviour of Concretes Exposed to Sea Water, Palermo, Italy, 24–26 May,* 1965, pp. 114–120.

[60] Senbetta, E. and Scholer, C., 'Absorptivity, a measure of curing quality as related to durability of concrete surfaces', *Proceedings, Second International Conference on the Durability of Building Materials and Components, September 14–16, 1981,* National Bureau of Standards, Gaithersburg, MD, pp. 153–159.

[61] Rehm, G., 'Korrosionschutz von Stahl in Beton' (The protection of steel in concrete against corrosion), *Betontechnische Berichte,* 1969, Beton-Verlag GmbH, Dusseldorf, 1970, pp. 57–65.

[62] Allen, R. T. L., 'Concrete in maritime works', Second Edition, *Publication 46.501,* Cement and Concrete Association, London, 1981.

[63] ACI, *Symposium on Concrete Construction in Aqueous Environments,*

ACI Publication SP8, American Concrete Institute, Detroit, 1964.

[64] ACI Committee 201, 'Guide to durable concrete', *ACI 201. 2R-77,* American Concrete Institute, Detroit, 1977.

[65] Idorn, G. M., 'Conditions for using concrete as a constructional material in Arctic harbours', *Proceedings from the First International Conference on Port and Ocean Engineering under Arctic Conditions,* Vol. II, 1971, pp. 1334–1342.

[66] Lyse, I., 'Mixing and pouring', RILEM, 'The behavior of concretes exposed to sea water', Topic II – General report, *RILEM Bulletin,* New Series No. 30, March, 1966, Paris, pp. 71–73.

[67] Mather, Bryant, 'Concrete in sea water', *Concrete International' Design and Construction,* **4,** (3), March 1982, pp. 28–34.

[68] Hansen, W. C., 'Twenty-year report on the long-time study of cement performance in concrete', *Research Department Bulletin 175,* Portland Cement Association, Research and Development Laboratories, Skokie, May 1965.

[69] RILEM-AIPCN, *International Symposium on Behaviour of Concrete Exposed to Sea Water, Palermo, Italy, May 24–28, 1965.*

[70] Ohama, Y., 'Comparison of properties with various polymer-modified mortars', *Synthetic Resins in Building Construction,* Vol. I, RILEM Symposium, Paris, September 4–6, 1967.

[71] Ohama, Y., 'Cement mortars modified by SB latexes with variable bound styrene', *Synthetic Resins in Building Construction,* Vol. I, RILEM Symposium, Paris, September 4–6, 1967.

[72] Ohama, Yoshihiko and Asahi, Hideji, 'Basic properties of paraffin-modified mortar', *Advances in Cement-Matrix Composites,* Editors: D. M. Roy, A. J. Majumdar, S. P. Shah and J. A. Manson, Materials Research Society, Proceedings, Symposium L, Boston, November 1980, pp. 259–260.

[73] Ohama, Y., 'Durability performance of polymer-modified mortars', *Proceedings, Second International Conference on the Durability of Building Materials and Components, September 14–16, 1981,* National Bureau of Standards, Gaithersburg, MD, pp. 242–248.

[74] De Vekey, R. C. and Majumdar, A. J. 'Durability of cement paste modified by polymer dispersions', *Supplementary Paper III-9,* The VI. International Congress on the Chemistry of Cement, Moscow, September, 1974.

[75] De Vekey, R. C. and Majumdar, A. J., 'Durability of cement pastes modified by polymer dispersions', *Materials and Structures-Research and Testing,* RILEM, 8 (46), Paris, July–August 1975, pp. 315–322.

[76] Bhatty, D. I., Dollimore, D., Gamlen, G. A. and Mangabhai, R. J., 'The mechanical properties of polymer modified cement paste cured under sea and tap water', *7th International Congress on the Chemistry of Cement,* Vol. IV, Edition Septima, Paris, 1981, pp. 363–367.

[77] Popovics, S., 'Polymer and Portland cement concrete combinations',

Proceedings of the 23rd Annual Arizona Conference on Roads and Streets, The University of Arizona, April, 1974, pp. 107–124.

[78] Popovics, S., 'Polymer cement concretes for field construction', *Journal of the Construction Division, ASCE,* 100 (CO3), *Proc. Paper 10806,* September 1974, pp. 469–487.

[79] Popovics, S., 'Cement-mortar experiments concerning the addition of water-dispersible epoxy of furfuryl alcohol systems', *Polymers in Concrete,* Proceedings of the First International Congress on Polymer Concretes, Paper 2.7, London, May 4–7, 1975, The Construction Press, Ltd., London, 1976, pp. 64–70.

[80] Popovics, S., 'Polymer pavement concrete for Arizona – Study II', *Research Report for the Arizona Department of Transportation, No. HPR 1–12 (154),* March 1976, 137 pp.

[81] Popovics, S. and Tamas, F., 'Investigation on Portland cement pastes and mortars modified by the addition of epoxy', *Polymers in Concrete,* International Symposium, *Publication SP58,* American Concrete Institute, Detroit, 1978, pp. 357–366.

Chapter 3

Exposure tests on concrete for offshore structures

K. M. BROOK and J. A. STILLWELL, Wimpey Laboratories Ltd., Hayes, Middlesex, UK

1 INTRODUCTION

In order to achieve a realistic estimate of the life of offshore concrete structures, the causes of deterioration of reinforced concrete must be investigated.

For structures in the marine environment, the most common cause of deterioration is corrosion of the reinforcement, with subsequent spalling of concrete. This can normally be prevented by the use of good quality concrete with adequate cover to the reinforcement. However, corrosion can occur if chlorides are able to penetrate to the surface of the reinforcement, either through cracks in the structure or through concrete of high permeability.

It is of interest to note that BS 6235, *Code of Practice for Fixed Offshore Structures* [1] states that research into the cover to reinforcement required for offshore structures is still incomplete.

As part of the 'Concrete in the Oceans' research programme, funded jointly by the Department of Energy and industrial contributors, the basic mechanisms of the corrosion of steel in concrete in the marine environment have been investigated. This work has been carried out by Wilkins and Lawrence [2] in a series of laboratory experiments.

The 'Exposure Tests' project on which this paper is based is also part of the 'Concrete in the Oceans' programme and it has investigated the conditions which cause the onset of corrosion and its rate of development. The parameters involved are the measurement of the rate of penetration of chloride into concrete and the rate at which corrosion of reinforcement occurs in large scale specimens exposed to a real sea environment. The conditions achieved are those typically experienced by structures in the North Sea.

Two exposure sites have been used, the first being a simulated splash zone site on a pier at Portland harbour, near Weymouth and the second being a deep water site at Loch Linnhe near Fort William where facilities are available to locate specimens in sea water at a depth of 140 m.

To determine the rate of penetration of chlorides, small unreinforced concrete prisms have been used, while the corrosion investigation has made use of pairs of reinforced concrete beams stressed back to back to induce cracking. The layout of reinforcement was designed such that cracks could be induced either transverse to the main bar or as a longitudinal crack along a main bar.

Phase 1 of the project, which is now complete, consisted of a total exposure period of two and a half years, during which chloride penetration measurements were made and the extent of reinforcement corrosion in cracked beams was investigated. The variables in the corrosion specimens included two grades of concrete and two depths of cover to reinforcement. The project has been continued into a second phase, to provide a total exposure period of five years and to include additional variables such as varying crack widths and the use of pulverised fuel ash in the concrete.

The work on fundamental corrosion and the exposure tests described in this paper have been complemented by the investigation into the durability of a concrete fort in the Thames Estuary [3].

2 TEST PROGRAMME

Five zones of marine environment have been used for the exposure of chloride penetration test specimens.

Zone A *Atmospheric*	Exposure in a natural saline atmosphere, but without direct contact with sea water.
Zone B *Splash*	Exposure just above mean high water spring level. The specimens are subject to wave action but not submerged in tidal sea water.
Zone C *Tidal*	Exposure at mean tide level with specimens frequently submerged in tidal water.
Zone D *Shallow immersion*	Exposure at a depth of about 10 m in sea water.
Zone E *Deep immersion*	Exposure at a depth of about 140 m in sea water.

Zones, A, B and C are located at Portland Harbour in Dorset and Zones D and E are located at Loch Linnhe in Scotland.

All five zones were used for the exposure of chloride penetration specimens, while corrosion investigation specimens were located only in Zones B and E.

For the chloride penetration investigation, four sets of test prisms were placed in all five exposure zones. Three sets from each zone have been tested at ages of six months, one year and two and a half years. The fourth set has been retained for testing in Phase 2.

Three sets of transverse crack beams were constructed for the investigation of corrosion and were placed in each of the exposure zones B and E. One set from each zone has been tested at one year and another set at two and a half years. The third set has been retained for testing in Phase 2.

Two sets of longitudinal crack beams were also constructed and placed in each of the exposure zones B and E. One set from each zone has been tested at one year and the second set has been retained for testing in Phase 2.

3 MATERIALS AND MIX DESIGN

Materials

The materials used in this project were as follows.

Cement	Ordinary Portland cement. Typical batch from Northfleet works of Blue Circle Industries.
Coarse aggregate:	Crushed granite (Mount Sorrel) 20 mm and 10 mm single size. From the Budden Quarry of Redland Roadstone Limited.
Fine aggregate:	Concreting sand. Zone 2 grading. From the Chertsey Pit of Hall Aggregates (Thames Valley) Limited.
Reinforcement:	'Torbar' cold worked high yield ribbed 25 mm diameter steel from Reinforcement Steel Services.

Mix design

Two grades of concrete were used and the mix proportions and concrete properties are given in Table 1.

Table 1

Grade of concrete	Standard	Low
Mixed proportions:		
Cement	1.00	1.00
20 mm aggregate	1.70	2.55
10 mm aggregate	0.80	1.20
Fine aggregate	1.50	2.25
Free water	0.45	0.70
Cement content	435 kg/m^3	305 kg/m^3
Workability: slump	50 mm	100 mm
Mean 28-day compressive strength	70 N/mm^2	35 N/mm^2

4 TEST SPECIMENS

Chloride penetration prisms

The chloride penetration prisms were blocks of plain concrete measuring 100 X 100 X 500 mm long with all faces except the ends sealed by a pitch/polyurethane material. In some prisms, a longitudinal crack was induced. Details of the specimens and the sampling procedure are given in Fig. 1.

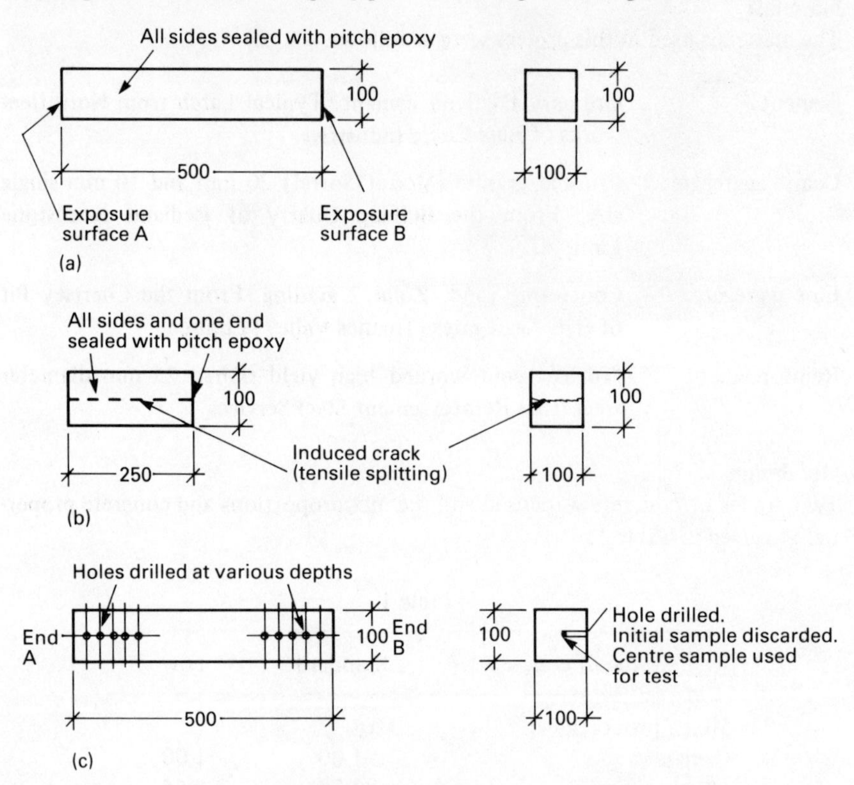

Fig. 1 – Details of chloride penetration test specimens.
(a) Standard double specimen (uncracked).
(b) Cracked single specimen.
(c) Method of drilling for test samples (chloride content).

Transverse crack beams

The transverse crack specimens were produced as pairs of 225 X 175 mm beams 1200 mm long, stressed back to back to induce cracking in the outer surfaces. The main reinforcement was a single 25 mm diameter bar along the length of the beam. Cover to the reinforcement was either 75 mm or 25 mm, the latter being achieved by including a recess in a beam, as shown by the details given in Fig. 2. The method of casting individual beams is shown in Fig. 3.

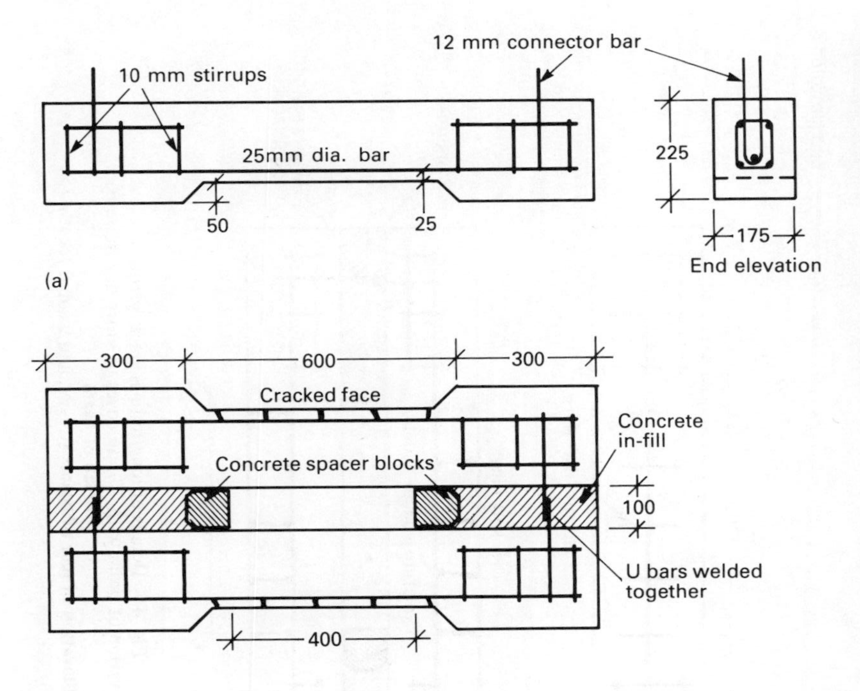

Fig. 2 – Details of transverse crack beams.
(a) Elevation of beam with 25 mm cover to longitudinal bar.
(b) Elevation of pair of beams stressed and welded together.

Fig. 3 – Casting individual beams.

Longitudinal crack beams

The longitudinal crack specimens were produced as pairs of 200 X 250 mm beams 1300 mm long, again stressed back to back to induce cracking. The main reinforcement was a single 25 mm diameter bar along the length of the beam, to which was welded a series of five short transverse bars, also 25 mm diameter. The cracks formed along these short bars.

The cover to reinforcement was made either 75 mm or 25 mm, using the same method as for the transverse crack beams, the details of which are shown in Fig. 4. The method of stressing a pair of beams to induce cracking is shown in Fig. 5.

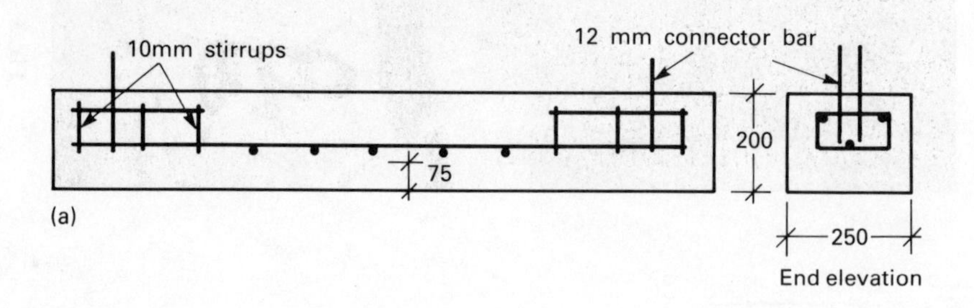

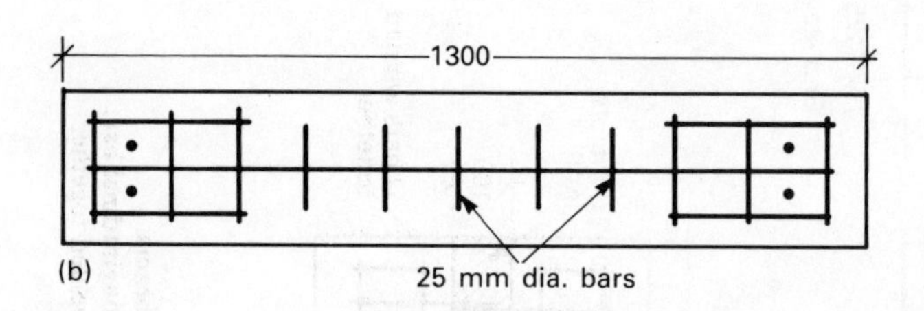

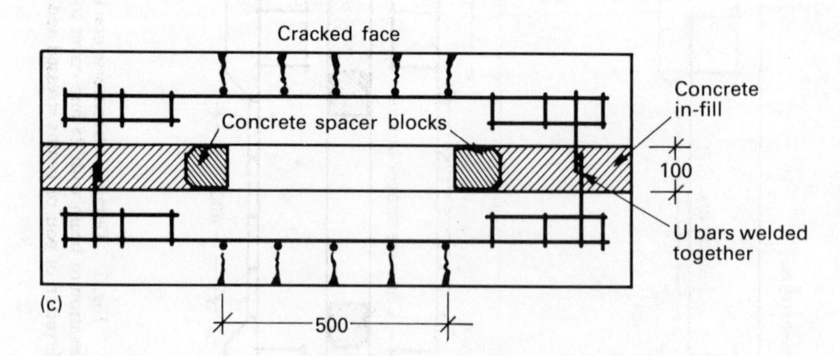

Fig. 4 – Details of longitudinal crack beams.
(a) Elevation of beam with 75 mm cover to short transverse bars.
(b) Plan of beam.
(c) Elevation of pair of beams stressed and welded together.

Fig. 5 – Stressing a pair of beams to induce cracking.

5 TEST FACILITIES

Splash zone facility at Portland

The requirement for a splash zone exposure facility was a site where the specimens would be alternately wetted and dried by salt water spray, without being immersed in water at any time. A site was chosen at Portland Harbour in Dorset, but it was found that owing to the shallow tidal range and the relative calmness of the weather on many occasions, it was not possible to achieve the required wetting conditions by natural means. Therefore, it was decided that the specimens should be subjected to an artificial spray on a regular daily basis. In addition to this they would also be subject to natural spray during severe weather conditions.

The specimens were grouped together in a vertical position within a steel frame on a stone jetty on the seaward side of the harbour breakwater and subjected to a spray of sea water supplied by an electrically operated pump controlled by a time switch. This was set to operate for a period of one and a half hours during the day in any twenty-four-hour period. The splash zone facility is shown in Fig. 6 with a detail of the control panel shown in Fig. 7.

Fig. 6 – Splash zone facility at Portland.

Fig. 7 – Control panel for spray system.

Deep water immersion zone facility at Loch Linnhe

It was necessary to find an exposure site providing sea water to a depth of about 150 m. Faced with very limited choice, the location chosen was Loch Linnhe, an open sea water loch near Fort William on the west coast of Scotland. The area was known to have sufficient depth and pier facilities were available. Investigation showed that composition of the loch water at depth was similar to North Sea conditions, with a satisfactory level of salinity and oxygen.

It was decided to support the specimens in steel cages from a pontoon moored in the loch at one of the positions where up to 150 m depth of water was available. Various methods of suspension were investigated and a system was devised whereby all the specimens were held in one large frame suspended by a fixed length of steel rope. This was joined at the bottom end to the main frame and at the top end to the pontoon by means of four-legged slings. Raising and lowering the specimens was achieved by the use of a floating gantry and winch. The frame containing the specimens is shown in Fig. 8, while Fig. 9 shows the pontoon moored in Loch Linnhe with a tug alongside.

Fig. 8 – Frame containing specimens at Loch Linnhe.

Fig. 9 – Pontoon moored in Loch Linnhe with tug alongside.

6 EXPOSURE DETAILS

Owing to various circumstances, in particular with reference to setting up the exposure sites and certain difficulties with the suspension system at Loch Linnhe, variations have occurred in the exposure of specimens, compared with the original intended programme. The actual exposure details were as follows.

Specimens were exposed in a saline atmosphere (Zone A) at Portland as originally proposed. However, specimens in the splash zone (Zone B) were exposed to a saline atmosphere for three months before being exposed to the saline spray for a short period each day. Also, the specimens in the tidal zone (Zone C) were exposed to a saline atmosphere for five months before being exposed to tidal immersion in sea water.

The shallow immersion specimens of Loch Linnhe (Zone D) were generally exposed in a 10 m depth of sea water over a three-year period, except for short periods at a depth of 2 m, totalling six months. The deep immersion specimens (Zone E) were generally exposed in a 140 m depth of sea water over a three-year period, except for short periods at a depth of 2 m, totalling nine months.

7 CHLORIDE PENETRATION

Levels of chloride penetration

The chloride content of concrete samples drilled at various depths from the exposed surface of the sealed prisms (see Fig. 1) was measured and, from the test results, comparisons were made of levels of chloride penetration in specimens from the different exposure zones, for both grades of concrete.

In the atmospheric exposure zone at Portland (Zone A) it was found that in the standard grade concrete, chloride penetration was limited to a depth of about 30 mm, while in the lower grade concrete this depth increased to 50 mm. The maximum level of absorption was 0.13% chloride (as Cl) by weight of concrete, with slightly higher values obtained from the low grade concrete compared with the higher grade. Neither grade of concrete showed any increase in chloride content after the initial six-month exposure period.

The depth of penetration of chloride into the specimens from the splash zone at Portland (Zone B) was similar to the values obtained from Zone A. However, there was a significant increase in the chloride content of the low grade concrete compared with the higher grade. Also included in this zone were pre-cracked specimens (see Fig. 2) and it was found that the chloride penetrated to a depth of at least 180 mm in the crack. Again, neither grade of concrete or type of specimen showed any increase in chloride content after the initial six-month exposure period.

The level and depth of penetration in the specimens from the tidal exposure zone at Portland (Zone C) was similar to the values obtained from Zone A.

Similar sets of specimens were exposed at Loch Linnhe and the pattern of chloride absorption in the specimens from the shallow immersion zone (Zone D) was similar to the Portland values, the penetration limited to a depth of about 30 mm in the standard grade concrete and 60 mm in the low grade. Generally, the levels of chloride absorption were higher than in the splash zone specimens, with significantly higher values obtained from the low grade concrete compared with the standard grade. There was a slight increase of chloride content in both grades of concrete after the initial six-month exposure period.

Finally, the level and depth of chloride penetration in the specimens from the deep exposure zone at Loch Linnhe (Zone E) was similar to the values obtained from the other exposure zones, with a slight increase in depth of penetration. Pre-cracked specimens were also included in this zone and the penetration of chlorides to a depth of at least 180 mm was recorded. Generally, there was some increase in chloride content after the initial exposure period.

Difference in chloride absorption for different exposure zones at the same age

A general comparison of the chloride contents obtained shows that although the actual levels vary there is a higher absorption of chloride into the lower grade concrete in all the exposure zones. Irrespective of grade of concrete, there was variation in levels of chloride absorption according to the type of exposure, the lowest values being obtained from Zones A and C, higher values from the splash Zone B and the highest values from the immersion Zones D and E. Therefore it can be concluded that there is a greater absorption of chloride when the concrete is fully immersed. There is also some evidence that the level of absorption tends to increase as the depth of immersion increases.

Difference in chloride absorption for different ages in the same exposure zone

In general, there is no particular evidence to suggest that there is a significant increase of chloride absorption with age in any exposure zone, except for a slight increase in the Loch Linnhe specimens. Even so this occurred without any increase in depth of penetration.

The test results suggest that the absorption of chloride takes place relatively quickly after the initial exposure, after which absorption and penetration either cease or continue at a very slow rate.

8 CORROSION INVESTIGATION

Preparation of specimens

In order to preserve the concrete adjacent to the reinforcement intact, it was decided that rather than break open the beams to expose the reinforcement, it would be more appropriate to cut cores from them at the crack positions. Each core could then be individually assessed and by splitting it open at the crack position, not only would the reinforcing bar be exposed for inspection, but the concrete surface adjacent to the steel and crack surface would also be available for examination.

100-mm diameter cores were cut from beams with cracks transverse to the single bar of reinforcement. However, for beams with cracks longitudinal to the short lengths of reinforcement placed transversely in the beam, 150 mm diameter cores were cut. This enabled the majority of the short transverse bar to be included in the core. Details of a typical core cut through a crack in a transverse crack beam are shown in Fig. 10, while similar details of a core cut from a longitudinal crack beam are given in Fig. 11.

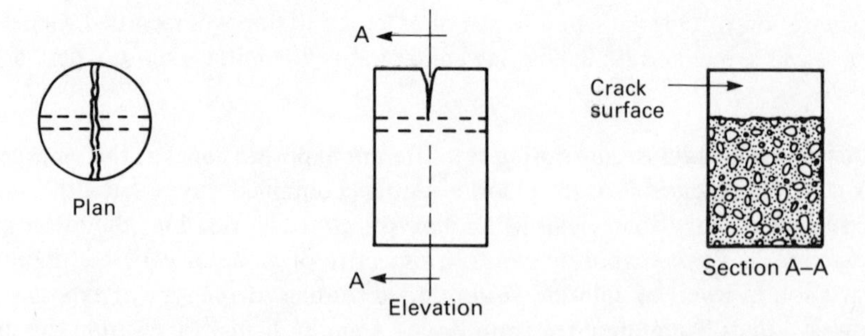

Fig. 10 – Typical core cut through crack in transverse crack beam.

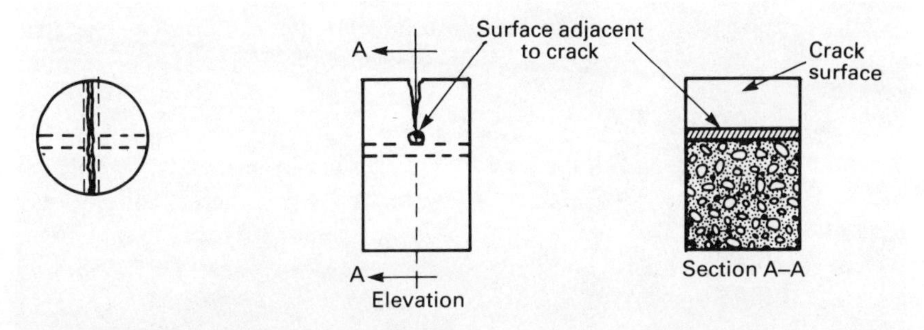

Fig. 11 – Typical core from longitudinal crack beam.

Assessment of reinforcement corrosion

A visual assessment of corrosion of the exposed reinforcement bars was made, the degree of which was defined according to a predetermined scale or grading. An inspection was also made of the lime bloom on the general surface area of the reinforcement and of the calcium carbonate deposit on the crack surface and on the steel at the base of the crack. Neither the one-year-old transverse and longitudinal crack specimens nor the two-and-a-half-year-old transverse crack specimens from Loch Linnhe showed any evidence of corrosion. In general, the reinforcement from the transverse crack specimens tended to have rather more lime bloom than the longitudinal crack specimens. This is possibly because in the former, the reinforcement remains covered by concrete except at the points of intersection with the crack, whereas the steel in the longitudinal crack specimens is exposed by the crack intersection along its full length.

As a result of the stressing carried out to induce cracking, bond between the bar and the concrete is broken at the point where the crack intersects the steel surface, thus the lime deposit tends to be increased in these areas. A typical example of the condition of the steel in a transverse crack beam from Loch Linnhe is shown in Fig. 12, while an example of steel in a longitudinal crack beam is given in Fig. 13.

The beams from the splash zone site at Portland showed there was significant corrosion on the reinforcement in the one-year-old longitudinal crack specimens and in most cases this was over the full length of exposed steel. There are two cases of fairly heavy corrosion in the two-and-a-half-year-old transverse crack specimens, which covered the majority of the surface of the reinforcement adjacent to the crack either side of the intersection. One of these cases was from a beam of low grade concrete with 75 mm cover while the other was from a beam of low grade concrete with 25 mm cover. Apart from the fact that both beams were from the splash zone, there does not seem to be any reason why these particular specimens corroded. In both cases, at least one other core was taken at a crack in the same beam which did not show any sign of active corrosion.

Fig. 12 – Typical example of condition of steel in transverse crack beam from Loch Linnhe.

Fig. 13 – Typical example of condition of steel in longitudinal crack beam from Loch Linnhe.

Fig. 14 – Typical example of condition of steel in transverse crack beam from Portland.

Fig. 15 – Typical example of condition of steel in longitudinal crack beam from Portland.

It is known that when the beams are stressed back to back and cracking occurs, there is a significant length of bar either side of the crack which becomes de-bonded. It may be that in these particular cases, the length of the bar which de-bonded was longer than normal and there was a significant gap between the bar and the concrete surface either side of the crack. This, together with a particular crack pattern, may have produced what was effectively an area of bare steel, which in the spray zone conditions would suffer corrosion. The first signs of corrosion in the transverse crack specimens had occured to a much smaller degree in some of the one-year-old specimens. A typical example of the condition of the steel in a transverse crack beam from Portland is given in Fig. 14 while an example of the steel from a longitudinal crack beam is given in Fig. 15.

Assessment of concrete properties

The strengths of several cores cut from the beam specimens were measured. They showed that the actual strength of the concrete in terms of equivalent cube strength was an average of about 82 N/mm^2 for standard grade and 37 N/mm^2 for low grade concrete at Loch Linnhe and 77 N/mm^2 for standard grade and 46 N/mm^2 for low grade concrete at Portland.

No significance can be attached to difference in core strengths between the splash and deep immersion zones, but it can be said that there is no reduction in strength due to either type of exposure.

A limited number of permeability tests were carried out giving mean values for permeability coefficient of around 1.2×10^{-11} m/s for standard grade concrete and 6.4×10^{-12} for low grade concrete at Portland. Equivalent values at Loch Linnhe were around 2.2×10^{-13} for standard grade and 3.3×10^{-11} for low grade concrete.

There were not sufficient test results to attach significance to the differences between the values obtained, but in general terms, they were typical for the grades of concrete being used.

9 GENERAL DISCUSSION OF TEST RESULTS

Chloride penetration

The results from measurement of chloride penetration into sealed prisms have been discussed in section 7. They show that there is a certain amount of variation in chloride content relative to the grade of the concrete and the conditions under which it was exposed. However, there is no evidence of any significant increase in chloride absorption after the initial exposure period of six months.

The average maximum chloride content by weight of sample for concrete exposed in the three zones above or at the water surface (i.e. atmosphere, splash, and tidal zones) varied according to the grade of concrete. The value of 0.15% chloride for the low grade concrete was about three times greater than the value of 0.05 measured in the standard grade material. There was a smaller difference between grades of concrete in the submerged zones, the lower grade having an average chloride content of about 0.25%, this being one and a half times the average value of 0.18% chloride in the standard grade. A difference in chloride content was also detected between specimens in the submerged zones and those in the surface zones. Submerged standard grade concrete absorbed on average about three times as much chloride as similar concrete in the surface zones, but for low grade concrete the difference was reduced to about twice as much.

Another aspect of chloride absorption which had been investigated is the depth of penetration into the prism. During the two-and-a-half-year exposure period, penetration of chloride into standard grade concrete exposed in the surface zones did not exceed a depth of 30 mm, with a slightly higher depth of penetration in the deep immersion zone. Penetration into the low grade concrete was rather higher, being in the range of 50 mm depth in the surface and 60 mm depth in the deep water.

All the chloride values discussed so far relate to sound concrete. Additional prisms were exposed which had cracks induced and in these the chloride penetrated down the crack very quickly. There was a slight decrease in chloride level with depth of crack, but differences in grade of concrete was not a significant factor. However, the exposure condition was significant and the general level of chloride at 75 mm depth was around 0.33% in deep water compared with around 0.20% in the splash zone. For the exposure period under consideration, these values are considerably greater than the chloride content of sound prisms measured at the same depth from the exposed surface. This suggests that 75 mm cover to the reinforcement is quite adequate to prevent chlorides penetrating to the steel surface in sound concrete, but if the concrete has cracks which are as deep as the cover provided chlorides will penetrate to the steel surface very quickly.

In addition to the prisms, measurements of chloride content were also made on samples of concrete drilled from cores cut from the exposure beams. In general, the difference between grades of concrete and exposure zones were of a similar pattern to the results from the penetration prisms, although the actual levels of chloride tended to be somewhat higher. This could be explained by the fact that unlike the prisms, none of the surfaces of the exposure beams were sealed. This allowed chloride to be absorbed from three different directions and in some cases, the length of the absorption path from the side of the beam was less than the depth from the cracked surface of the beam at which the sample was taken.

Corrosion of reinforcement

The results of the investigation into the rate of corrosion of the reinforcement in the test beams have been discussed in section 8. They show that so far significant corrosion has been limited to those beams which have cracks longitudinal to the reinforcement which have been exposed in the splash zone. Although the extent of corrosion is significant compared with the other situations, even in its worst case, the depth of corrosion is very small in comparison with the 25 mm diameter of the reinforcing bar. Unless the rate of corrosion increases it would take at least ten years before the bar area was reduced a significant amount. However, the only results so far for longitudinal crack beams relates to an exposure period of only one year. After a longer exposure period, the pattern of corrosion may well change, although it must be remembered that any corrosion obtained so far has been in cracked specimens from the splash zone. It is unlikely that cracks in this area on an actual structure will remain untreated.

There was no corrosion of reinforcement in the transverse crack beams after one year's exposure in the splash zone and after two-and-a-half years' exposure there were only two significant examples in which corrosion extended 40 to 50 mm along the bar surface, either side of the crack position. It was not possible to find a common factor to relate this to any of the variables such as concrete grade, cover to reinforcement or crack width. However, it appears that for a crack width no greater than 0.4 mm, it is possible for corrosion to occur at the point where a transverse crack intercepts the reinforcing bar in a beam exposed in the splash zone.

None of the specimens retrieved from the deep exposure zone during Phase 1 of the project showed any signs of corrosion. Therefore it seems reasonable to conclude that the risk of corrosion of reinforcement in cracked concrete is far greater than the splash zone than when it is submerged.

The period of exposure covered by Phase 1 of this project has so far not produced any major differences in the rate of corrosion, relative to the variables being investigated. However, it is possible that after a longer period of exposure more significant information on the effect of the different parameters may be obtained.

Concrete properties

The results of tests on the properties of the concrete have been discussed in Section 8. The compressive strength of the concrete has been measured and shows no significant reduction after exposure at either Loch Linnhe or Portland for the two-and-a-half-year period. The test results show that compared with the original 28-day cube strengths, there has been the normal increase of strength with age.

Some measurements of coefficient permeability were also made and although there was some variation in the values obtained from different samples they were all within the expected range of 10^{-11} to 10^{-13} m/s which is normally obtained from medium to high grade concrete.

From the limited number of tests carried out so far, there has been no evidence of any chemical attack on the concrete and it is reasonable to assume that there is unlikely to be any significant deterioration of the concrete over a long period.

Any test results and discussion within this paper refer only to the Phase 1 exposure period and it is anticipated that when the five-year exposure period is completed on Phase 2, a wider range of more significant information will be available.

10 SUMMARY

This paper has described the work carried out in the 'Exposure Tests' project of Phase 1 of the 'Concrete in the Oceans' programme. In order to determine the factors which cause the onset and development of the corrosion of steel in reinforced concrete, large-scale specimens have been exposed to a sea environment under conditions typically experienced by North Sea structures.

Small, unreinforced concrete prisms have been used to determine the rate of penetration of chlorides. The corrosion investigation has made use of pairs of reinforced concrete beams stressed back to back to induce cracking. During Phase 1 of the project, specimens were exposed for a period of two-and-a-half years. The variables in the corrosion specimens included two grades of concrete and two depths of cover to the reinforcement.

The results of the Phase 1 investigation showed that in general chloride penetration is limited to depth of about 30 mm in the standard grade concrete and about 60 mm in the lower grade concrete without any significant increase of either chloride content or depth of penetration after the initial six-month exposure period.

Significant corrosion of reinforcements has occurred only in those specimens with cracks longitudinal to the reinforcement in the splash zone at Portland. There were isolated occurrences of corrosion in beams with cracks transverse to the reinforcement in the splash zone after an exposure period of two and a half years. There was no evidence of corrosion in any of the specimens exposed in the deep water zone at Loch Linnhe.

The project is being continued in Phase 2 of the 'Concrete in the Oceans' programme, which provides a total exposure period of five years for specimens transferred from Phase 1 and has allowed additional variables to be investigated.

11 ACKNOWLEDGEMENT

The authors are grateful to the Management Committee of the 'Concrete in the Oceans' programme for permission to publish the information in this paper.

REFERENCES

[1] BS 6235:1982, *Code of Practice for Fixed Offshore Structures*, British Standards Institution.

[2] Wilkins, N. J. M. and Lawrence, P. F., Fundamental mechanisms of corrosion of steel reinforcement in concrete immersed in sea water, *Concrete in the Oceans Technical Report No. 6*, C & CA, 1980.

[3] Taylor Woodrow Research Laboratories, Marine durability survey of the Tongue Sands Tower. *Concrete in the Oceans Technical Report No. 5*, C & CA, 1980.

Chapter 4

Corrosion of steel reinforcement in offshore concrete – Experience from the Concrete-in-the-Oceans Programme

M. B. LEEMING, CIRIA/UEG, Westminster, London, UK

1. INTRODUCTION

The Concrete-in-the-Oceans programme of research was started in 1976 and is funded jointly by the offshore industry and the Department of Energy with coordination by CIRIA/UEG. There were eight major projects within Phase I costing £416,000 and all but one of the final reports which were originally confidential to the contributors are now available on open sale (see Table 1). A second phase of the programme began in 1980 and at present comprises 22 projects (see Table 2) covering three main areas: corrosion, fatigue, and the structural aspects of concrete offshore. Phase II, funded by the organisations listed in Table 3, is estimated to cost £700,000 over four years. The results of this Phase are confidential to the contributors until the final reports are published for general sale, one year after their circulation to the contributors.

Table 1 – Concrete-in-the-Oceans technical reports (Phase I).

1. Cracking and Corrosion. A. W. Beeby (Cement & Concrete Association).
2. Modes of Failure of Concrete Platforms. D. J. Dowrick (Ove Arup & Partners).
3. Effects of Temperature Gradients on Walls of Oil Storage Structures. J. L. Clarke and R. M. Symmonds (Cement & Concrete Association).
4. Behaviour of Concrete Caisson and Tower Members. P. E. Regan and Y. D. Hamadi (Polytechnic of Central London).
5. Marine Durability Survey of the Tongue Sands Tower. Taylor Woodrow Research Laboratories.
6. Fundamental Mechanisms of Corrosion of Steel Reinforcements in Concrete Immersed in Seawater. N. J. M. Wilkins and P. F. Lawrence (Harwell Corrosion Service).
7. Fatigue Strength of Reinforced Concrete in Seawater. W. S. Paterson, M. J. Dill and R. Newby (John Laing R & D).
8. Exposure Tests on Reinforced Concrete. K. M. Brook and J. Stillwell (Wimpey Laboratories Ltd).

(Note' This report is still confidential to the contributors).

Table 2 – List of projects in Phase II, Concrete-in-the-Oceans.

Project	Title	Contractors
1	Crack and Corrosion Criteria	
	Fundamental mechanisms	Harwell Corrosion Service,
	Location of corrosion cells	Det norske Veritas,
	Efficiency of cover	Taylor Woodrow Research Laboratories
	Exposure tests	Wimpey Laboratories Ltd,
		Underwater Trials Ltd,
		& Scottish Marine Biological Association.
2	Implosion Strength	Lloyds Register of Shipping
3	Punching shear and Impact Resistance	Wimpey Laboratories Ltd.
4	Temperature Gradients	Cement and Concrete Association.
5	Fatigue and Corrrosion Fatigue	
	Long term & further work	John Laing R & D Ltd,
	Fatigue review	Gifford & Partners,
	Bare bars	British Steel Corporation
	Reverse bending	Glasgow University.
6	Survey of Existing Structures	Taylor Woodrow Research Laboratories
	Development of Inspection Techniques	Ove Arup & Partners.

Table 3 – List of contributors to Phase II.

UK Department of Energy	Sir William Halcrow & Partners
Ove Arup & Partners	Harris & Partners
Britoil Ltd	John Laing Construction Ltd
BP International Ltd	Lloyds Register of Shipping
Building Design Partnership	Sir Robert McAlpine & Sons Ltd
Cement & Concrete Association	Pell Frischmann & Partners
Central Electricity Generating Board	Reinforcement Steel Services
Det norske Veritas	Messrs Sandberg
Esso Exploration & Production UK	Shell UK Exploration & Production Ltd
Freeman Fox & Partners	Taylor Woodrow Construction Ltd
Sir Alexander Gibb & Partners	Total Oil Marine Ltd
E. W. H. Gifford & Partners	Transport & Road Research Laboratory
University of Glasgow	Wimpey Construction (UK) Ltd

A number of projects within the Concrete-in-the-Oceans programme cover different aspects of the problem of corrosion and have been carried out by several research contractors (see Table 2). Technical advice on the conduct of the projects has been provided by the Corrosion Steering Group which is made up of the research contractors, of Contributors' representatives and invited experts. This process allows each project to benefit from the experience gained in the others and provides correlation between experiments in the laboratory and in the sea. In both phases of the programme, 461 corrosion specimens have been made and tested in the five testing sites. In addition the fatigue projects are also generating information relevant to the corrosion process.

This paper relates the current concepts of the corrosion of reinforcement in concrete to the environmental conditions in the North Sea. It then represents the experience from the Concrete-in-the-Oceans programme in achieving similar environments for testing purposes combined with suitable specimen sizes and testing methods. The experiments studying the basic corrosion process employ monitoring and testing methods that are being further developed for the inspection of existing structures so that a more accurate understanding of their condition can be obtained.

2. THE CORROSION PROCESS

There are two distinct periods in the life of a structure from the corrosion view point: t_0, the time from construction to initiation of corrosion and t_1, the time from initiation to the occurrence of unacceptable damage which may be unsightly spalling or structural inadequacy. Exposure tests have shown that, where cracked concrete is exposed to a marine environment, t_0 is small compared with the design life and that the time t_1 is important and is dependent on the rate of corrosion.

The action of casting reinforcement into concrete places the steel in a passive condition and as long as this state pertains corrosion does not occur. However, carbonation and the penetration of chlorides into the concrete depassivate the steel so that corrosion may ensue. The majority of concrete practice today is aimed at maintaining the reinforcement in a passive condition so that anodic sites are not set up and a corrosion cell cannot form.

This strategy has stood the test of time for land-based structures in relatively dry conditions where the main depassivating mechanism is carbonation. For marine concrete, research into the rate of chloride penetration and the level at which depassivation takes place has led to the specification of a sufficient thickness of concrete cover to the reinforcement so as to avoid depassivation within an acceptable life for the structure. However, experience from Phase I of the Concrete-in-the-Oceans programme has shown that chlorides penetrate rapidly down cracks when concrete is immersed in sea water. This is true for quite narrow cracks within the design range and it is evident that design measures to control the width of cracks do not have a significant effect in controlling the initiation of corrosion. While it is prudent to take all possible measures to avoid the initiation of corrosion, there is considerable benefit in studying methods of slowing down the rate of corrosion when initiation cannot be avoided.

Beeby [1] has suggested that corrosion resistance is a function of bar diameter, cover thickness and some measure of concrete strength. This must be particularly true when spalling is the failure criterion. Spalling is the structural failure of the cover zone when a sufficient thickness of rust has accumulated to burst off the concrete. The above factors will certainly play a part but to them should be added a factor which takes account of the geometry of the surface of the concrete in relation to the bar. This factor will allow for spalling at a corner, as a wedge from a single bar or as laminations from closely spaced bars. These 'structural' aspects probably have a strong influence by extending the total time to unacceptable damage (t_1) for a given corrosion rate but do not necessarily directly control that rate. Research in the Concrete-in-the-Oceans programme is now directed towards investigating what factors particularly influence the rate of corrosion recognising that this controls the design life rather than initiation.

Early on in Phase II, a desk study entitled 'The Location of Corrosion Cells' was set up. This project, carried out by Det norsk Veritas, was an appraisal of current design practice and recent research on the resistance of concrete marine structures to corrosion of the reinforcement. It identified possible locations of potential corrosion cells that may affect the safety and serviceability of these structures during their lifetime. It provided recommendations for organising inspection programmes and for further research on other projects. Mr Roland's paper to this conference, 'The relevance of present day design criteria for corrosion resistance of marine reinforced concrete structures', relates to this aspect of the programme.

Theoretical models for the corrosion cell [2, 3, 4, 5] have been proposed and in studying these it soon became clear that the process is very complex; none of these models can be predictive at present. Their benefit lies, however, in indicating the more important relationships of one part of the cell to another. It is clear from these models that the cathodic part of the cell has a strong influence in controlling the rate of corrosion. Oxygen is necessary for the cathodic process and therefore the rate of oxygen diffusion through sound concrete to the steel surface is an important factor. Saturated concrete inhibits this diffusion but also provides low resistivity which is ideal for the electrical circuit. The balance of these two factors to give the highest corrosion rate has yet to be determined so that this situation can be avoided in practice. However, in the splash zone of a marine structure there is no steady state and alternate wetting and drying must lead to higher corrosion rates as oxygen, which penetrates during the dry period, is used in the corrosion process when the concrete is wet.

The corrosion models also show that the relative sizes of cathodic to anodic area can influence the rate of corrosion. As sea water provides a good electrolyte, macro corrosion cells can form in marine concrete leading to severe localised corrosion. On the other hand a wide distribution of cracks can perhaps limit the cathodic area available to each anode. Tests in Concrete-in-the-Oceans Phase I demonstrated that more significant corrosion occurs where cracks intersect voids at the steel surface, where they run along the bar and where the concrete is porous [6] and this higher level of corrosion appears to be independent of crack width. The effect of the anode/cathode ratio is being investigated in Phase II of the programme and also in Norway [7].

It is normally assumed that corrosion of reinforcement embedded in concrete demonstrates itself by spalling of the concrete. This phenomenon is due to the expansive nature of the corrosion products which can produce large forces without much loss of steel area. Experiments carried out simulating a large cathodic area (polarised to −200 mV Ag/AgCl) have resulted in significant localised corrosion leading to a 40% loss of bar area in twelve months which was not accompanied by severe spalling. The expansive red rust products are formed by a secondary reaction of $2Fe_2(OH)_2$ with oxygen from the air. In conditions where immersion limits oxygen availability other rust products can form which are soft and less expansive. These products can diffuse away from the steel surface, along the crack and could be washed away by wave action. Whether these findings are significant with regard to the visual inspection of concrete underwater is a subject that is being investigated in the programme of additional work that is being set up within Concrete-in-the-Oceans.

Little research had been done prior to Phase I of the Concrete-in-the-Oceans programme on the effects on reinforced concrete of deep immersion in sea water. In Phase I concrete beams were exposed at a depth of 140 metres in sea water in Loch Linnhe in Scotland and this work is being continued in Phase II to give a final exposure period of 5 years. The results so far have confirmed that significant corrosion has not occurred at this depth even though the reinforcement is rapidly depassivated by chloride ingress. This is thought to be due to a lack of oxygen combined with the blocking of the finer cracks close to the steel surface. As the North Sea is saturated with oxygen at all depths, a shallower immersion is expected to show similar results.

Underwater concrete has generally been found to be less permeable than concrete above water, probably as a result of the blocking of pores in the concrete by materials like brucite and aragonite formed from chemical reactions with sea water [8, 9]. This low permeability concrete provides better protection for the reinforcement and reduces the access of oxygen to the corrosion cell. The action of sea water on concrete leading to the blocking of cracks was clearly illustrated in the project investigating the effect of fatigue loading on reinforced concrete immersed in sea water. Under the fluctuating loads, the sea water is forced in and out of the cracks leaving behind hard deposits. The crack blocking results in a smaller deflection range, reducing also the stress range which has a beneficial effect on fatigue life. The same mechanism may well occur in static cracks but to a lesser extent due to reduced movement of water in and out of the cracks.

3. THE ENVIRONMENT

The splash zone is accepted to be the area most vulnerable to corrosion. There are two extremes. The lower tidal or wave zone is subject to frequent wetting and drying either on a 12-hourly cycle or more frequently where the area is subject to the minimum wave height. Under these conditions the concrete does not have time to dry properly and remains saturated. At the upper end of the splash zone the concrete is subject to a much more prolonged cycle of wetting and drying and in periods of calm weather the concrete can dry out allowing diffusion of oxygen. Surface levels of chloride have been found to be highest in this area although the depth of penetration is little different from immersed concrete.

The important environmental factors relating to the North Sea are as Table 4. There can be other factors which can cause local changes in the environment. Hot oil in risers may affect the sea water temperature locally. Marine growth on the surface of the concrete may have some effect on the diffusion of oxygen and may keep the concrete in a moister condition. Inside oil storage cells anaerobic conditions are known to exist [10] in the water phase leading possibly to sulphate attack of the concrete. Some work has been done on this subject [11, 12, 13] in connection with sewage disposal works.

Table 4

		Surface	At depth (40 M)
Temperature	Summer	4–7°C	6–8°C
	Winter	12–16°C	6–8°C
Salinity		34.4 to 35.4 parts per thousand	
Dissolved oxygen		Saturated at all depths 8.5–10 p.p.m. depending on temperature	
pH		7.9–8.1 (little variation with depth)	

4. THE TEST SITES

The Concrete-in-the-Oceans programme covers both the fundamental mechanisms of corrosion and real sea exposure of concrete specimens. Loch Linnhe on the west coast of Scotland was chosen as the deep sea exposure site as it provided the greatest depth of sea water in a reasonably sheltered location that could be found around the shores of Britain. At a depth of 140 metres the following parameters have been measured:

Temperature	7°C (winter)–13°C (summer)
Salinity	32.3 parts per thousand
Dissolved oxygen	9.3 p.p.m.
pH	8.15

These figures show that at this facility the sea water is slightly less saline than the North Sea but that it is saturated with oxygen.

The exposure zones are located at Portland Harbour in Dorset. This is a man-made harbour with no major freshwater inlets and provides a secure site as it is located on a government establishment. The splash zone specimens were placed on the harbour breakwater. However, the natural environment of this reasonably exposed location could not be relied upon to provide the sort of spray conditions experienced in the North Sea so a spray system was installed to pump fresh sea water over the specimens for two periods of 45 minutes per day. The much more southerly coastal location provides a slightly warmer climate than the North Sea but, as corrosion progresses faster at higher temperatures, the exposure is in this respect a little more severe.

The Harwell Corrosion Service which has been carrying out the fundamental studies into the corrosion processes in both phases of the programme have a testing facility situated on the same breakwater at Portland. Smaller specimens have been placed in a tank where fresh seawater is pumped through, going to waste don the other side of the breakwater. Two other projects involving the immersion of concrete in sea water have obtained supplies of sea water from the same site where it has been transported by tankers inland for use in the testing. The water is stored in tanks and pumped round the specimens. The pH, dissolved oxygen, and the temperature are regularly monitored and, when they fall below acceptable values, fresh supplies are obtained.

The first of the two inland testing sites is located at Elstree where Laing R & D are testing reinforced concrete beams under cyclic loading. The beams are placed in the fatigue rig in troughs through which the sea water is circulated. The second facility is located at Taylor Woodrow's Laboratory at Southall, Middlesex. The project there is investigating the factors within the cover zone controlling the corrosion rate and the testing is carried out in a tank which provides shallow immersed, tidal, splash and marine atmospheric zones. This is achieved by arranging the specimens at three levels on racks. The immersed specimens are placed in the second balancing tank which holds the sea water for most of the time. For three hours in every twelve the main tank is flooded to cover the lowest level of specimens to simulate the tidal zone. For five minutes at the end of this cycle the water level is raised above the next level of specimens to simulate the splash or wave zone. The water is then pumped back into the balancing tank to begin the next cycle. The upper level of specimens is never covered but air is bubbled through the water remaining in the bottom of the tank so that they remain in a slat-laden atmosphere with a relative humidity of approximately 96%.

5. THE TEST SPECIMENS

The policy of the programme has been to use, wherever possible, concrete mixes, bar sizes, cover thicknesses and concrete strengths which are typical of offshore construction practice. A granite aggregate concrete mix was adopted with a cement content of 435 kg/cubic metre and strengths in the region of 60 n/mm^2. A lower grade mix was also used in Phase I in the exposure project but, as the quality of the concrete did not appear to have a significant effect on the amount of corrosion in the short term, it was not used in Phase II so that more important parameters, such as a 20% PFA mix, could be investigated. For the reinforcement, 25 mm diameter Torbar has been used in most cases.

Cover thickness is a variable in a number of projects. Typically 25 mm and 75 mm covers are being examined to see what influence these have on the degree of corrosion. Crack width had been investigated in several projects. Phase I included widths up to 0.6 mm. In this range no correlation between crack width and time to initiation of corrosion was established. Beeby in his report [14] stated that 'cracks of the width covered in exposure tests (up to 1.5 mm) will be likely to induce corrosion where bars intersected them, but the amount of corrosion occurring at the cracks over the design life of the structure will not be significantly influenced by the width of the cracks, only their presence'. In Phase II of the programme larger crack widths, typical of construction joints and structural defects, are being investigated. The details of the numbers of specimens and the parameters being investigated in the programme are summarised in Table 5.

The real sea exposure of large beam specimens has been carried out by Wimpey Laboratories and is described in their paper to this conference, 'Exposure tests on concrete for offshore structures', K. M. Brook and J. A. Stillwell. The objective of the project is to compare the durability of reinforced concretes with different crack widths, crack orientations, covers and types of concrete. Pairs of beams are placed back to back and stressed to hold open the cracks, see Fig. 1. A number of beams from Phase I have been retained in Phase II to achieve an exposure period of 5 years.

In the same project plain concrete prisms, 100 × 100 × 200 mm, have been used in the chloride penetration studies. These had all faces except the two ends sealed by a pitch/polyurethane material and were exposed in all five exposure zones, atmospheric, splash, tidal, shallow immersion and deep immersion as well as on an offshore platform.

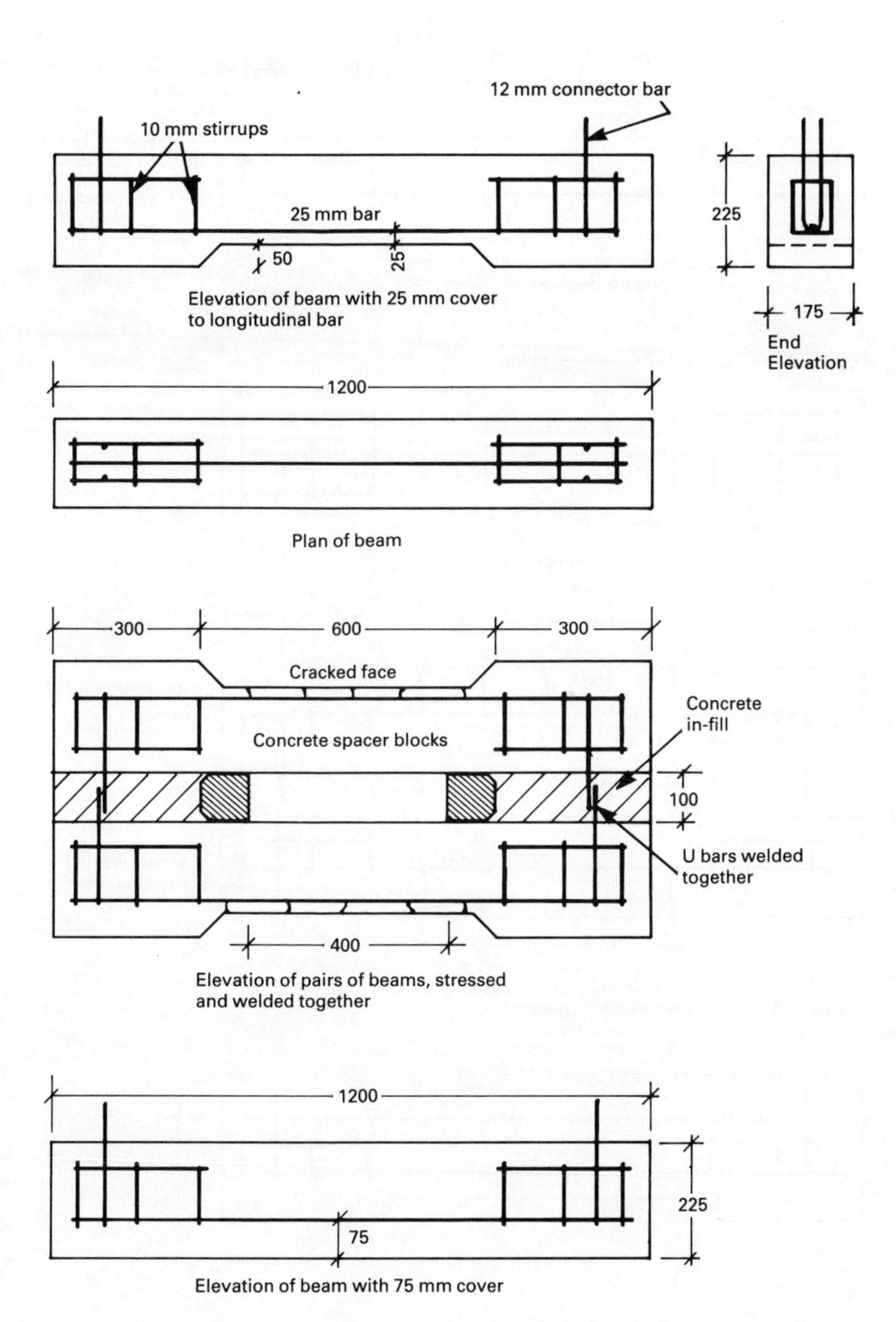

Fig. 1 – Exposure beams (continued next page)

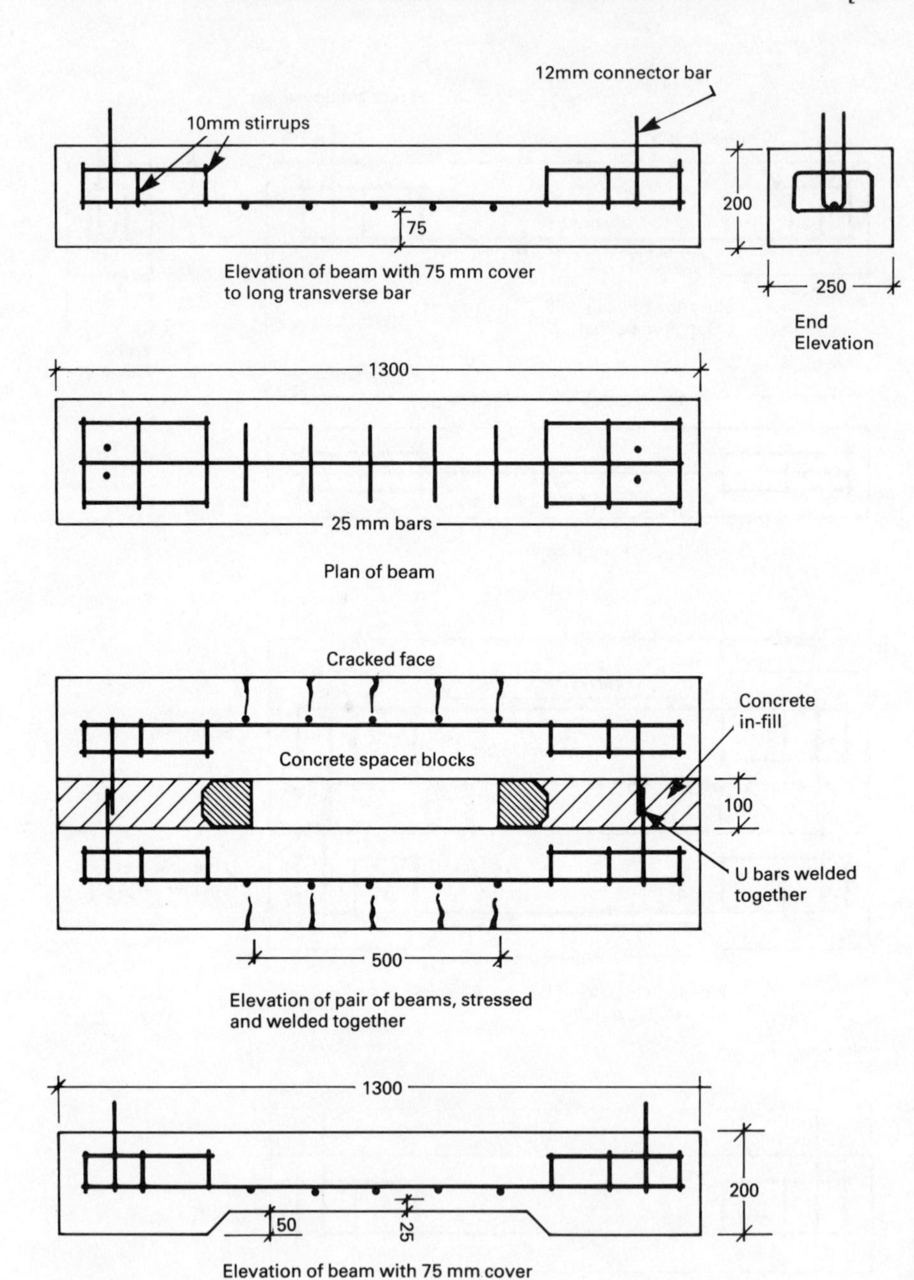

Fig. 1 – Exposure beams (continued from previous page)

The fundamental mechanisms of corrosion are being studied by the Harwell Corrosion Service which is the subject of a paper to this conference by N. J. M. Wilkins and P. F. Lawrence, 'The corrosion of steel reinforcements in concrete immersed in sea water'. The project set out to investigate electrochemically the mechanisms by which steel reinforcement in concrete immersed in sea water is protected from corrosion, the reasons why corrosion of the steel may eventually occur and the factors determining the distribution and severity of the corrosion. The cylindrical specimens used in Phase I, illustrated in Fig. 2, were cracked by tensile loading or were slotted by a saw cut to expose the steel to the sea water.

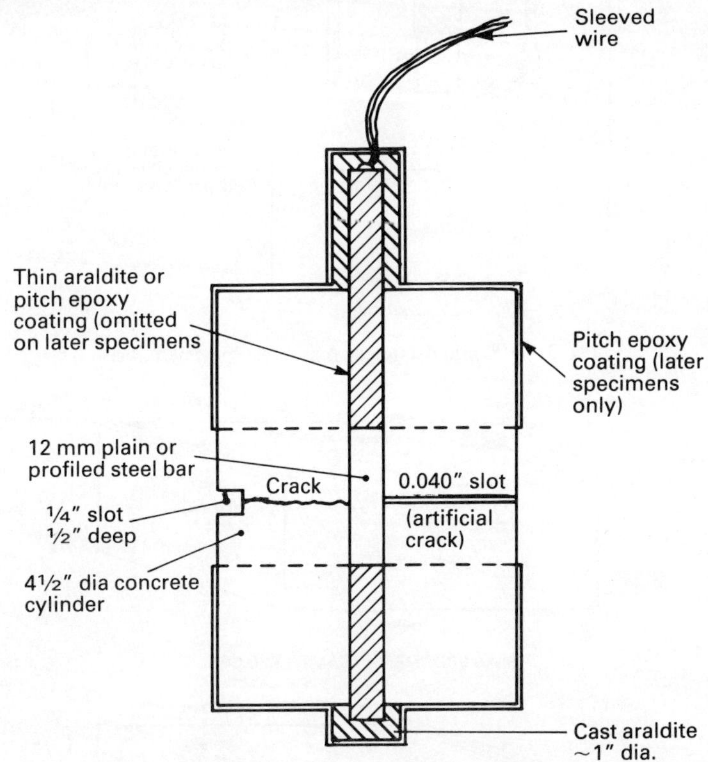

Fig. 2 – 100 mm diameter cyclindrical specimen (Phase I).

Cylindrical specimens are also being used in Phase II of the progamme in order to determine the rate of cathodic reactions on steel in concrete. These are illustrated in Fig. 3. The effect of voids is also being investigated. In the same project the effect of the width of cracks on the passivity of embedded steel is being investigated by the use of slab specimens illustrated in Fig. 4. The slabs are held in steel frames which were used to crack the concrete and subsequently to maintain the crack open.

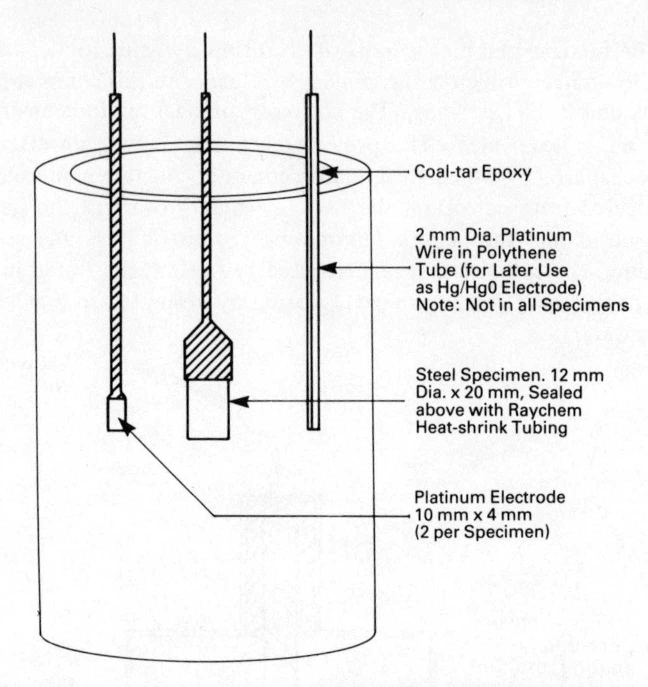

Fig. 3 – 100 mm diameter cyclindrical specimen (Phase II).

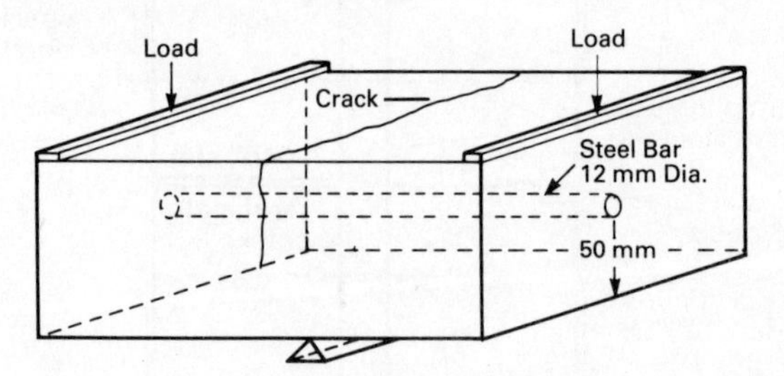

SLAB SPECIMEN – TRANSVERSE CRACK

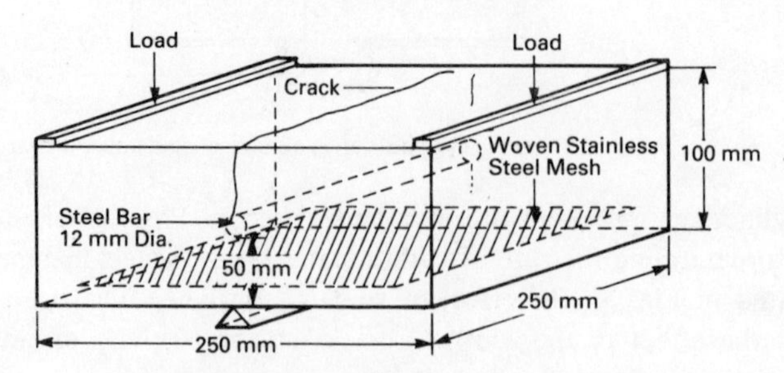

SLAB SPECIMEN – LONGITUDINAL CRACK

Fig. 4 – Exposure slabs.

Identical slabs are also being used as part of an experiment to investigate the effect of the ratio of relative sizes of the anode to the cathode in the exposure project. Here the reinforcement in the small cracked slabs are electrically connected to the reinforcement in larger beams that are uncracked. The coupled specimens are located in Portland Harbour with approximately half the specimens maintained on staging in the tidal range and the remainder immersed in the sea below the platform. The connections are made via a junction box on the jetty above so that pairs of specimens are placed in tidal/tidal, immersed/immersed, tidal/immersed and immersed/tidal exposures. Corrosion currents over a tidal cycle are also being monitored to study the effects of alternate wetting and drying.

The 'Efficiency of Cover' project being carried out by Taylor Woodrow Research Laboratories also uses slab specimens but of a different design as illustrated in Fig. 5. The larger size has been used so that cores can be taken at various times to investigate such parameters as permeability and oxygen diffusion.

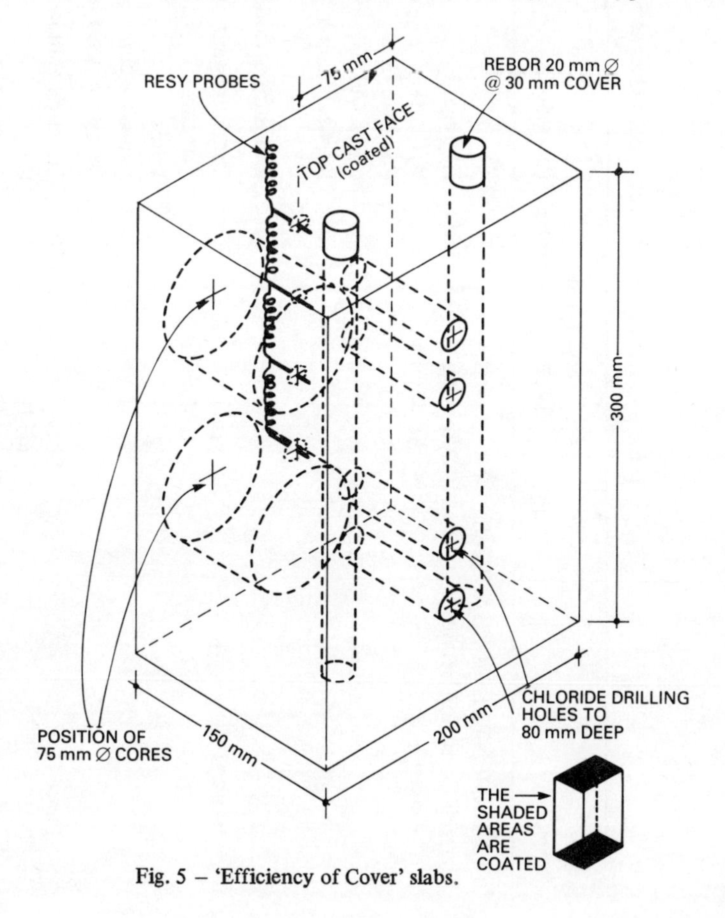

Fig. 5 – 'Efficiency of Cover' slabs.

Table 5 – Details of Concrete-in-the-Oceans Specimens

Project	Specimens Type	Specimens Number	Parameters	
FUNDAMENTAL MECHANISMS				
Harwell Corrosion Service	Cylinders	88	Aggregate	Granite and gravel
Phase I	100 mm dia.		W/C ratio	0.4 to 0.7
	Fig. 2		Covers	25 mm and 50 mm
			Crack widths	0.3, 0.6, 1.0 mm and 1 mm sawn slot
			Exposure zones	Immersed at 1 atmos and 500 psi and tidal
			Exposure period	3 months, 6 months and 3.5 years
Phase II	Cylinders	25	Concrete grade	65N/mm²
	100 mm dia		Cements	OPC and 20% PFA replacement
	Fig. 3.		Voids	
			Exposure zone	Shallow immersion
			Exposure periods	2.5 years
Phase II	Slabs	30	Concrete grade	65 N/mm² OPC
	Fig. 4		Cover	40 mm
			Crack widths	0.3, 0.6, 1.0 and 2.0 mm
			Crack orientation	Longitudinal and tranverse
			Exposure zone	Shallow immersion
			Exposure period	17 months, 2 years and 2.5 years

Wimpey Laboratories Ltd	Beams	86	Concrete grade	60 N/mm² and 35 N/mm² (phase I)
Phase I and II	Fig. 1		Cements	OPC and 20% PFA replacement
			Covers	25 mm and 75 mm
			Crack widths	Uncracked, 0.3 mm, 1.0 mm and voids
			Crack orientations	Longitudinal and transverse
			Exposure zone	Splash and deep (140 M) immersion
			Exposure period	12 months, 2.5 years and 5 years
Coupled specimens	Slabs and	32	Concrete grade	65 N/mm² OPC
Phase II	beams		Covers: – slabs	40 mm
	Figs. 4 and 1		beams	25 mm and 75 mm
			Crack widths (slabs)	0.3 mm and 1.0 mm
			Crack orientations	Transverse
			Exposure zone	Tidal and shallow immersion
			Exposure period	2 years
Chloride penetration	Prisms	74	Concrete grade	65 N/mm² and 35 N/mm²
Phase I and II	100 × 100		Cements	OPC and PFA
	× 200 MM		Cracks	Cracked and uncracked
			Exposure zones	Atmospheric, splash, tidal, shallow and deep immersion
			Exposure periods	6 months, 1 year, 2.5 years and 5 years
EFFICIENCY OF COVER				
Taylor Woodrow	Slabs	108	W/C ratio	0.4, 0.5 and 0.6
Phase II	Fig. 5.		Curing	Air, fog and seawater
			Cements	7% C_3A, 14% C_3A and 20% PFA replacement
			Exposure zones	Atmosphere, splash, tidal and shallow immersion
			Exposure period	2 years

6. THE PARAMETERS BEING MONITORED

In addition to strength tests, the taking of cores and assessment of the extent of corrosion, the following parameters are measured in the various projects. The electrical potential of the reinforcement is taken as a routine in all corrosion projects in order to determine the condition of the reinforcement. The potentials of the deep water specimens in Loch Linnhe are being measured automatically with respect to a zinc half cell by two independent recorders, one placed at depth with the specimens and the other being located on the supporting pontoon. These recorders have been supplied by the Scottish Marine Biological Association.

Other recording methods and different half cells have been used to suit individual investigators so the conversion factors, given in Table 6, have been used to bring the results to a common standard with respect to a Ag/AgCl half cell. The programme as a result has amassed a vast amount of data on the potentials of steel embedded in concrete which is being evaluated so that a better understanding of this parameter can be obtained.

Table 6 – Half-cell conversion factors (after L. L. Shreir).

Half-cell	Std half-cell EMF @ 25°C	Conversion to	Millivolts
Zinc in art. sea water	−0.81 Volts	Zinc in sea water	−10
		SCE	−1050
		Silver/silver chloride	−1060
		Copper/copper sulphate	−1120
Zinc in sea water	−0.80 Volts	Zinc in art. sea water	+10
		SCE	−1040
		Silver/silver chloride	−1050
		Copper/copper sulphate	−1130
SCE	+0.241 Volts	Zinc in art. sea water	+1050
		Zinc in sea water	+1040
		Silver/silver chloride	−10
		Copper/copper sulphate	−80
Silver/silver chloride in sea water	+0.25 Volts	Zinc in art. sea water	+1060
		Zinc in sea water	+1050
		SCE	+10
		Copper/copper sulphate	−70
Copper/copper sulphate	+0.318 Volts	Zinc in art. sea water	+1130
		Zinc in sea water	+1120
		SCE	+80
		Silver/silver chloride	+70

The resistivity of concrete has been measured in various projects, such as surveys of existing structures and laboratory experiments. However, there are several methods of measurement of this parameter ranging from *in situ* methods [15] to those carried out on newly cast specimens [16, 17]. There is no standard method for carrying out these measurements but the four-probe method developed by Taylor Woodrow in the Tongue Sands Survey [18] has generally been used in *in situ* tests.

Chloride penetration is measured by drilling the concrete and analysing the dust from various levels by titration. Within the programme it is possible to compare the results of chloride penetration in various exposures including both laboratory and real sea environments, for different ages of concrete and at various depths of exposure.

The permeability of concrete has been measured in a number of projects so that the change with time and exposure can be noted by comparing results from newly cast concrete at various ages with those from surveys of old structures with various marine exposures. No standard BS or ASTM test exists for measuring this parameter, so use is made of a triaxial cell apparatus under constant head conditions in one project.

Oxygen diffusion measurements have been taken on a few slices from 100 mm diameter concrete cores. This technique is in the development stage so that there are too few results to draw any firm conclusions.

7. INSPECTION

A survey of the Tongue Sands Fort was carried out in Phase I [18] and in Phase II three further surveys were carried out: of the Mulberry Harbour Units in Portland Harbour, the Royal Sovereign Lighthouse and Shoreham Harbour in Sussex. The objective of these surveys was to relate the durability of reinforced concrete structures in a marine environment to the effects of age and construction practice in different exposures. These projects were successful in many ways but the value of the information gained was limited by the survey methods available. It became clear that the true relationship of the corrosion of the reinforcement to the parameters controlling the process could not be ascertained without some destruction of the structure to find out the extent of corrosion hidden by a concrete cover. Destruction was not desirable for the structures surveyed nor would it be for most structures in service, so it was decided to devote the remaining funds allocated to the 'Survey of existing structures' project to developing non-destructive inspection procedures.

The various inspection methods that were evaluated in the programme are shown in Table 7 and are grouped by type and arranged in increasing degree of destruction. These methods are well reviewed in reference [19]. Ultrasonic methods are fairly well developed but whilst they give a good indication of the quality of the concrete cover they can at the best only indicate the likely causes of corrosion and can give no indication of its extent or severity. A. C Impedance [20] and noise measurements [21] are promising methods being developed by others so it was decided to concentrate on half cell potential and resistivity methods, of which there was already considerable experience in the programme. Both these methods are based on the electrical nature of the corrosion cell and seemed in combination to offer the best opportunity for development. It seems unlikely that one single tool will ever provide a complete picture and that several methods will need to be used together. The measurement of resistivity has shown itself to be useful in indicating the quality of the concrete cover and in this context is perhaps as good if not better than ultrasonic methods.

Table 7 – Inspection techniques

Degree of Destruction (↓)	Method	Technique
	Visual	–Photographic
		–Video
		–Defect classification
		–Training
	Impact	–Chains
		–Schmidt hammer
		–Windsor probe
Degree	Ultrasonic	–Pundit
		–Gamma radiography
		–X-ray photography
of	Electrical	–Covermeter
		–Half cell potentials
Destruction		–Resistivity
		–A.C. impedance
	Drillings	–Chemical analysis
		–Chloride penetration
	Cores	–Strength
		–Permeability
		–Carbonation

There has been considerable experience of taking half cell potential measurements within the Concrete-in-the-Oceans programme and elsewhere, but the interpretation of the resulting corrosion potential is less easy. The present Van Daveer criteria [22] only provide wide bands of potential related to the probability that the steel is corroding based on experience of bridge decks in America. The absolute value of individual readings are of less significance; the pattern of potential differences over a structure giving a much better indication of its condition. With modern data gathering methods, a large number of readings on a close grid can more easily be taken to allow the construction of electropotential contour maps which give a much clearer indication of likely corrosion sites.

Visual survey methods are most commonly used in inspection because they are the obvious first step in any detailed survey but also for want of any other reliable methods. While they are easy and quick to carry out, it needs considerable experience for a proper evaluation of the results. A detailed classification of defects is presently lacking and it is hoped that some progress towards such a document can be made in the additional work programme. Its objective will be to facilitate clear communication between inspector, evaluating engineer and certifying authority.

8. SUMMARY

The Concrete-in-the-Oceans programme of research provides an opportunity to look at the corrosion problem from different aspects and the active participation of industry is ensuring that the results are related to practice offshore.

Current concrete practice is aimed at maintaining the passivity of the reinforcement; yet research being carried out within the programme seems to indicate that the reinforcement rapidly depassivates in a marine environment and the corrosion rate in practice is controlled by the cathodic reaction. Oxygen diffusion and the ratio of anodic to cathodic areas are the important parameters.

The programme has as far as possible reproduced the offshore environment and the materials used in offshore construction. The experience gained during the experiments in testing and monitoring is now being used for developing inspection techniques so that the condition of existing structures can be more accurately determined.

REFERENCES

[1] Beeby, A. W. Cover to reinforcement and corrosion protection. *Int. Symp. Behaviour of Offshore Concrete Structures, Brest.* Paper I. 3. Oct. 1080.

[2] Bazant, Z. P. Physical model for steel corrosion in concrete sea structures – Theory. *Proc. ASCE Struct. Div.*, **105** (ST6). Paper 14651. June 1979.

[3] Bazant, Z. P. Physical model for steel corrosion in concrete sea structures – Application. *Proc. ASCE Struct. Div.*, **105** (ST6). Paper 14652. June 1979.

[4] Wu, S. T. and Clifton, J. R. Analysis and modelling of corrosion of steel in prestressed concrete. *Eng. Lab. Nat. Bureau of Standards. Washington, DC.* Nov. 1980. NBSIR-81-2390.

[5] Gronvold, F. O. A model of the initiation and growth of localised corrosion of steel reinforcement in concrete. *2nd Int. Seminar on Electrochemistry and Corrosion of Steel in Concrete, Copenhagen,* May 1982.

[6] Wilkins, N. J. M. and Lawrence, P. F. Fundamental mechanisms of corrosion of steel reinforcements in concrete immersed in seawater. *Concrete in the Oceans Technical Report No. 6,* C & CA, 1980.

[7] Vennesland, O. and Gjorv, O. Effect of cracks in submerged concrete structures on steel corrosion. *Corrosion 81, Toronto,* April 1981. Paper 50.

[8] Conjeaud, M. L. Mechansisms of seawater attack on cement mortar. *Conf. on Performance of Concrete in Marine Environment, New Brunswick;* A.C.I. Pub. SP65-3. Aug. 1980.

[9] Regourd, M. Physico-chemical studies of cement paste, mortar, and concretes exposed to seawater. *Conf. on Performance of Concrete in Marine Environment, New Brunswick.* A.C.I. Pub. SP65-4. Aug. 1980.

[10] Charlton, R. M. North Sea Platform Operations. *Proc. Instn. Civ. Engrs, Part 1,* 1981, **70,** pp. 616–619.

[11] Forrester, J. A. An unusual example of concrete corrosion, induced by sulphur bacteria in a sewer. *C & CA Technical Report TRA/320.* July 1959.

[12] Thistlethwaite, D. K. B. (Editor). The Control of Sulphides in Sewerage Systems. *Butterworth.* 1972.

[13] Harrison, W. H., Teychenne, D. C., Forrester, J. A. and Spratt, B. H. Long-term investigation of sulphate resistance of concrete at Northwick Park – Field testing and initial testing of concrete. *CIRIA Technical Note No. 57.* July 1974.

[14] Beeby, A. W. Cracking and corrosion. *Concrete in the Oceans Technical Report No. 1.* C & CA. 1978.

[15] Wilkins, N. J. M. Resistivity of Concrete. *Materials Development Division. AERE Harwell.* Jan. 1982. AERE M-3232.

[16] Gjorv, O., Vennesland, O. and El-Busaidy, A. Electrical resistivity of concrete in the oceans. *O.T.C.* May 1977. OTC 2803. p. 581.

[17] McCarter, W. J., Forde, M. C. and Whittington, H. W. Resistivity characteristics of concrete. *Proc. Instn. Civ. Engrs. Part 2,* Mar. 1981. Paper 8397.

[18] Taylor Woodrow Research Laboratories. Marine durability survey of the Tongue Sands Tower. *Concrete in the Oceans Technical Report No. 5.* C & CA. 1980.

[19] Vassie, P. R. A survey of site tests for the assessment of corrosion in reinforced concrete. *TRRL Laboratory Report 953.*

[20] John, D. G., Searson, P. C. and Dawson, J. L. Use of a.c. impedance technique in studies on steel in concrete in immersed conditions. *Br. Corros. Journal,* **16,** (2), 1981.

[21] Hladky, K. and Dawson, J. L. The measurement of corrosion using electrochemical L/F noise. *Corrosion Science,* **22** (3), pp. 231–237, 1982.

[22] Van Deever, J. R. Techniques for evaluating reinforced concrete bridge decks. *ACI Journal.* Dec. 1975. Title No. 72-47C.

Chapter 5

Failure analysis of the collapse of the Berlin Congress Hall

B. ISECKE, Bundesanstalt für Materialprüfung, 1000 Berlin 45, Federal Republic of Germany

INTRODUCTION: DESCRIPTION OF THE BUILDING

In recent years various failures of prestressed concrete structures due to HISCC have arisen [1]. The most serious failure of this type occurred in May 1980 when the southern outer roof of the Berlin Congress Hall collapsed.

The Berlin Congress Hall was built in 1957. Fig. 1 shows the building shortly after completion. The roof of the auditorium consists of inner and outer sections, separated by an oval ring beam, 40 cm thick and 2–8 m wide, resting on the outer wall (Fig. 2). The rims of the north and south roofs projecting out over the auditorium consist of hollow reinforced concrete arches tensioned at two anchorpoints on the east and west side. The outer roof sections consist of 24 separate panels approximately 2,125 m wide and 7 cm thick. Each panel has 2 to 4 tension elements, altogether 82 tension elements in each half of the roof. Each tension element contains between 7 and 10 prestressed rods made of quenched and tempered high tensile steel St 145/160 in an oval duct. These elements serve to anchor the outer reinforced arch to the ring beam.

Fig. 3 shows a section through the outer roof, the ring beam, panels and arch were cast separately. A suitable joint at the ring beam was supposed to allow movement of the panels against the ring beam. It consisted of a step on which the panels rest. Fig. 4 shows an enlargement of Fig. 3 in the region of this joint. Bituminous felt was placed on the floor of the step between the panels and the ring beam. This was possibly intended just to separate the panels and the ring beam or perhaps to ease movement of the panels when tightening the tension elements [2]. The method of assembly was as follows: before removing the wooden formers of the arch the tension elements were partly stressed, thus lifting the arch out of its shell. Next, the joints between the panel and the ring beam and also between the panel and the arch were filled with mortar. When this had dried out the tension elements were fully tightened to 55% of maximum stress. Finally the ducts were injected full of grout.

Fig. 1 – The Berlin Congress Hall.

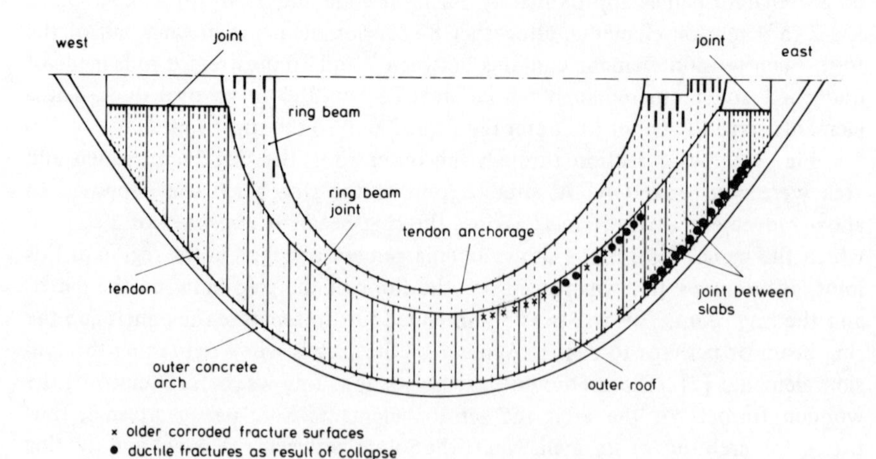

Fig. 2 – Southern outer roof of the Berlin Congress Hall (top view).

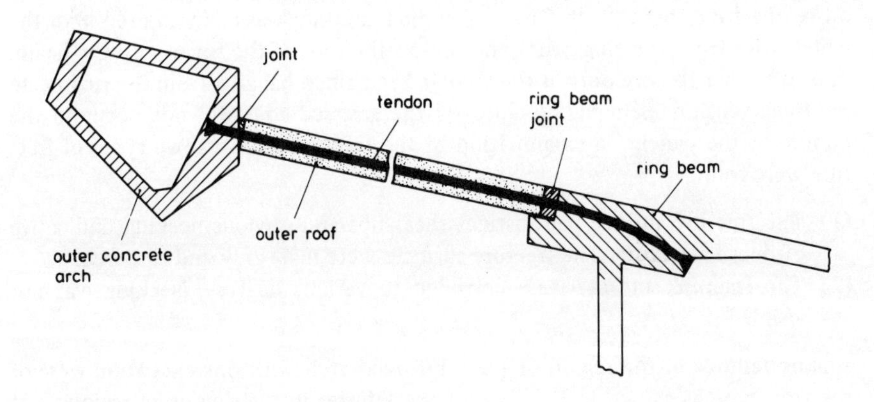

Fig. 3 – Cross-sectional view of the southern outer roof of the Berlin Congress Hall.

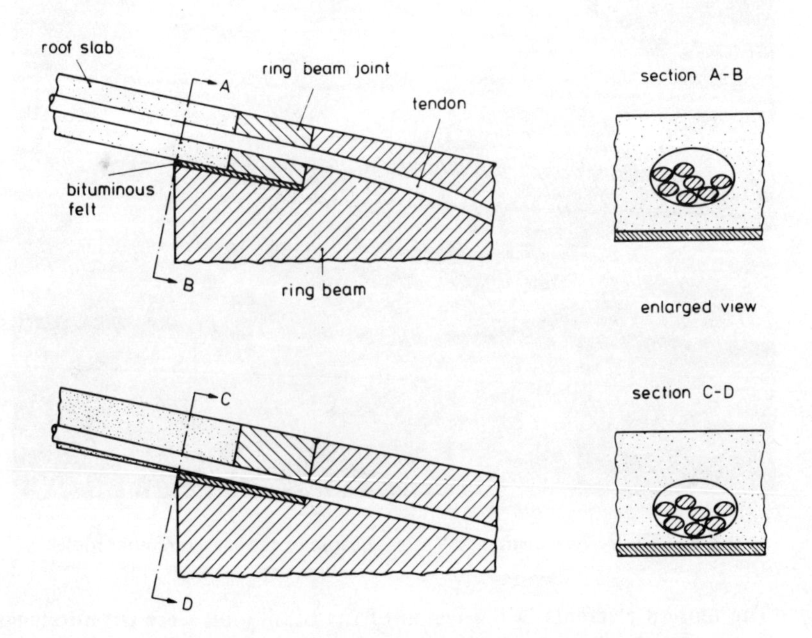

Fig. 4 – Construction in the area of the ring beam joint.

OBSERVATIONS AFTER FAILURE

On 21 May 1980 the southern outer roof collapsed (Fig. 5). The outer roof fell on to the foyer beneath it. One person died, another was badly injured. 8 of the panels tore from the ring beam and lay on the roof of the foyer with the beam. The other panels tore only at the arch and remained hanging from the ring beam by their tension elements. Failure of the stressed rods did not occur in the middle of the panel. On examination of the fracture surfaces two types of fracture were found:

(1) The fracture surface had distinct shear lips and obvious necking, indicative of ductile fracture. The fracture surfaces were matt-grey and rust-free.
(2) The fracture surfaces were corroded to various degrees. Necking was not visible.

All the failures in the region of the reinforced arch with one exception were of the first type as shown in Fig. 2. 8 of the failures in the ring beam region (7 of them adjacent) were corroded fractures of the second type, while in one tension element in this area both corroded and ductile fractures were found. All other torn rods in the ring beam joint were of the first type.

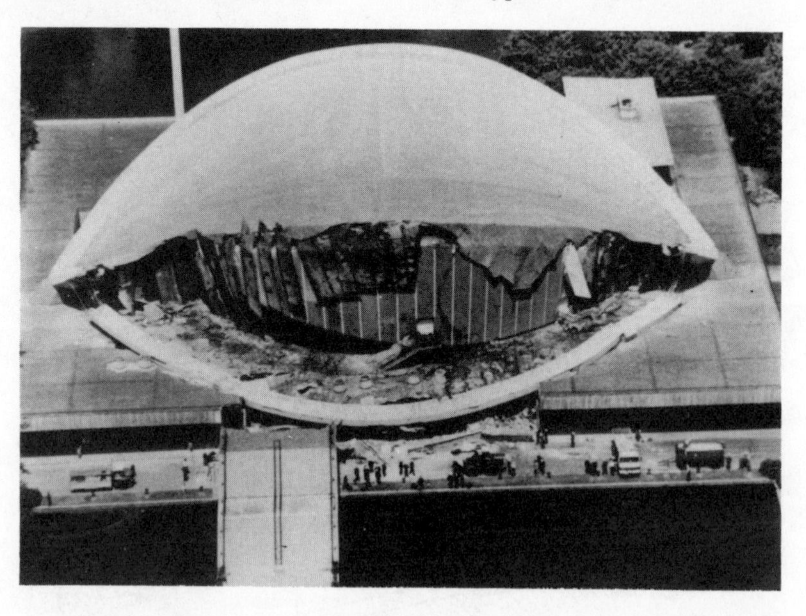

Fig. 5 – The Berlin Congress Hall after collapse of the southern outer roof.

The tension elements in the region of the beam joint were cut into lengths 0.5 m from the fracture for further investigation. Fig. 6 shows one of the ducts with prestressed rods from this area. It is obvious that the duct has suffered extreme corrosion damage preferentially on the underside. The prestressed rods in the bottom section of the duct have consequently rusted the most. Remnants of the bituminous felt laid in the step of the ring beam were found adhering to the undersides of several ducts.

Fig. 6 – Prestressed steel tendons and duct after fracture.

This indicates that these ducts were resting on the bituminous felt and hence that their undersides were not covered by cement when the joint was filled. These conditions are illustrated in the lower part of Fig. 3.

The injected grout filling of the ducts in the joint region was also studied. The degree of filling varied. Fig. 7 shows a section through a tension element whose prestressed rods underwent ductile fracture. The duct is in this case uniformly filled with grout. By contrast it can be seen in Fig. 8 that the the rods lie to the bottom, a result of the rods being pulled down during tensioning. Most of the corroded fractures occurred in elements in this condition. The undersides of the ducts were always rusted to a greater degree than the upper sides, as can be seen clearly in Fig. 9 – the duct shown from the joint region is almost fully rusted away underneath, while the top, although attacked is not yet rusted through correspondingly, the fracture surfaces of the rods lying to the bottom of the duct were the most corroded and probably fractured before the upper rods.

Since all the corrosion fractures occurred in the eastern end of the ring beam joint, various cement samples were taken there for analysis. In contrast to the western side it appeared that 5 different filling materials had been used in this region. The material in contact with the ducts was sand-rich and crumbly with high porosity – ideal conditions for steel corrosion.

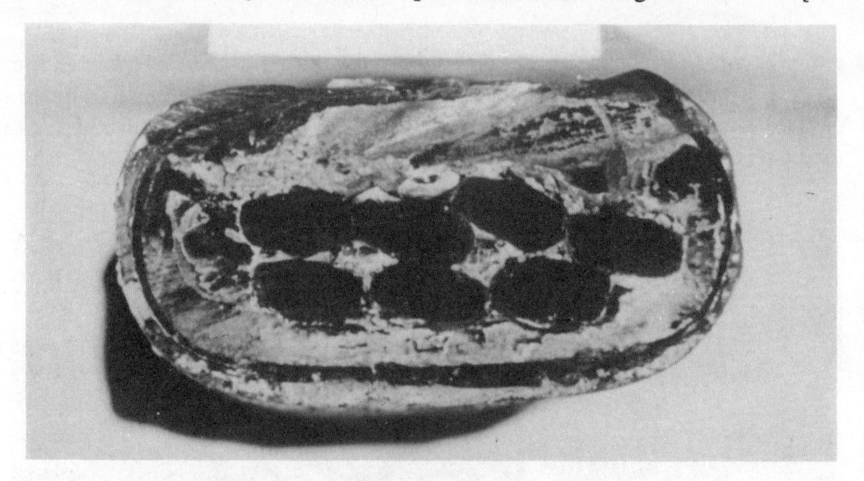

Fig. 7 – State of grout injection in a duct.

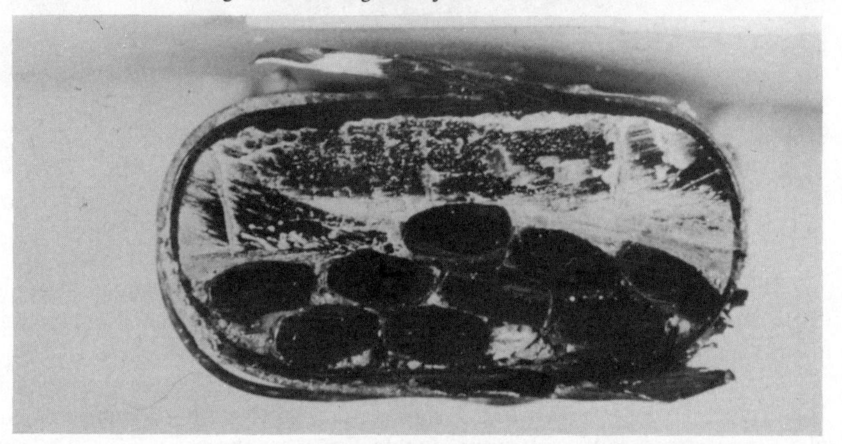

Fig. 8 – State of grout injection with steel tendons on the bottom of the duct.

Fig. 9 – Corroded duct (view from below).

The actual load on the tension elements did not correspond to those specified for the construction of the roof. The loads due to the dead weight, snow and wind as well as movement of the beam resulted in concentrated bending stresses at the ring and the arch, leading to changes of curvature in the tension elements at the anchor points. The overloads occurring in the system were taken up by flexing at these points, thereby breaking up the concrete around them and theoretically loading the rods beyond the yield point.

FRACTURE INVESTIGATION

As was previously described the fractures fell into two groups. Fresh matt-grey fractures from the connection of the panels to the arch had prominent necking and shear lips when viewed through a stereo microscope (Fig. 10). These fractures only occurred as the roof collapsed and are irrelevant to the investigation of the failure.

Fig. 10 – Ductile fracture of a quenched and tempered prestressed steel.

Fig. 11 shows one of the badly rusted fractures from the joint region. In contrast to the type 1 fractures there is no visible necking in these fractures. The samples were de-rusted and afterwards viewed under a scanning electron microscope. Some fractures were so badly corroded that even after de-rusting evaluation was impossible. Other, less corroded fractures showed that failure had begun by intergranular fracture (Fig. 12). In contrast to the dimple morph-

Fig. 11 – Corroded fracture surface.

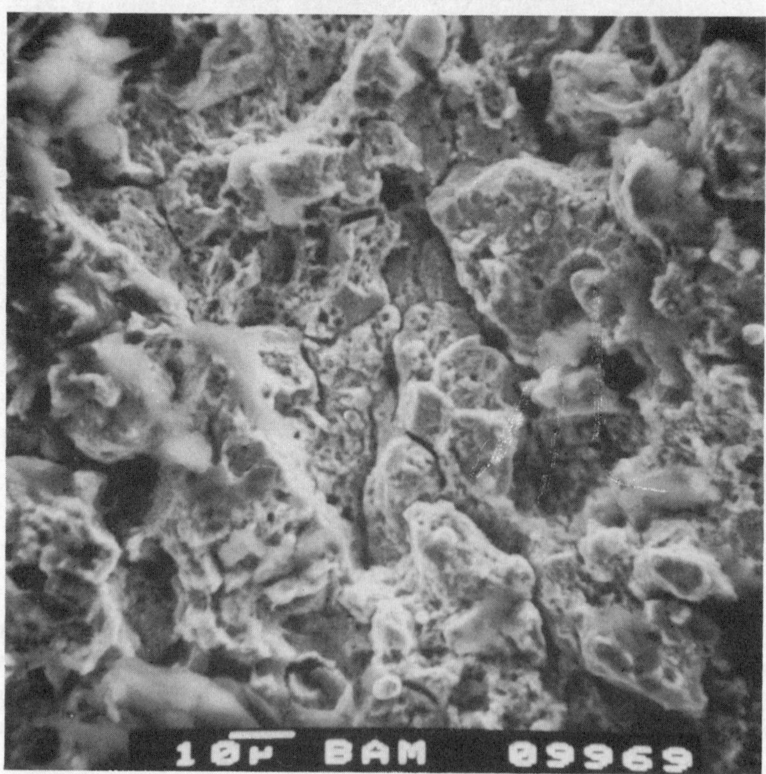

Fig. 12 – Dissolved areas along the former autenite grain boundaries.

ology of a ductile fracture, the effect of heavy grain boundary corrosion and dissolution of former austenite grain boundary regions is visible. This fracture appearance is typical of hydrogen induced stress corrosion cracking in this kind of steel, comparable to hydrogen-induced fractures produced in the laboratory in steels of the same kind after hydrogen absorption. It was also noted that the fracture surfaces contained micro- and macroscopic cracking and the surfaces of the rods were deeply pitted (Fig. 13).

Fig. 13 – Fracture surface with a macroscopic crack.

DEVELOPMENT OF THE FAILURE

The collapse of the southern outer roof and the arch occurred due to extreme weakening of the anchoring of the ring beam. This resulted from the total failure of 8 tension elements and the partial failure of 2, themselves caused by previous failures of the tension rods. This was made worse by the fact that 7 of the elements, which had been fracturing steadily for some time, lay directly next to one another in one section of the eastern roof. As a result of successive fractures in the tension elements the load capacity of the remaining elements was exceeded, the south arch became unstable and collapsed.

REASONS FOR FAILURE

Conditions likely to cause corrosion of the prestressed steels due to structural conditions and finishing of the ring beam joint. Bending stresses near the anchoring points caused cracks to appear in the concrete panels as far as the tension elements and partial opening of the ring beam joint. Deformation of the elements due to excessive bending at the ring beam led to increased stress of the elements. Since the concrete and mortar in the ring beam joint in the area of the already failed elements were porous and permeable they provided insufficient corrosion protection for the steels. Together with the mechanical considerations, moisture permeated down through the thickness of the joint.

Some of the tension elements which had already failed lay with their ducts directly in the bituminous felt, laid in the step of the ring beam before cementing in the roof panels (Fig. 4, bottom). As a result of these construction failures they were insufficiently surrounded by concrete, if at all. Even if high quality joint material had been used protection against corrosion would have been lacking, moisture could still have permeated the structure even from below. The backward slope of the step in the ring beam aggravated this by trapping moisture in the joint.

Since the tension rods lay mainly on the bottom of the ducts, with no grout as protection around them the corrosion gradually destroying them transferred quickly to the tension rods. In connection with this, the geometric conditions of the gap between tension element and floor of the joint and also between the tension rods were such that the hydrogen evolved during corrosion reaction could be absorbed by the steel. There, as the SEM inspection showed, the tension rods were embrittled by hydrogen.

SUMMARY

During the last few years there have been a lot of failures of prestressed concrete constructions because of hydrogen-induced stress corrosion cracking. The most important failure in that field happened in May 1980 when the Berlin Congress Hall collapsed. The roof of this exceptional piece of architecture consisted of an inner and an outer part. The outer part of the roof was characterized by a reinforced concrete arch which was hollow. 82 tendons which were installed in the 7 cm thick roof sheeting served as the anchorage of the arch against an inner ring beam. The outer surface of the roof sheeting was divided into 24 slabs which were 2 metres wide each. The slabs contained 2 to 4 tendons with 7 to 10 prestressed wires ribbed made of annealed steel (St 145/160).

On 21 May 1980 the southern outer roof collapsed. The slabs in the southeast part broke away from the ring beam. The slabs in the southwest part were fractured at the outer arch and remained hanging on the ring beam held by the tendons.

Normal stereo microscope and electron scan microscope analysis of the fracture surfaces of the broken prestressed wires were made in order to find the cause of the failure. The fracture structures could be divided into two groups. Fractures from the transition area from the slabs to the outer arch showed ductile character and emerged as secondary failures of the collapse. In contrast to these fractures, the fracture surfaces of the broken prestressed wires at the ring beam in southeast part of the roof were more or less corroded. Here only very few newly ductilely broken wires were found. The majority of fractures had already happened before the collapse. They showed macroscopic embrittlement (restricted cross-sectional area of contraction) the microfractographically dissolved areas along the former austenite grain boundaries as well as numerous inner cracks.

Hydrogen-induced stress corrosion cracking of the prestressed wires in the area of the ring beam joint was identified to be the cause of the collapse. The surrounding conditions in this part of the building caused by the constructional finishing led very soon to the formation of a corrosion environment. This environment attacked and gradually destroyed the encasing tubes by corrosion. Because the prestressed steels were lying on the bottom of the duct there was no grout in this space. So the corrosion resistance was very low and allowed hydrogen-induced stress corrosion cracking of the prestressed wires.

REFERENCES

[1] Nürnberger, U., Analysis and evaluation of failures of prestressed steels. *Forschung Straßenbau und Straßenverkehrstechnik, bulletin 308,* Bundesminister für Verkehr , Bonn–Bad Godesberg (1980).

[2] Fleckner, S., *Beton und Stahlbetonbau,* **52** (1957), pp. 233–236.

Chapter 6

Deterioration of marine concrete structures with special emphasis on corrosion of steel and its remedies

M. A. AZIZ, M. A. MANSUR, National University of Singapore, Kent Ridge, Singapore

ABSTRACT

Concrete, both conventionally reinforced and prestressed, is being used extensively in the construction of various types of structures in marine environment. Long-term performance of marine concrete structures requires a careful procedure to be followed in both design and construction stages. Selection of materials, mix design, proper detailing of reinforcement, appropriate construction technique and a strict control programme are the essential parameters to produce a durable marine concrete structure. This paper presents the various factors affecting the durability of marine concrete structures and the mechanism of deterioration of concrete and corrosion of steel reinforcement by various aggressive sea salts. Results of laboratory and field investigations of some aspects of concrete deterioration are also reported. Preventive measures to be adopted in the design and construction stages are presented in an orderly form.

INTRODUCTION

Conventional reinforced concrete has been extensively used in marine environment for constructing port facilities like jetties and wharves, and for protective structures like embankments and sea walls. Prestressed concrete is now being used for the construction of general marine structures and for offshore drilling, production and shipping facilities. All these marine concrete structures involve huge investments of money and all concerned rightfully demand that such structures be strong, durable, free from deterioration and excessive maintenance requirements – because there is little possibility of applying remedial measures to these structures after they have been built and put into service.

A marine structure may be analysed for anticipated external loads by any one of the sophisticated methods currently available and be designed for strength with adequate margin of safety. However, its long-term performance depends largely on the conditions of exposure. Inadequate knowledge of the aggressive marine environment may lead to serious economic losses and even disasters to the owners of such structures because repair works are difficult and very expensive and even the success of such measures is somewhat uncertain.

Some studies have been made of the various factors involved in the durability of concrete structures under the marine environment over the past few years [1–8]. These studies have revealed some very important facts. But still it remains to be a dynamic subject of further study and research.

This paper approaches the subject of durability of marine concrete structures from a different angle, endeavouring to identify the possible aggressive elements in marine environment and their mechanism of attack on concrete structures. It also highlights the appropriate measures to be adopted in order to produce a durable structure.

THE STRUCTURE AND THE MARINE ENVIRONMENT

According to the conditions of exposure, a marine concrete structure may, in general, be divided into three distinct zones: embedded, submerged and atmospheric as shown in Fig. 1. The top and the bottom zones of the structure are subjected to atmospheric and underground exposure conditions, respectively while the central zone, i.e. the portion in between the mud line and the high-water level, is exposed to the marine environment. The present discussion will be limited to this central zone of the structure only. The marine environment consists of various active physical, chemical and biological agents. Sea water contains chlorides, sulphates, combined alkalis with magnesium, and oxygen and carbon dioxide absorbed from the atmosphere or produced by the sea flora and fauna. It also contains suspended solids such as sand, silt and ice. Moreover, wave actions, cross currents and tidal effects are present in the sea water.

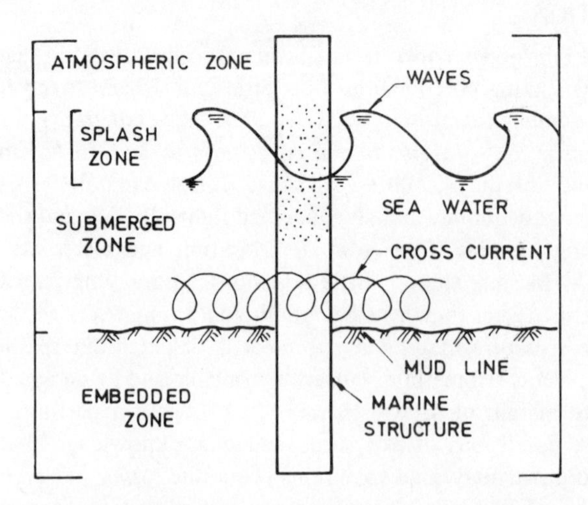

Fig. 1 – Various zones of a marine concrete structure.

The structure subjected to the above environmental conditions consists mainly of concrete and steel. The principal constituents of concrete are cement, coarse aggregates, sand, water and admixtures, if any. The composition of these various constituents of concrete together with the environmental exposure conditions plays a vital role in the long-term performance of marine concrete structures (both reinforced and prestressed).

AGENTS PRODUCING DISRUPTION

Disruption of a marine concrete structure is caused by the various aggressive agents present in the environment as well as in the concrete and steel, and by the interaction among them. These agents may be broadly classified into two categories – external and internal.

The external agents are those present mainly in the environment. Therefore, these are peculiar to the particular conditions the structure is exposed to. In marine environment, the aggressive agents that affect the durability of concrete adversely may be listed as follows:

(a) Various salts like sulphates, chlorides, magnesium, oxygen, carbon dioxide and acids, etc. present in sea water.
(b) Freezing and thawing.
(c) Water movements such as waves, currents and tides.
(d) Suspended solids like sand, silt and ice causing abrasion and impact.
(e) Various sea flora and fauna.
(f) Overloads from storm waves, impact from ocean-going vessels, etc. which may produce cracks in the structure.
(g) Spilled oils and aggressive chemicals stored in the structure (special cases).

Internal agents are inherent to the structure and the type of materials used, and are, therefore, independent of the conditions of exposure. They include the following:

(a) Unsound and reactive aggregates.
(b) High alkali and tricalcium aluminate content of cement.
(c) High sulphide content of cement, water and aggregates.
(d) Insufficient quantity of cement in making concrete.
(e) Inadequate cover to reinforcement.
(f) Previous concrete.
(g) Dissimilar metals used as reinforcement.
(h) Honeycomb, bleed holes, etc.
(i) In the case of prestressed concrete – inadequate grouting of ducts, and inadequate sealing of anchorages or tendon ends.

MECHANISM OF DETERIORATION

Deterioration of marine concrete structures takes place mainly by (a) disintegration of the concrete, and (b) corrosion of reinforcing or prestressing steel and their end anchorages.

Disintegration of Concrete

The majority of cases of serious disintegration are due to chemical attack on highly permeable concrete. The common forms of chemical attack are the leaching out of cement and the action of sulphates and acidic waters.

Sulphate solutions present in sea water react with hydrated cement paste and produce insoluble salts. These salts have a considerably larger volume than the compounds they replace. The end products of the reaction thus lead to progressive cracking and spalling of the concrete.

Carbon dioxide and various types of acids that may be present in sea water or produced by microorganisms can dissolve calcium hydroxide, thus causing surface erosion.

Cement containing more than 8% tricalcium aluminate can lead to the replacement of calcium compounds in the hydrated cement by magnesium compounds from the sea water. This results in serious loss of strength.

High chloride concentration in the concrete mix or in water accelerate corrosion, particularly when the concrete is of low quality with high water–cement ratio.

In addition to the chemical action, crystallization of various sea salts may take place in the pores of the concrete due to capillary intrusion of saline water and alternate wetting and drying conditions. Gradual accumulation of these salts may result in the disruption of concrete owing to the bursting pressure exerted by the salt crystals. This form of attack is more severe in concrete between the tide marks, i.e. in the splash zone (Fig. 1).

Unsound and reactive aggregates in concrete may expand due to prolonged immersion in sea water producing cracking and general disintegration.

Deterioration of concrete in marine environment is aggravated by waves, currents and tides, which continuously replenish the aggressive agents to the affected area. Incessant water movements coupled with abrasion caused by the suspended solids also tend to scour away the softer parts of the concrete and thus expose fresh surfaces to further attack.

Corrosion of Steel

The corrosion of embedded steel in concrete structures is mainly due to aggressive salts. Concrete usually provides embedded steel with a high degree of protection against corrosion. This is due to the fact that the liquid phase of concrete normally has a pH value in excess of 12.5 (ranging between 12.5 to 13.2). In this alkaline environment, a thin protective coating $\gamma - Fe_2O_3$ is formed on the steel surface. This protective coating remains intact at high pH values (around 13). The protection afforded to steel by high pH value persists while concrete sets and hardens and remains intact so long as the concrete environment surrounding the steel does not change significantly. Once the hardened concrete has dried to an equilibrium moisture content, the combined protection of steel depends entirely on the ability of the concrete to maintain the appropriate alkaline environment. However, the protective film of $\gamma-Fe_2O_3$ on the steel surface is usually disrupted when the pH value is reduced by ingression of sea salts [8–10] due to capillary action, cracking and spalling of concrete and poor construction practices. The disruption of protective cover leads to corrosive attack on the reinforcement. Once this attack commences, corrosion products start to build up on the steel surface. Gradual accumulation of these products results in the development of disrupting tensile forces which may rupture the surrounding concrete and expose the steel bars to corrosive element (Fig. 2).

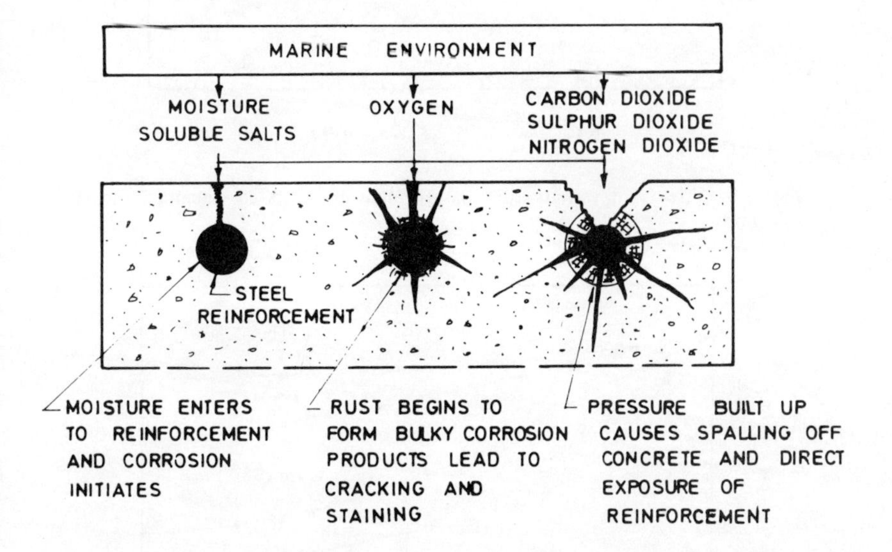

Fig. 2 – Mechanism of corrosion of steel reinforcement in marine concrete structures.

LABORATORY AND FIELD INVESTIGATIONS

In the present study, laboratory experiments were conducted to investigate the effect of various sea salts on the pH value of the concrete phase surrounding the steel reinforcement. The change in this value is a measure of the susceptibility of reinforcing bars to corrosion. Varying concentrations of different sea salts were added to both saturated calcium hydroxide solutions and hydrated cement extracts, and the change in pH values was recorded.

The results of these tests are presented in Figs. 3 and 4. From these figures, it is observed that of the various salts present in sea water magnesium sulphate has the most pronounced effects on the alkalinity of both saturated calcium hydroxide solutions and hydrated cement extracts. This behaviour can be attributed to the low solubility product of magnesium hydroxide [$Mg(OH)_2$].

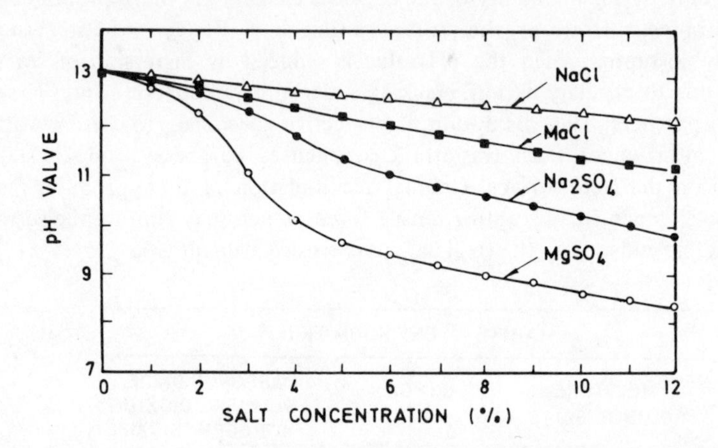

Fig. 3 – Effects of various salts on pH value of saturated calcium hydroxide solutions.

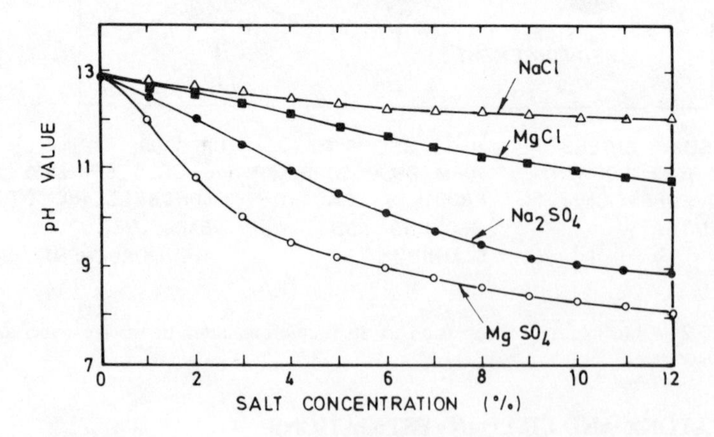

Fig. 4 – Effects of various salts on pH value of hydrated cement extracts.

It can be seen from Figs. 3 and 4 that an addition of only 2% of magnesium sulphate ($MgSO_4$) has decreased the pH value of saturated $Ca(OH)_2$ solution from 12.8 to 10.7 but the same dosage of $MgSO_4$ on hydrated cement extracts has reduced the pH value from 13.1 to 12.3. The reduction of pH values by other salts of the same dosage is not very significant; while further increase of $MgSO_4$ up to 4% lowered the pH value of $Ca(OH)_2$ solution and hydrated cement extract from 12.8 to 9.5 and 13.1 to 10.1 respectively. Thereafter, only a slow rate of decrease of pH value is observed up to 10% of $MgSO_4$ solution in both the cases. The effect of sodium sulphate (Na_2SO_4) is found to be much less pronounced in lowering the pH values compared to that of $MgSO_4$, but quite significant as compared to those of magnesium chloride (MgCl) and sodium chloride (NaCl). The effect of sodium chloride is found to be negligible. These experimental findings also confirm the results of similar studies carried out by Gjorv and Vennesland [10].

Some field investigations were also carried out to study the effects of sea salts on marine concrete structures in Singapore. These investigations have revealed that large amounts of salt can penetrate deep into the concrete. A chloride concentration of 5% to 12% by weight of free observed water was found in 25 to 50 mm thick concrete covers removed from old marine structures above and below the fluctuating sea water level. Chloride concentration of 10% to 15% was found at a distance of 50 to 100 mm below the concrete surface within tidal zones. Magnesium content of 12 to 15 times the original amount in the cement paste had been observed at a depth of 75 to 100 mm below the concrete surface. The steel reinforcement was found to be corroded, concrete cover spalled off exposing the steel reinforcement to further attack by sea water as shown in Fig. 5. The average pH value of concrete surrounding the corroded steel was found to be 9.0.

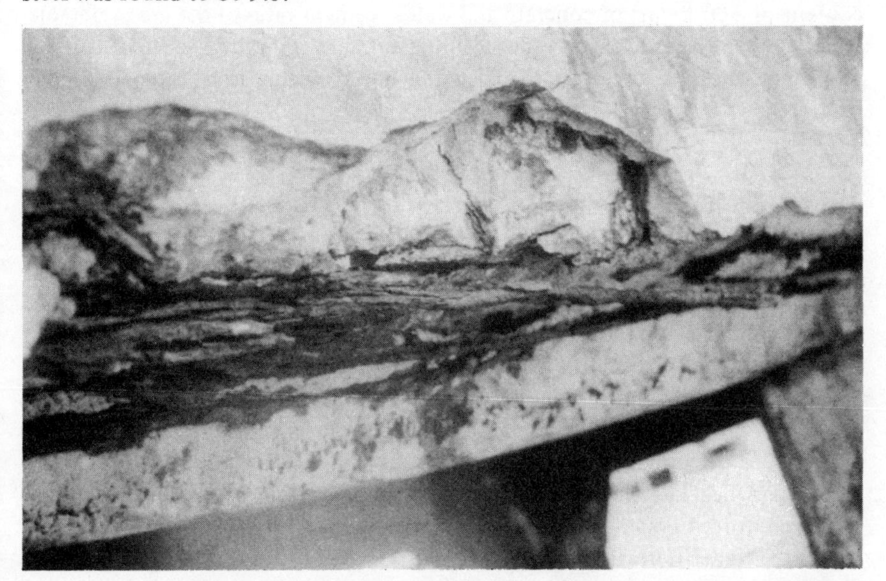

Fig. 5 – Highly deteriorated concrete structure exposed to marine environment.

PREVENTIVE MEASURES

Various preventive measures to be adopted in order to obtain a durable marine concrete structure are summarized below.

Structural Design and Detailing

A thorough check for serviceability limit state of cracking must be made to limit the surface crack width to 0.2 mm concrete cover over prestressing tendons and main reinforcement should be at least 60 mm and that over secondary reinforcement be 40 mm. Special care must be taken in detailing the joints to avoid congestion of reinforcement. Sharp corners should be avoided.

Selection of Constituent Materials

Ordinary Portland cement with 5% to 6% C_3A content and an alkali content of not more than 0.65% should be used. Aggregates should be strong, durable, free from chlorides and non-alkali-reactive. Mixing and curing water must be of potable water quality. Suitable admixtures (water-reducing, retarding or fluidizing) should be free from any chlorides. Steel should be checked for any contamination.

Mix Design

Aggregates should be properly graded to obtain a dense concrete. A cement content of 300 kg/m^3 of concrete and water–cement ratio of 0.4 are preferable. Water content may even be reduced using suitable admixtures. Use of suitable retarding agents is advisable to delay the initial setting time for better compaction.

Construction Technique

Construction materials should be preserved to avoid any contamination. Formworks must be tight specially at corners to prevent mortar leakage. Concrete should be properly mixed and placed, and thoroughly compacted. Finished surface should be smooth and free from honeycomb, and any bleed-holes if present must be filled with epoxy. Extreme care must be exercised in preparing construction joints and the number of joints should be kept to a minimum. The quality of materials and construction must be monitored at every stage by experienced personnel.

Experience has shown that prestressed concrete is preferable to conventional reinforced concrete in the marine environment, because it can be designed crack-free under normal service loading. It is also believed that precast construction is in general less vulnerable to attack than *in situ* casting.

CONCLUSIONS

The successful performance of a marine concrete structure depends to a great extent on its durability against the aggressive marine environment. An understanding of the aggressive elements of the environment and the mechanism of their attack on concrete structures are essential to develop a right course of action in providing structures to best withstand the aggression. Most of the problems of deterioration of concrete structures could perhaps be eliminated if appropriate measures are taken in the selection of materials, mix designs, reinforcement detailing, construction technique and quality control.

REFERENCES

[1] Gjorv, O. E. (1971). Long-time durability of concrete in seawater. *ACI Journal,* **68,** (1), pp. 60–67.

[2] Biczok, I. (1972). *Concrete Corrosion – Concrete Protection.* Budapest, Akademiai Kiado.

[3] Browne, R. D. (1973). The performance of concrete structures in marine environment. *Proc. Symp. Corr. Mar. Envr.,* Inst. Mar. Engrs., London, pp. 50–57.

[4] Bury, M. R. C. and Domone, P. L. (1974). The role of research in the design of concrete offshore structures. *Proc. Conf. Offshore Technology, Dallas.* pp. 155–168.

[5] Berman, H. A. (1975). Sodium chloride, corrosion of reinforcing steel and pH of calcium hydroxide solution. *ACI Journal,* **72** (4), pp. 150–157.

[6] Gerwick, B. C. Jr. (1975). Practical methods of ensuring durability of prestressed concrete ocean structures. *ACI Publication SP-46: Durability of Concrete,* pp. 317–324.

[7] Aziz, M. A., Ramaswamy, S. D. and Roy, S. K. (1978). Some aspects of concrete corrosion and preventive measures. *Proc. Int. Conf. Mat. Constr. Dev. Count., Bangkok,* Vol. 1, pp. 427–441.

[8] Yeomans, S. R. and Cook, D. J. (1978). Corrosion of steel reinforcement in concrete – causes and prevention. *Proc. Int. Conf. Mat. Constr. Dev. Count, Bangkok,* Vol. 1, pp. 443–458.

[9] Shalon, R. and Raphael, M. (1959). Influence of seawater on corrosion of reinforcement. *ACI Journal,* **55,** pp. 1251–1268.

[10] Gjorv, O. E. and Vennesland, O. (1976). Sea salts and alkalinity of concrete. *ACI Journal,* **73,** pp. 512–516.

Chapter 7

The influence of concrete quality of carbonation in Middle Eastern conditions – A preliminary study

K. W. J. TREADAWAY, Building Research Establishment, Watford, UK
G. MACMILLAN, P. HAWKINS, Ministry of Works, Power and Water, Bahrain
C. FONTENAY, Haji Hassan Ready Mix Co, Bahrain

INTRODUCTION

Reinforced concrete is one of the most important structural materials used in the construction industry worldwide. Generally speaking it has excellent structural and durability performance but there are examples of early deterioration due to a number of factors one of the most important of which is reinforcement corrosion [1]. The expansive rust scale formed as reinforcement corrodes applies bursting forces to the concrete cover which leads to cracking in that cover and, if unattended, subsequent spalling. This can have serious implications both from a point of view of expensive maintenance requirements and, in the longer term, from the point of view of loss of structural integrity if no action is taken to restrain the corrosion progress. While reinforcement corrosion occurs in poor quality concrete throughout the world more severe problems are encountered in hot climates, particularly in coastal locations such as the Arabian Gulf area. In this area the rate of deterioration is of great concern [2]. Serious reinforcement corrosion has occurred within a short period from construction leading to virtual re-building or even demolition within ten years in some examples. Such rates of deterioration are much higher than have occurred in more temperate climates and an important question which needs to be answered before modification can be made to specifications or practice is the cause of this deterioration.

Regardless of exposure condition steel will only corrode in concrete when it becomes depassivated. Depassivation can occur through one of two major routes, loss of alkalinity in the concrete or attack on the steel by aggressive ions, or by a combination of both these factors.

Loss of alkalinity results from reaction of the hydrating cement matrix with acidic components in the atmosphere, and, to some extent, leaching of hydroxyl ion from the concrete. When the pH of the matrix drops below a value in the range 9.5 to 10 steel can corrode given adequate supplies of moisture (to provide the electrolytic path) and oxygen (to stimulate the cathodic reaction). Likewise passivity can be destroyed by aggressive ions. Chloride is by far the most aggressive ion in relation to steel corrosion and its presence in sufficient quantitites in concrete can readily destroy passivity leading to rapid and often serious corrosion.

IMPLICATIONS TO THE CONCRETE

Concrete being a porous medium will allow the ingress of atmospheric gases. These will contain carbon dioxide which will readily react with the alkalis present so reducing the pH. However, the extent of ingress will be controlled, amongst other factors, by permeability of the concrete matrix and of the aggregate. It is important therefore to minimize permeability to prevent excessive atmospheric ingress and consequently alkalinity reduction.

While carbonation is one important route for the loss of protection to the reinforcement another is the presence of chloride in the concrete. It can be introduced in concrete in a number of ways but in the Middle East if chlorides are present in the concrete they have usually been introduced as contaminants in the aggregate, in the mix water or as a result of exposure to chloride-laden environments. Of these routes that via the aggregate is the most common. As well as with chlorides, Middle Eastern aggregates can be contaminated with sulphate and this gives rise to the additional problem of containing any potential sulphate attack on the concrete. Tricalcium aluminate is the component in the cement which can combine chemically with the sulphate to form expansive calcium sulpho aluminate hydrates but it also has a similar combining ability with chloride and in this case the amount of free chloride in the matrix is reduced to a value sufficient to provide equilibrium with that in the solid combination [3]. It is this free chloride which is available to stimulate the corrosion reactions. However, competitive reactions between chloride and sulphate for the tricalcium aluminate can occur and as a consequence it is possible that higher free-chloride concentrations in the cement matrix porewater can occur if sulphate preferentially complexes with the tricalcium aluminate. The amount of tricalcium aluminate in the cement can, in its own right, significantly influence the potential to chloride-induced corrosion.

JOINT EFFECT OF CARBONATION AND CHLORIDE

From the foregoing it can be seen that two main factors create conditions in which reinforcement corrosion is likely to occur:

(a) the presence of sufficient quantitites of chloride in the concrete.
(b) carbonation to the full depth of cover.

It is important to realise that in addition to their individual effects the combination of carbonation and chloride attack needs to be considered. Full carbonation of the cover to the steel and even a low-level of contamination of chloride will lead to an increase in corrosion over that when carbonated concrete is uncontaminated by chloride. This may be of critical importance when it is remembered that the recommended maximum chloride limits in Codes of Practice in developed countries, which are widely applied and often achieved in the Middle East, are based on the situation where concrete surrounding the steel is uncarbonated. In general the timescales for the two influences of chloride-induced corrosion and carbonation are likely to be different. On the one hand high levels of chloride in the concrete could pose an immediate risk of stimulating reinforcement corrosion. On the other, in European conditions, rates of carbonation in good quality dense aggregate concrete should be slow and so carbonation-induced reinforcement corrosion should occur much later than that due to chloride. While ranges of chloride in relation to timescales to cracking have been published for both UK [4] and Middle Eastern conditions [5], data on carbonation rates is less well characterised, although in a major survey of deterioration of concrete structures in Middle Eastern conditions the rate of carbonation was observed to be higher at higher temperatures [6]. Furthermore in early work by the Public Works Department of the Ministry of Works, Power and Water, Bahrain, significant carbonation was determined at an early stage in the design life of some structures. As a result of this work a decision was made to investigate, in controlled experiments, the development of carbonation in concretes manufactured from Middle Eastern materials in Middle Eastern conditions. This paper outlines the results of the original survey and continues to describe the laboratory experiments which have been commenced to measure the influence of a number of variables on rates of carbonation in concrete.

SURVEY OF EXISTING STRUCTURES

Various surveys of structures have been carried out, all commissioned by the Ministry of Works, Power and Water of the State of Bahrain, to determine the causes of reinforcement corrosion which have led to premature failure. Since 1978 these surveys have included the determination of carbonation depths using phenolphthalein as a pH indicator. The surveys have been carried out on both sound and deteriorated structures and these are annotated in the results illustrated in Table 1.

It will be seen from Table 1 that many of the structures tested were built in the decade of the 1970s when a building boom occurred in Bahrain. During that period great demands were placed on materials supply and as a consequence there are many examples of the use of chloride-contaminated aggregates for the manufacture of concrete. The presence of chloride has created additional corrosive conditions to those resulting from carbonation and has resulted in early and serious corrosion failures. Nevertheless carbonation in its own right can cause serious problems and it will be seen from Table 1 that the rate of carbonation in Bahrain structures is, in many cases, considerably higher than would have been anticipated, to the extent that the depth of cover to the steel may be, and in some cases has been, exceeded by the depth of carbonation within the design life of the building. Although cause for concern is indicated, because of the wide scatter of results even from mixes which were intended to be identical, this concern could not be quantified. It does appear however, that the potential for corrosion due to loss of alkalinity will be a problem in many structures and it will occur in the medium rather than the long-term. These implications are of major importance to the economy of Bahrain and have led to the setting-up of the programme described in this paper.

Table 1 – Data from measurement of depths of carbonation in buildings and structures.

Type of structure	Element	Test date	Age of test (years)	Carbonation depth (mm)	
Health Authority building	Roof slab	1978	12	70	
School	Foundation	1979	1.75	50	
Housing Complex Block (1)	Floor slab	1980	7	Upper Surface	lower
				3	16
				8	7
				5	8
				0	NR
				3	NR
	Column	1980	7	7.5	
				10	
				15	Average
				13	19.25
				25	
Block (2)	Floor slab	1980	7	0	
				0	
				0	
	Column	1980		20	Average
				20	18.3
				15	
				upper	lower
Block (3)	Floor slab	1980	7	15–20	6–8
				0.5	1–4
	Column	1980	7	20	
				13	
				20	Average
				10	15
				12	
				15	

Type of structure	Element	Test date	Age of test (years)	Carbonation depth (mm)	
Heavy industrial structure					
(a) main stock	Foundations	1979	5–6	15	
				10–13	
				16	
	Machine plinths	1979	5–6	16	
				19	
				11	
				12–13	
				15–16	Average
				18–22	14.8
				15–20	
				14–18	
				15	
				15–19	
				upper	lower
(b) subsidiary block	Floor slab	1980	6–7	7–10	10–15
	Column	1980	6–7	20–25	
	,,	,,	,,	5–10	
	,,	,,	,,	20–25	
	,,	,,	,,	5–20	
	,,	,,	,,	5–25	
				surface	
				upper	lower
Precast slabs		1980	4 months	1	0
			5 months	2	3
			9 months	3	3
			11 months	1	0
			17 months	2	2
			17 months	5	0
			23 months	7	0
			27 months	7	4
			30 months	9	8
			36 months	5	2
			37 months	7	2
			43 months	4	5

NR: Not reported.

EXPERIMENTAL APPROACH

The objectives of the preliminary study were defined as follows:

(1) To establish a method of observing the rate of carbonation of concrete specimens.

(2) To determine the rate of carbonation of concretes used as standard mixes on public sector projects.

(3) To determine relative rates of carbonation of uncured and 28-day moist-cured concrete.

EXPERIMENTAL PROCEDURES

The preliminary study of carbonation described in this paper is based on measurement of the progress of carbonation into 100 X 100 X 500 mm concrete beams. Following casting and curing the beams are exposed and at intervals returned to the laboratory for sectional splitting and subsequent examination of depth of penetration of carbonation. The remainder of the beam is then resited on the exposure site.

The mix designs used for the casting of beams are based on those prepared by a major ready mixed concrete supplier in Bahrain for the Public Works Affairs specification:

	Minimum cement content	Maximum w/c ratio	Maximum normal aggregate use
Class I	330	0.50	20 mm Grade 30
Class II	370	0.45	20 mm Grade 45
Class III	220	–	20 mm Grade 15

The casting programme was extended with mixes with higher as well as lower w/c ratios and cement contents. An air-entraining agent was incorporated in additional mixes of Class I and II. Materials details are given in Appendix 1 to this chapter.

The mixing, batching and casting procedure was as follows. The pre-mix weights of coarse and fine aggregates, cement and water for each mix were recorded and the admixture introduced measured by volume. Samples of each material were taken from the weigh belt or when entering the mixer for all relevant physical and chemical testing.

The concrete was batched in a 3 m^3 punch-card-operated turbo mixer, water being added in the mixer until the required workability was obtained. Two batches were prepared and discharged into a truck mixer, the truck being weighed prior to and following loading as a double-check on yield.

A 150 l sample of the 6 m^3 concrete was taken for beam and cube preparation. This sample was measured for slump, unit weight, air content and temperature (see details in Table 2). Table 2 also gives details of the cement content, water to cement ratio and combined grading. Correction of the w/c ratio due to bleeding during compaction and initial curing had, of necessity, to be omitted.

The cube and beam moulds were filled in layers of 50 mm while placed on a vibrating table. The concrete sample was remixed regularly by scoop to avoid segregation during the casting operation. After vibration, and trowelling of the

Table 2 – Mix details

Mix number	Air entrained	Cement content kg/m³	Water:cement	Combined grading % age passing BS sieve size										Flakiness index		Slump mm	Air content %	Fresh unit weight kg/m³	Concrete temp °C	(Modified BS 4550) SO_3 in aggs. and water % by weight of concrete	SO_3 % in cement (BS 4550)	Total SO_3 % by weight of cement (BS 4550)	NaCl % by weight of concrete (BS 812)	NaCl % by weight of cement (BS 812)
				75 μm	150 μm	300 μm	600 μm	1.18 mm	2.36 mm	5 mm	10 mm	20 mm	28 mm	20 mm	10 mm									
7		420	0.38	0.6	3	9	23	31	36	46	66	98	100	12	20	70	2.2	2422	29	0.071	1.58	1.99	0.036	0.21
5		370	0.44	0.6	4	11	26	33	38	49	64	98	100	10	21	60	2.1	2417	30	0.073	1.72	2.20	0.038	0.24
1		368	0.48	0.7	4	11	24	32	36	46	68	99	100	10	22	100	1.9	2409	29	0.076	1.49	1.98	0.048	0.31
11		331	0.49	0.4	3	12	25	35	39	48	68	99	100	13	19	75	2.5	2416	24	0.117	1.67	2.51	0.035	0.25
2		330	0.52	0.8	3	10	24	32	37	49	67	99	100	12	23	75	1.6	2449	29	0.082	1.68	2.28	0.037	0.27
3		221	0.72	0.8	4	12	27	35	40	52	70	99	100	15	21	55	1.6	2404	27	0.084	1.80	2.72	0.033	0.36
10		135	1.20	0.6	3	9	23	33	38	48	66	99	100	14	20	70	2.1	2368	26	0.071	1.66	2.90	0.032	0.55
9		132	1.42	0.8	3	10	27	36	41	53	72	99	100	16	29	50	1.1	2375	25	0.088	1.75	3.32	0.044	0.79
6		129	1.43	0.7	3	10	27	36	41	50	66	99	100	9	17	60	1.9	2385	–	0.104	1.54	3.42	0.076	0.84
8	AE	373	0.41	0.7	3	9	22	29	33	43	64	98	100	12	19	80	5.2	2345	29	0.081	1.63	2.14	0.028	0.18
4	AE	331	0.44	0.7	3	10	23	30	35	45	64	95	100	9	24	85	5.8	2383	29	0.067	1.82	2.29	0.029	0.20

concrete surface of the specimens they were divided into groups of three different curing regimes and immediately carefully removed to the curing sites. The three curing regimes were:

(1) *Reference:* the cubes and beams were cured according to BS 1881 for laboratory-made specimens in tropical climates.
(2) *Site reference curing:* the specimens were cured outdoors covered with wet hessian and polythene until demoulding the following day. The specimens were then left outdoors wrapped in wet hessian and polythene sheet until 28 days. On the 28th day the beams were unwrapped and left exposed.
(3) *Non-cured specimens:* the specimens were placed outdoors uncovered immediately after casting and demoulded the following day, they were subsequently left exposed.

Cubes from curing regimes 2 and 3 were weighed in air 24 hours before compression testing and then water cured until testing according to BS 1881. For each curing regime one cube was tested at 7 days and two at 28 days.

At the time of commencement of exposure each beam was placed in the open 0.25 m above ground level supported on two points. The exposure site was situated 2.5 km from the nearest coast line, approximately 7 m above sea level.

At the completion of each exposure period the test beam was removed from the exposure site and weighed. A transverse section of the concrete between 30 and 50 mm thick was then split from the beam by crushing under point loading at the appropriate section thickness. After fracture the transverse section was air blown to remove dust particles and larger loose material was removed by hand. The section was then marked, the freshly fractured surface sprayed with phenolphthalein solution and then photographed. Carbonation measurements were taken from the edge of the specimen to the edge of the darkest mauve colouration, measurements being made at regular intervals around the perimeter of the section and a diagram of the carbonation depth (with measurements) being recorded on a proforma. The remainder of the beam was reweighed and then replaced on exposure. The reported depth of carbonation (k) was determined as a weighted average to the nearest millimetre of the depth at the two sides and the bottom of each specimen, the trowelled surface being ignored. The peak depth of carbonation (k^1) was also recorded as in the paper by Meyer [7].

RESULTS AND DISCUSSION

Cube densities and compressive strengths are shown in Table 3. The density of the non-cured cube at 27 days is based on the cube weight at 27 days and the cube volume measured at 28 days.

Table 3

Mi Mix no.	Cem.– cont.	W/c ratio	Refer-ence	Non-cured	27 days dry density	28 days SSD-density	7 days strength	28 days strength	Weighted mean depth of carbonation mm at 3½–5 months	at 6 months	at 9 months	Beam density actual 9-months calculation
4	420	0.38	x		–	2456	54.5	65.0	<1	<1	<1	2386
				x	2408	2445	45.5	50.0	2.5	3.0	3.0	2388
5	370	0.44	x		–	2478	50.0	59.5	<1	<1	<1	2400
				x	2440	2471	43.5	50.0	2.0	3.0	3.0	2405
1	368	0.48	x		–	2448	46.0	56.0	<1	<1	1.0	2367
				x	2405	2445	39.0	40.5	3.5	4.5	4.0	2387
11	331	0.49	x		–	2436	33.0	49.5	1.0	2.0	4.0	2344
				x	2396	2425	32.0	40.5	2.5	3.0	5.0	2374
2	330	0.52	x		–	2465	46.0	54.0	1.5	2.0	3.0	2374
				x	2421	2466	35.0	37.0	4.0	4.5	5.0	2406
3	221	0.72	x		–	2441	26.0	32.0	4.0	5.0	7.0	2329
				x	2341	2421	21.5	20.0	6.5	8.0	10.0	2329
10	135	1.20	x		–	2424	10.0	14.0	6.0	9.5	12.0	2294
				x	2330	2423	9.0	10.5	9.5	12.0	15.0	2322
9	132	1.42	x		–	2414	10.5	14.5	6.5	9.0	11.0	2273
				x	2359	2452	9.5	10.5	11.0	13.5	16.0	2353
6	129	1.43	x		–	2447	9.0	12.5	9.5	11.5	15.0	2317
				x	2346	2437	9.5	10.0	11.5	14.5	18.0	2335
8 AE	373	0.41	x		–	2378	44.0	50.0	<1	1.0	1.0	2300
				x	2353	2390	36.0	39.5	3.0	4.0	5.0	2330
4 AE	331	0.48	x		–	2385	36.5	44.5	1.0	2.0	2.0	2302
				x	2336	2375	29.5	30.5	4.0	4.5	5.0	2326

The actual beam density at 9 months has been determined on basis of the 27 or 28 days cube density and the weight change of the beam from 28 days to 9 months.

The depth of carbonation has been measured at 3½–5, 6 and 9 months; Table 3 shows the results for the reference and non-cured beams. The results of the site reference beams are omitted, as they had the same (within 1 mm) carbonation at 9 months relative to the reference beams.

As a result of the uniformity of the measured depth of carbonation of each of the four beams from the different curing regimes at 3½–5 months and at 6 months (see Table 4), the results have been presented as the mean of four beams. Also, the examination at 9 months was carried out on a single beam from each curing regime.

Table 4 – Depth of carbonation (mm) at 6 months for mixes 5 and 9.

	Reference				Site reference				Non-cured			
Beam no.	1	2	3	4	5	6	7	8	9	10	11	12
Mix 5	<1	<1	<1	<1	<1	<1	<1	<1	3	3	3	3
Mix 9	9	8	9	9	9	10	10	9	14	13	14	12

The relationships between depth of carbonation at 9 months and cement content and water cement ratio, respectively, are shown in Figs. 1 and 2.

The rate of carbonation for a range of the mixes is shown in Figs. 3 and 4 for reference and non-cured beams, respectively.

The depth of carbonation versus the cube compressive strength is given in Fig. 5.

In the current series of experiments the relationship between average carbonation depth (k) and water/cement ratio of cement content are illustrated in Figs. 1 and 2 respectively. The maximum average depth of carbonation was 18 mm after nine months exposure, with a peak depth (k^1) of 20 mm. After the same period of exposure some beams have shown only traces ($k<1$ mm and $k^1 = 3$ mm) of carbonation.

Although, as stated above, the k range at nine months is from <1 mm to 18 mm, in general ($k^1 - k$) is in the range 1 mm to 4 mm, the maximum ($k^1 - k$) value recorded being 6 mm (in 2 of 198 observations).

Whilst, in an experimental study of carbonation rates of different concretes, k forms the basis of comparison, it can be seen from the above that k^1 is significantly different from k and must also be considered in the context of concretes containing reinforcing steel, where even a localised loss of passivation can lead to corrosion.

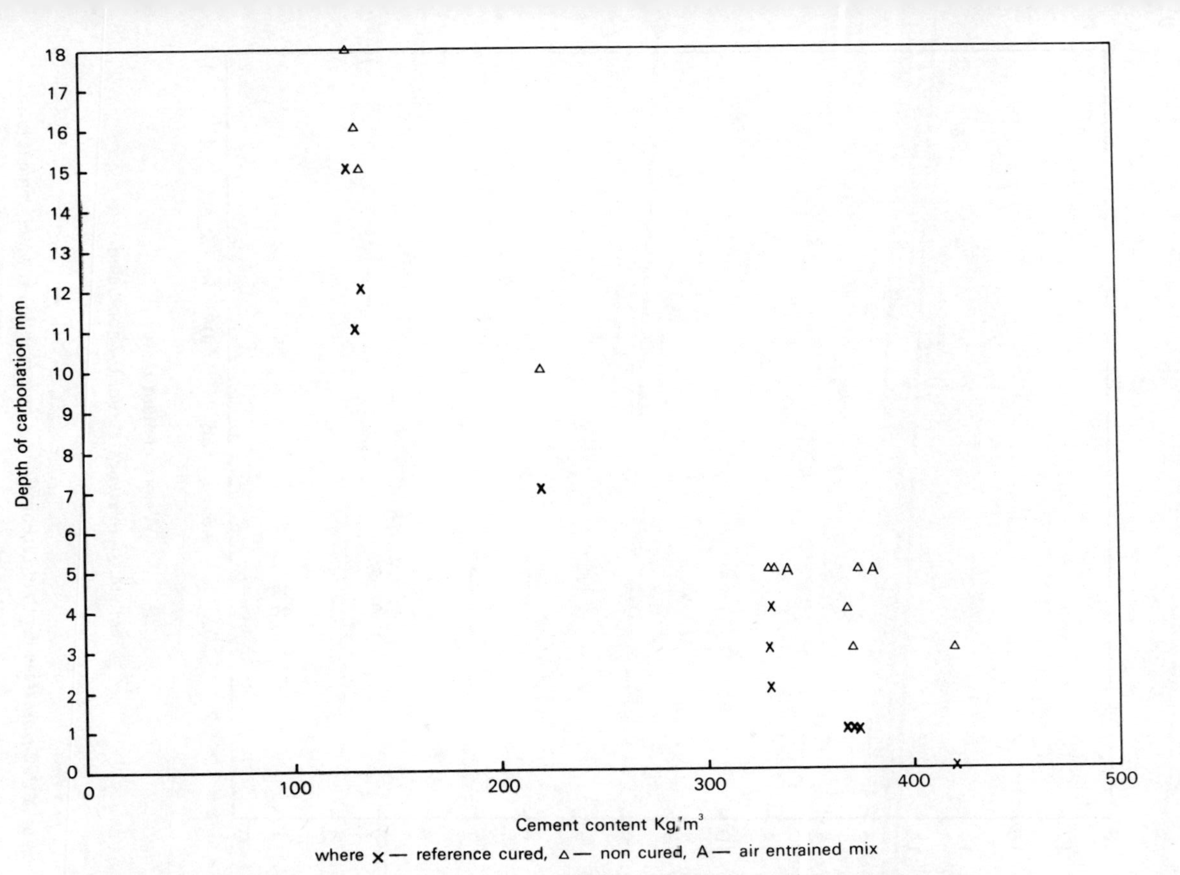

Fig. 1 – Average depth of carbonation (at nine months) v. cement content.

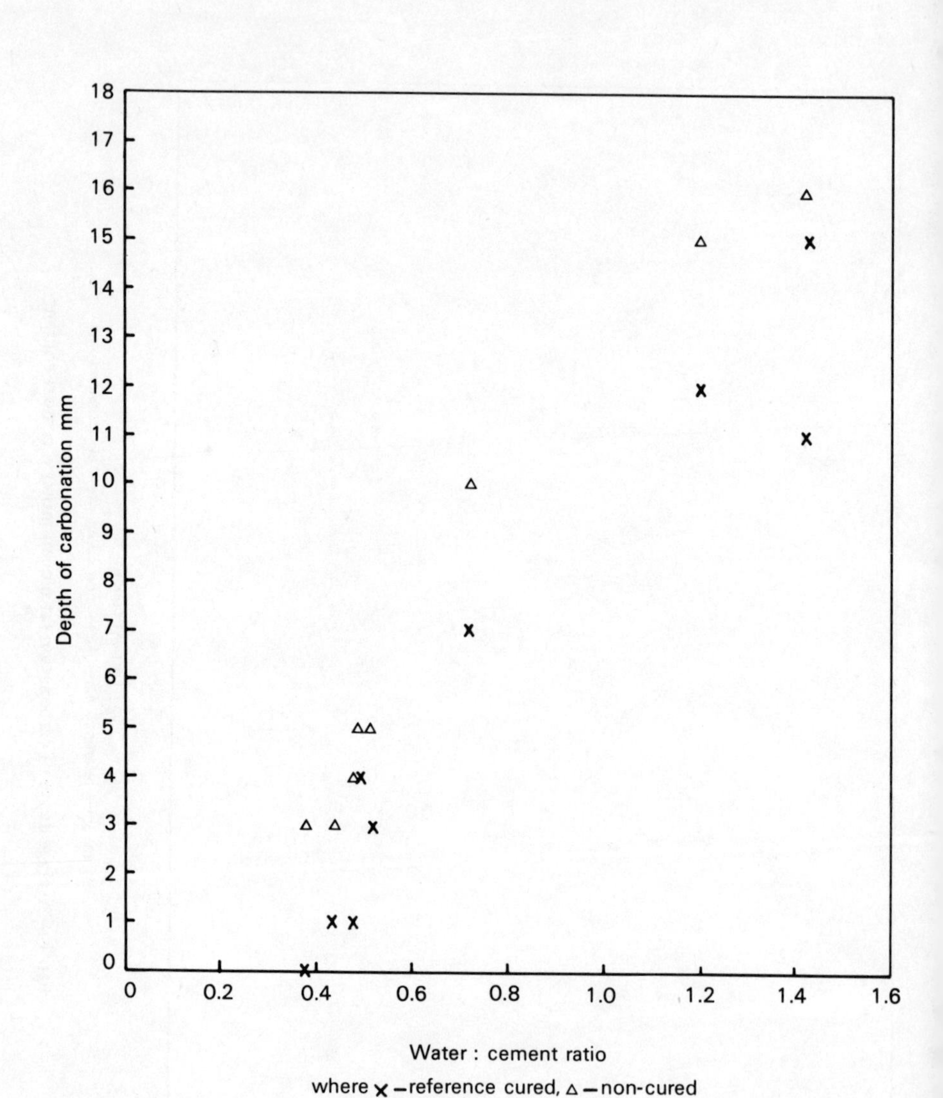

Fig. 2 – Average depth of carbonation (at nine months) v. water/cement ratio.

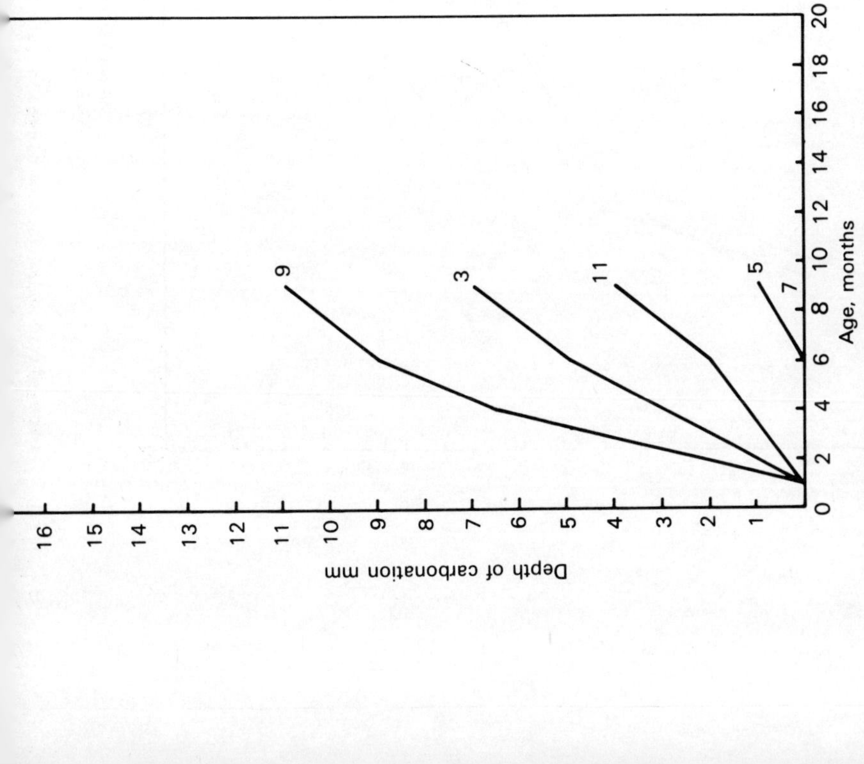

Fig. 3 – Development of carbonation with time for cured beams.

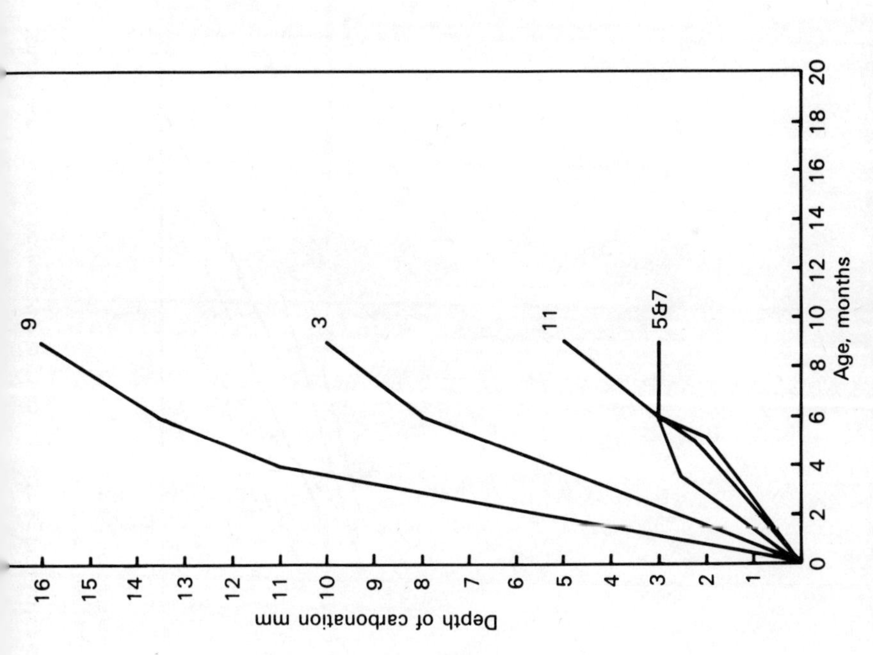

Fig. 4 – Development of carbonation with time for non-cured beams.

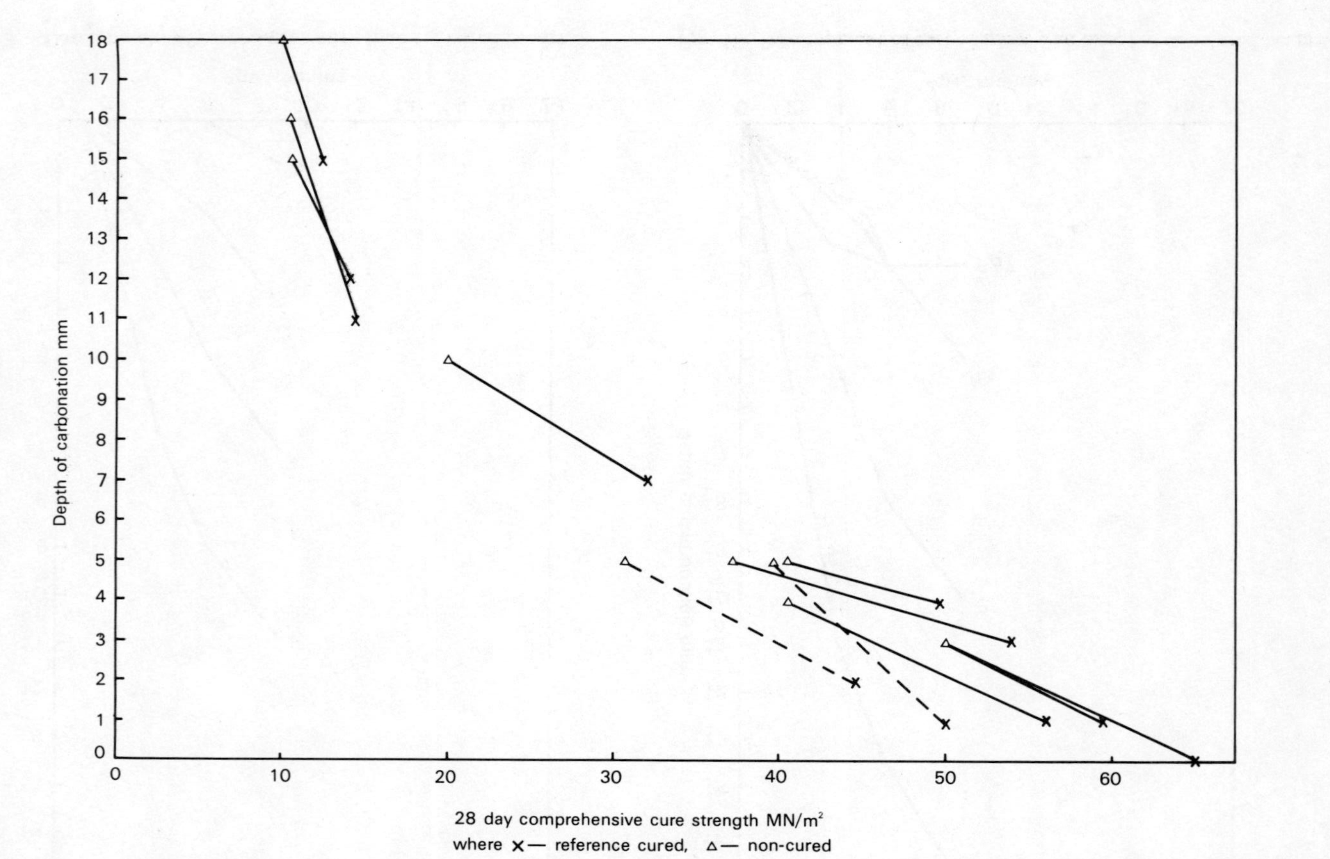

Fig. 5 – Depth of carbonation (at nine months) v. cube compressive strength (28 days) for cured and non-cured concrete specimens.

The results from non-air-entrained concretes (Fig. 5) suggest, over a nine-month period, a similar relationship for the strength and carbonation relationship between cured and uncured states. It should be noted, however, that these specimens have been well compacted and every effort has been made to ensure a minimum of inhomogeneity. Furthermore external ambient temperatures were similar to the laboratory control temperature at the time of casting and curing.

The test results indicate that by maintaining cement content and workability an air-entraining admixture can be incorporated in the concrete mix without significantly affecting the rate of carbonation (Fig. 5). The air-entrained mix has a lower capillary porosity due to the reduced water/cement ratio. Thus the apparent comparison in carbonation rate with non air-entrained mixes is only possible if part of the transport phenomena necessary for carbonation can take place through the entrained air pores.

The ratio of non-cured compressive strength to reference compressive strength at 25 days, for all water/cement ratios, ranges from 0.63 to 0.84 with an average of 0.75; the minimum difference between the compressive strengths of cubes within a mix is 2.5 N/mm^2 and the maximum 17 N/mm^2.

Irrespective of the water/cement ratio lack of curing increases carbonation by approximately 3 mm at 9 months testing age. For the higher water/cement ratio mixes lack of curing has apparently proportionally the same effect on the depth of carbonation as on the strength, whereas for the mixes with low water/cement ratios the proportional effect is much greater for the carbonation depth than for the strength. By properly curing a concrete with a low water/cement ratio it is shown that it is possible to eliminate measurable amounts of carbonation at least within a nine-month period. On the other hand, lack of curing seems to give at least 3 mm depth of carbonation at the same test time.

All the mixes were cast during December 1981 and January 1982, a period which was cold, and sometimes, rainy in Bahrain. Greater depths of carbonation for non-cured concrete would be expected for concrete cast in the summer-time. Therefore, the present observations are only valid for winter casting in Bahrain.

The carbonation rates observed in concrete during a 9-month exposure period do not follow the relationship cited by Lawrence [8] in his review of mechanisms of transport processes in concrete, the Middle Eastern data showing higher rates than would be expected from the application of the Smolczyk relationship. Thus from the preliminary data doubt is cast on some of the current procedures applied in the Middle East to avoid reinforcement corrosion within the specified design life, since these may influence the carbonation of the surrounding concrete.

In order to determine the longer-term effects the examination of this series of beams will continue until 1986. Furthermore a series of beams was cast during September 1982 to study the following aspects of carbonation:

(1) The effect of varying water/cement ratio at fixed cement contents.
(2) The effect of varying cement content at fixed water/cement ratios.
(3) A comparison of carbonation rates between beams cast in Bahrain and exposed in both England and Bahrain and vice versa.
(4) The effect of casting and curing in summer conditions.

Additional beam castings are planned and these will allow studies of the following.

(1) The effect of different cement types (BS 12, BS 4027, BFSC, pozzolan).
(2) A more detailed study of the effect of curing, to include both curing period and delay before the onset of curing.
(3) The effect of curing compounds.
(4) The effect of surface coatings and decorate finishes.
(5) The effect of admixtures.

CONCLUSIONS

The method of multiple splitting of beam specimens has proved to be practically feasible and appears to have reduced some of the inherent lack of homogeneity problems experienced in multi-specimen experiments.

The following conclusions refer only to the mixes studied, cast in Middle East winter conditions:

(1) The rate of carbonation is a monotonic function of water/cement ratio and of cement content.
(2) Effective (28 days wet) curing reduces carbonation by an average of 3 mm relative to totally uncured concrete.
(3) An air-entraining admixture can be incorporated in a concrete mix without significantly affecting the rate of carbonation.
(4) All carbonation rates observed in structural concretes are higher than those indicated by the relationship cited by Lawrence [8].
(5) The rates of carbonation observed would suggest that concretes meeting the requirement of CP 110 for severe/very severe exposure are required for the exposure conditions of this experiment.

ACKNOWLEDGEMENT

This paper forms part of the joint research carried out at the Building Research Establishment, Department of the Environment and the Ministry of Works, Power and Water, Bahrain, and is published by permission of the Minister of Works, Power and Water and the Director, BRE.

REFERENCES

[1] Midgley, H. C., Figg, J. W. and McLean, M. J., *Concrete,* **7**, 24–26 (1973).
[2] Pollock, G. J., Kay E. A. and Fookes, P. G., *Concrete,* **15**, 12–18 (1981).
[3] Lea, F. M., *The Chemistry of Cement and Concrete,* 3rd edn (Edward Arnold, London, 1971).
[4] 'The durability of steel in concrete: Part 2 Diagnosis and assessment of corrosion – cracked concrete', *Building Research Establishment Digest 264,* Garston, 1982.
[5] Kay, E. A., Fookes, P. G. and Pollock, G. J., *Concrete,* **15**, 11, 22–28 (1981).
[6] Fookes, P. G., Pollock, G. J. and Kay, E. A., *Concrete,* **15**, 9, 12–19 (1981).
[7] Meyer, A., *Proc. 5th Int. Cong. on Chemistry of Cements,* Part III, pp. 394–401, Cement Association of Japan, Tokyo, 1969.
[8] Lawrence, C. D., *Cement and Concrete Association Technical Report 544,* Cement and Concrete Association, Slough, 1981.

APPENDIX 1 MATERIALS

The coarse and fine aggregates used for the manufacture of beams and cubes for this project are representative of the materials used throughout Bahrain for concrete with reasonable quality requirements. Likewise the cement is used for more than 50% of all concrete manufactured in Bahrain. A water-reducing admixture is normally used for all ready-mixed concrete supplied on the Island.

The coarse aggregate is a crushed dense grey limestone imported from the United Arab Emirates. The saturated surface dry specific gravity normally falls between 2.68 and 2.70 and the absorption of 0.4%. The coarse aggregate is screened to BS 20 mm and 10 mm single sizes and is washed in order to remove dust and to be at least in a saturated, surface dry condition when used. The flakiness indices are given in Table 2.

Marine sands from different local sources are blended and washed with the undersizes from the coarse aggregate for making a homogeneous blend of fine material with the grading complying with BS zone 2 sand. The washing also removes 60–80% of the chlorides from the sand resulting in a chloride level of between 0.05 and 0.1% chloride as sodium chloride. A significant part of the sand is shell fragments. The saturated surface dry specific gravity is normally 2.65 and the absorption 1–1.5%.

The sulphate resisting cement (BS 4027, ASTM C150, Type V) has a Blaine fineness of 3100 $cm^2\ g^{-1}$, its initial and final set (Vicat) is 3 h 15 m and 4 h 30 m, respectively, and the strength is approximately 23 N/mm^2 at 7 days (ASTM C109). A typical Bogue composition is:

C_3S : 59%
C_2S : 20%
C_3A : 1.8%
C_4AF : 13.9%

The sulphate content of the cement is reported in Table 2.

The mixing water has been treated in a reverse osmosis plant, whereby the total dissolved solids is reduced to 300–400 ppm. A water-reducing admixture (ASTM C494 Type A) is incorporated in all mixes, the dosage recommended by the manufacturer being used. For two mixes in this programme a vinsol resin-based air-entraining admixture (ASTM C260) has been used. All mix details are given in Table 2.

Part 2
MECHANISMS

Chapter 8

The corrosion of steel reinforcements in concrete immersed in seawater

N. J. M. WILKINS and **P. F. LAWRENCE,** AERE Harwell, Oxon, UK

ABSTRACT

Good quality concrete can provide embedded steel with long-term protection against the marine environment. This is particularly true of concrete fully immersed in sea water when the cathodic reaction (ultimately oxygen reduction) is severely restricted; such a situation is indicated by very negative potentials in the range −800 to −1100 mV with reference to silver/silver chloride. The steel is then actively corroding, but the corrosion is spread over the largest possible area of the reinforcement and the actual corrosion rates, limited by the rate of oxygen reduction, are extremely small.

In conditions where cathodic oxygen is more readily available, such as in concrete at or near the tideline or possibly in concrete exposed to air on one side, as is a hollow immersed concrete structure, potentials in the range +50 to −300 mV are observed. In these circumstances the corrosion of bare steel or steel embedded in a highly porous or cracked concrete can be accelerated by galvanic coupling to a 'cathode' of reinforcing steel in sound concrete. Local corrosion damage will depend on the relative sizes of corroding and 'cathodic' areas.

Corrosion of embedded steel is not necessarily associated with chloride penetration and cracks extending to the steel surface may not result in significant corrosion. The extent of corrosion damage depends on a number of factors including crack geometry, void content of the concrete, relative areas of cracked and sound concrete, and the precise chemical and physical properties of the concrete mix and its continuing long-term reactions with the sea water. Other factors such as sea water flow or turbulence, tidal range and marine fouling may also have some influence. In anaerobic conditions corrosion by sulphate-reducing bacteria has been observed on steel in highly porous concrete.

This work has been undertaken as part of the Concrete-in-the-Oceans programme funded by the Department of Energy and the Offshore Industries, coordinated by CIRIA/UEG.

INTRODUCTION

Steel is usually very effectively protected from corrosion when embedded in Portland cement concrete and reinforced concrete structures which are still in good condition after approaching 100 years exposure to various environments bear testimony to the durability that can be achieved. Nevertheless, under certain conditions, deterioration of steel-reinforced structures can be a serious problem and a great deal of research has been directed, particularly in the past 10–15 years, to understanding the processes which lead to corrosion of embedded steel. A large section in the recent C.E.B. state-of-the-art report on Durability of Concrete Structures [1] is devoted to corrosion of reinforcement and a recent article in Nature [2] gives a comprehensive account of current information and understanding of the mechanisms of corrosion of steel in concrete. Both documents emphasise the electrochemical processes involved and the complex interrelations between the physical and chemical properties of the concrete and its environment, which determine the long-term behaviour of the system.

Corrosion of steel reinforcement has formed an important part of the Concrete-in-the-Oceans research programme which, since 1976, have been seeking to further our understanding of the special problems of design, construction and durability of concrete oil production platforms and other marine concrete structures. This paper is not a detailed or comprehensive account of the Concrete-in-the-Oceans corrosion research, but it draws on results obtained by the authors as part of this programme [3] to discuss the factors which could, in certain circumstances, lead to corrosion of steel reinforcement in immersed parts of a marine concrete structure.

ELECTROCHEMICAL CORROSION OF STEEL IN CONCRETE

The corrosion of steel in concrete is an electrochemical process in which the moist concrete forms the electrolyte. Let us consider the reactions involved.

The Anodic Reaction

The reactions of iron in aqueous solutions can be described by a potential-pH diagram [4] (Fig. 1). Only four primary anodic reactions need to be considered:

$$Fe \rightarrow Fe^{2+} + 2e' \quad (1)$$

$$Fe + 2H_2O \rightarrow Fe(OH)_2 + 2H^+ + 2e' \quad (2)$$

$$3Fe + 4H_2O \rightarrow Fe_3O_4 + 8H^+ + 8e' \quad (3)$$

$$Fe + 2H_2O \rightarrow FeO(OH)^- + 3H^+ + 2e' \quad (4)$$

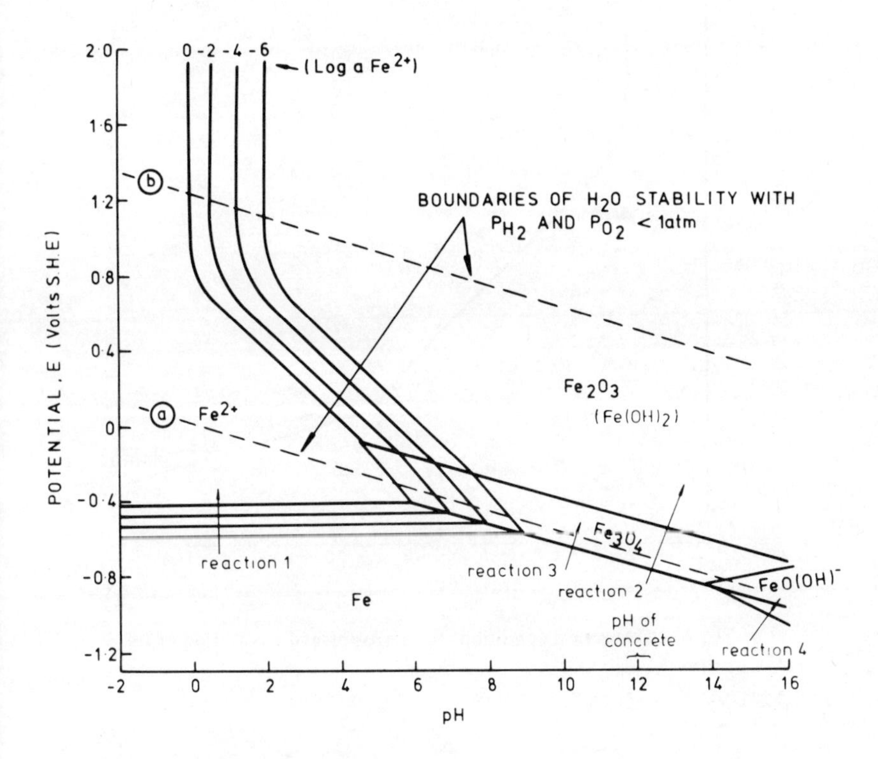

Fig. 1 – Potential–pH equilibrium diagram for the system Fe-H_2O at 25°C (after Pourbaix [4]).

A simplified potential–pH diagram (Fig. 2) can be divided into regions representing immunity, corrosion (reactions 1 and 4) and passivity (reactions 2 and 3, which can form a protective layer of corrosion product on the surface). The basis of the passivity of iron and steel in an alkaline concrete environment is apparent. Page [5] suggests that passivity is maintained in concrete by a lime layer in intimate contact with the steel surface, which stabilises the pH in the passive range.

Penetration of chloride to the steel surfaces does not necessarily destroy passivity. Pourbaix [6] presents a modified potential–pH diagram (Fig. 3) which indicates that even with very high chloride concentrations, a zone of perfect passivity remains, and it has been shown that corrosion of steel in concrete exposed to a high chloride environment can be prevented by polarising to a potential within this zone [7, 8].

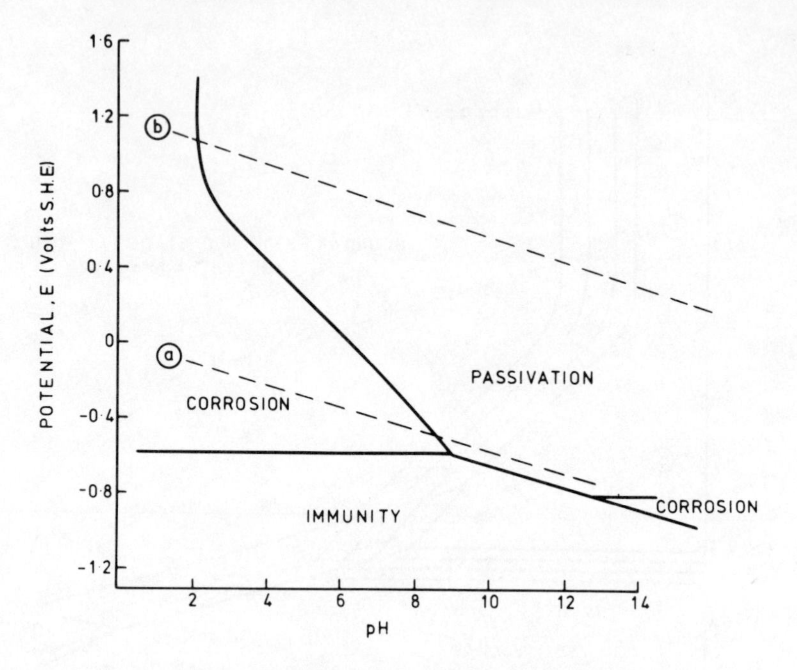

Fig. 2 – Theoretical conditions for corrosion and passivation of iron.

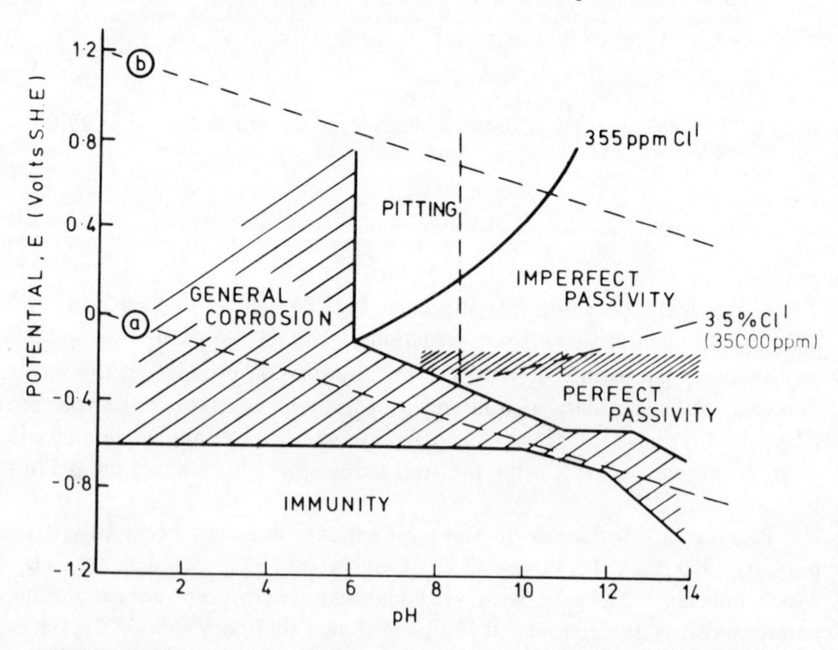

Fig. 3 – Influence of chloride on corrosion and passivation of iron (after Pourbaix [6]).

The Cathodic Reaction

The two principal cathodic reactions available for the corrosion of steel in concrete are indicated on the potential–pH diagrams:

$$\text{Line (a) } 2H^+ + 2e' \rightarrow H_2 \tag{5}$$

$$\text{Line (b) } \tfrac{1}{2}O_2 + H_2O + 2e' \rightarrow 2OH' \tag{6}$$

However, oxygen reduction is recognised as the cathodic reaction in most cases of significant corrosion of steel reinforcement.

The Composite Reaction

Whatever the environment, unless the cement paste can dry out completely, corrosion activity can be related to differences in potential within the structure. If passivity breaks down locally in part of the reinforcement, this will become anodic to the passive steel, and if oxygen is available at the passive steel surface, a current will flow from these passive regions to the anode. Most cases of severe corrosion of reinforcement occur in the atmosphere where oxygen to depolarise this macro-cell is plentiful.

In air, as a result of current flow in the macro-cell, a potential difference (iR) can be measured from one point to another on the surface of the concrete (Fig. 4(a)). However, the cathode area is limited, since the resistance path through the concrete from the anode to remote cathodic sites is high. In fully immersed conditions, the seawater has a very much higher conductivity than the concrete, and no significant potential differences will be measurable at the concrete surface. Since the resistance path is short-circuited by the seawater, it is possible for an anode to draw upon a much larger cathode area (Fig. 4(b)), and the steel in a large submerged structure could carry a significant current to a small anode; even though in this case the cathode is inefficient compared with concrete in air, due to limited access of oxygen.

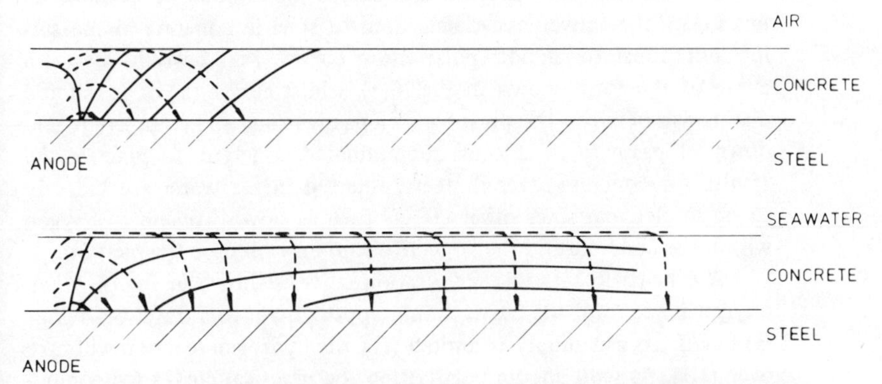

Fig. 4 – Current flow and potential distribution, (a) concrete in air, (b) concrete in sea water. Solid line, equipotential lines; dashed line, current flow.

Electrochemical Measurements

Three basic measurements can be made.

(i) *Change of potential with time.* This is not in itself an indicator of corrosion, since change to a more negative (active) potential could be due to increased corrosion or a limited cathodic reaction (Fig. 5).

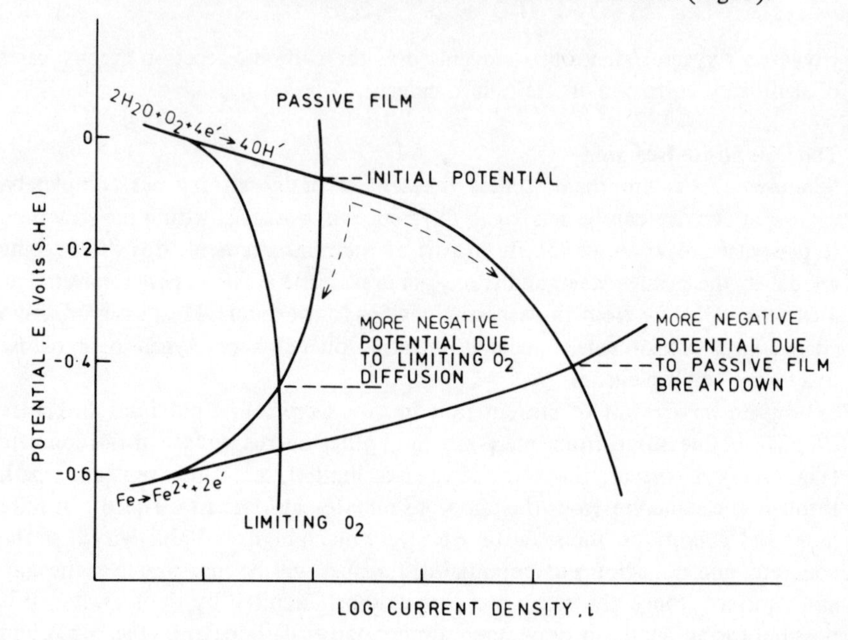

Fig. 5 – Movement of potential due to changes of anodic and cathodic polarisation.

(ii) *Anodic and cathodic polarisation curves.* It should be possible to investigate the active/passive behaviour of steel in concrete by measuring 'potentiostatic' anodic polarisation curves. Fig. 2 would suggest a curve of the form shown in Fig. 6(a), whilst curves (b) or (c) should distinguish between the presence of a passive zone and complete breakdown of passivity in chloride-contaminated concrete. In practice the results for concrete-covered steel immersed in sea water are unlikely to be so clear-cut since other effects such as slow diffusion of oxygen within the concrete and the large '*iR* drop' may obscure the picture.

A schematic cathodic polarisation curve is shown in Fig. 7. When oxygen is plentiful the curve should approximate to a Tafel behaviour (I) but if oxygen supply is limited (II), the hydrogen reaction will take over (III). As with anodic polarisation the practical curves for composite specimens are likely to be much modified by other factors.

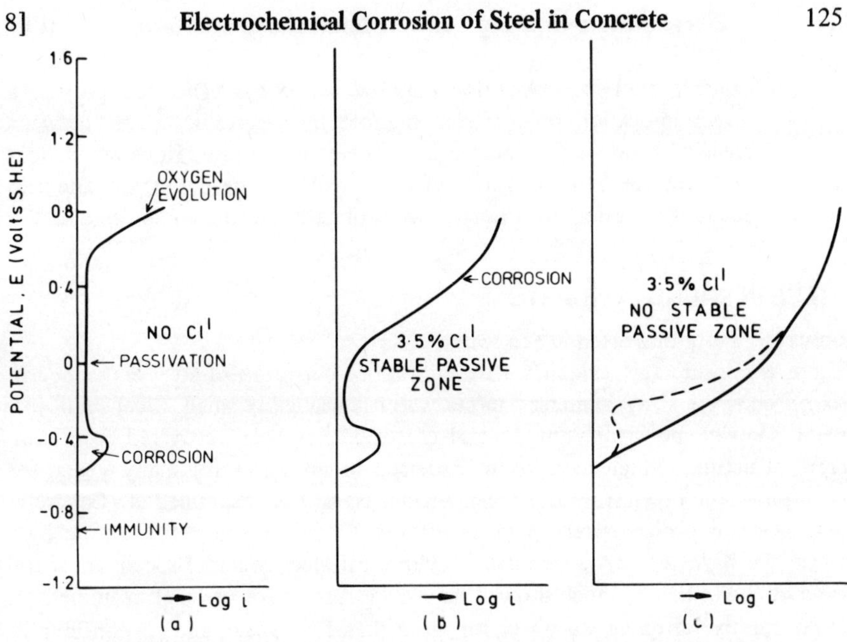

Fig. 6 – Anodic polarisation curves for steel in concrete (schematic).

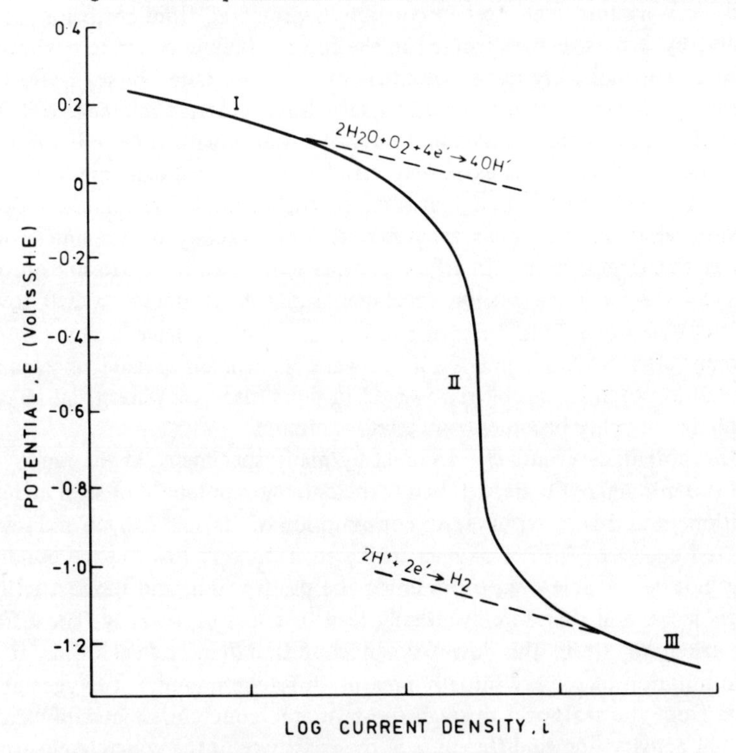

Fig. 7 – Cathodic polarisation curve for steel in concrete (schematic).

(iii) *Current flow between two coupled specimens.* This can provide a direct indication of corrosion induced by a galvanic couple between areas of reinforced concrete in different conditions. However, it does not measure 'self-corrosion' of either part of the couple or take into account the effect of different ratios of cathodic and anodic area.

STEEL IN SOUND CONCRETE

Concrete Fully Immersed in Sea water

There is considerable evidence that the rate of corrosion of steel embedded in *sound* concrete fully immersed in sea water is negligibly small. Steel reinforcement showing no significant general corrosion has been recovered from concrete structures which have been immersed in sea water for many years; for example a core taken from the Tongue Sands Naval Fort examined in a Concrete-in-the-Oceans project after 35 years off the Thames Estuary [9]. In our own work, we have recently examined a cylindrical steel/concrete specimen of the type shown in Fig. 8, which showed no visible trace of corrosion after immersion in constantly refreshed sea water for more than four years, and we confidently expect that similar conditions exist in specimens over seven years old.

We can assume that steel in concrete is protected from corrosion, at least initially, by a passive film formed in the highly alkaline concrete environment and specimens normally show potentials on first immersion in sea water in the range 0 to −100 mV with respect to a silver/silver chloride reference cell. These potentials usually tend towards more negative values with time, but the rate at which the potential falls is extremely variable, ranging for example, from −100 mV to over −800 mV in one year for apparently identical specimens in flowing sea water (Fig. 9). Potentials generally fall more rapidly on specimens where voids at the steel/concrete interface provide local sites for corrosion on otherwise passive steel. Occasionally, specimens show no tendency to drift towards more negative values, and potentials as high as −200 mV have been recorded on specimens after 5 years immersed in sea water. On the other hand, an even more rapid fall in potential has been observed in pressurised sea water (Fig. 10) when the concrete quickly becomes completely saturated.

The potentials eventually attained by many specimens, in the range −800 to −1100 mV Ag/AgCl, are well below the corrosion potential of steel in aerated conditions, and this is attributed to consumption of residual oxygen and severely restricted access of further oxygen to the steel. In very low oxygen conditions it may not be possible even to maintain the passive film, and the restriction of oxygen access will therefore eventually lead to a loss of passivity. The different times taken to attain this low oxygen condition may reflect different rates of consumption of oxygen initially present, different amounts of oxygen able to diffuse from the sea water through the saturated concrete, or possibly variable local cell activity due to differences in the resistivity of the concrete electrolyte.

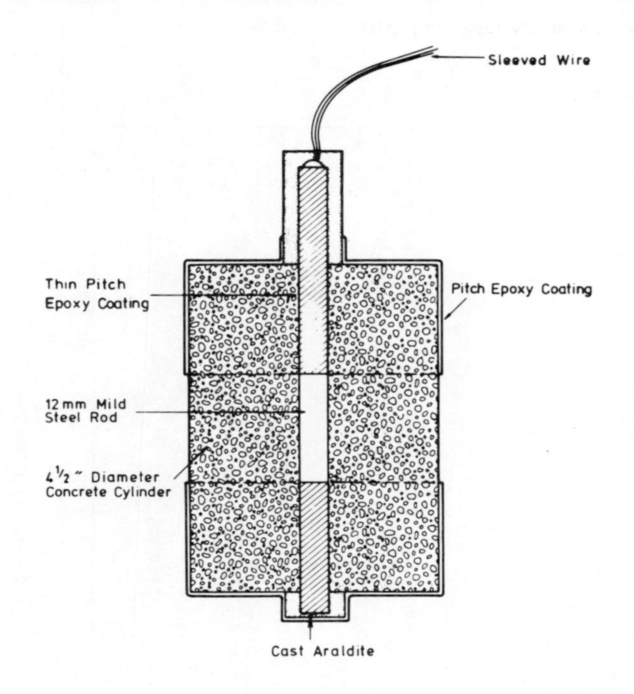

Fig. 8 – Test specimen – reinforcing steel in concrete.

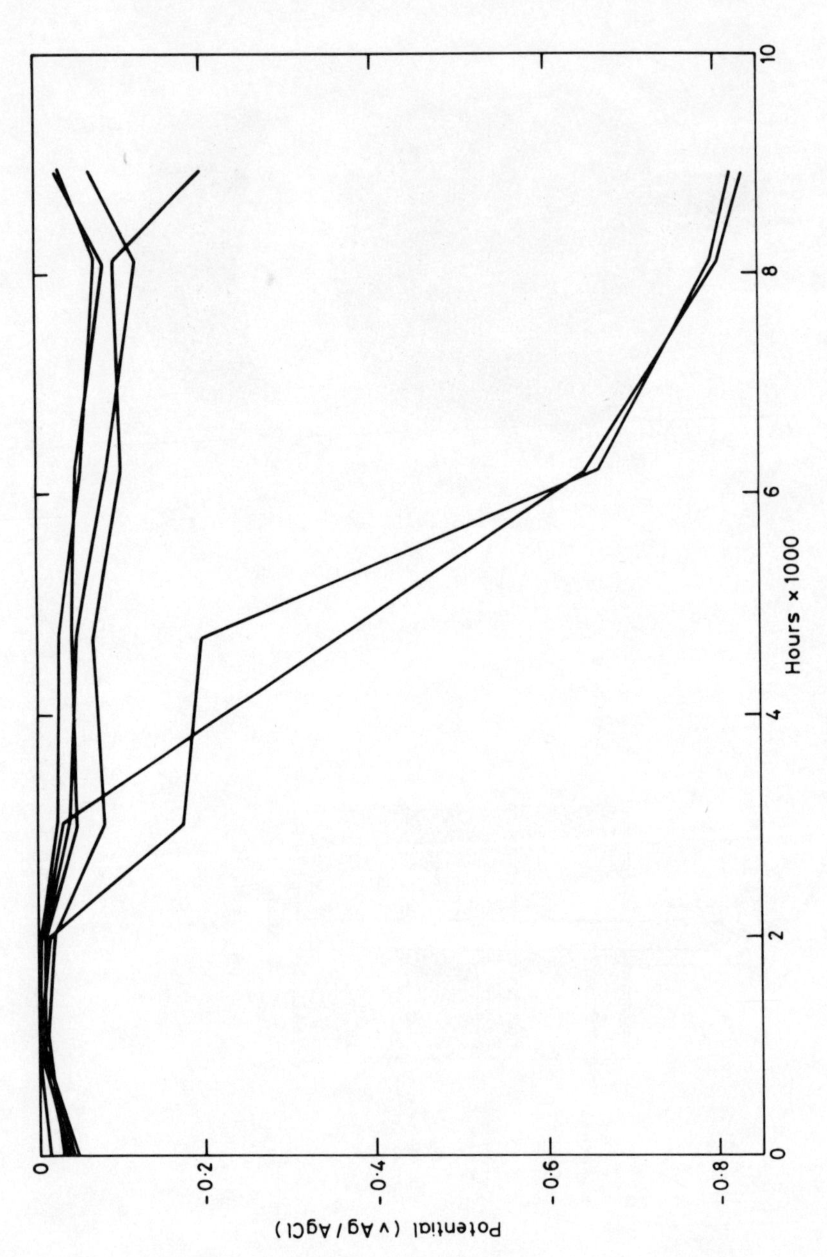

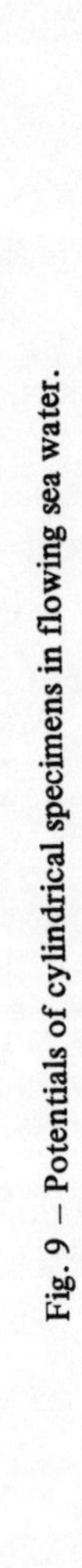
Fig. 9 – Potentials of cylindrical specimens in flowing sea water.

In the case of pressurised sea water the higher conductivity of concrete fully saturated with sea water, by allowing increased local cell activity, might tend to accelerate consumption of oxygen and cause the more rapid fall in potential usually observed in these conditions.

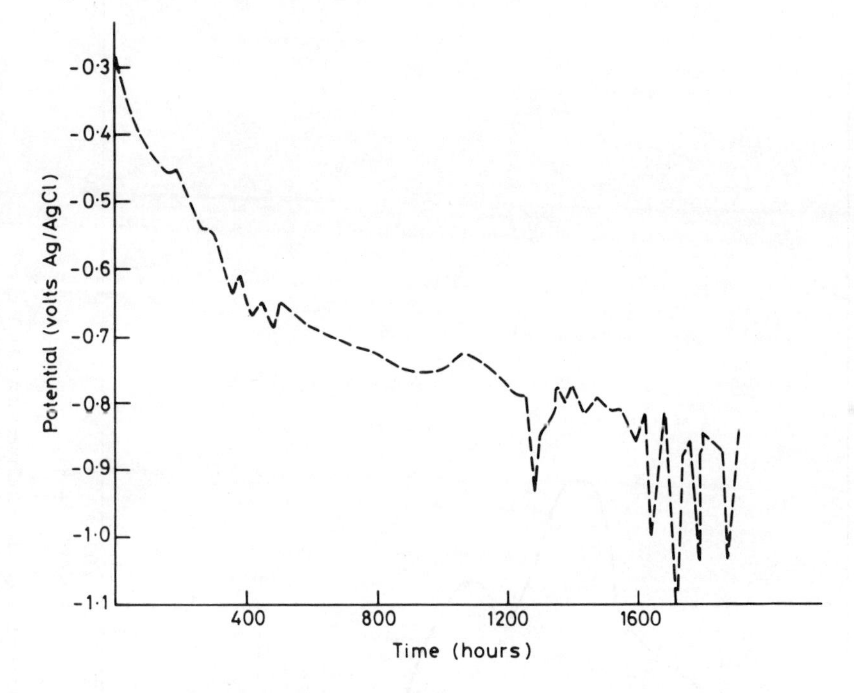

Fig. 10 – Potential of cylindrical specimens in flowing sea water at 500 psi.

It is important to note that removal of oxygen by corrosion does not necessarily imply an irreversible breakdown of the passivity imparted to steel by the alkaline concrete environment, even though in some cases chloride may have already diffused to the steel/concrete interface. Anodic polarisation curves produced at a scan rate of 10 mV per minute indicate repassivation at potentials above about −600 mV (Fig. 11). These are, of course not equilibrium curves, and the '*iR* drop' will also introduce an error of several tens of millivolts at the highest (repassivation) current. Nevertheless the trends are clear, and the 'passive current' gives at least an upper limit to the steel corrosion rate. In the example shown in Fig. 11 the 'limiting' current of $\sim 10\ \mu A$ applies to a steel surface of $\sim 20\ cm^2$, i.e. a current density of $0.5\ \mu A\ cm^{-2}$ equivalent to $< 10^{-3}$ cm per year corrosion.

It appears then that if potentials lie in the range −800 to −1100 mV Ag/AgCl the steel is corroding actively, but the corrosion is spread over the largest possible area of reinforcement and the actual corrosion rates will be

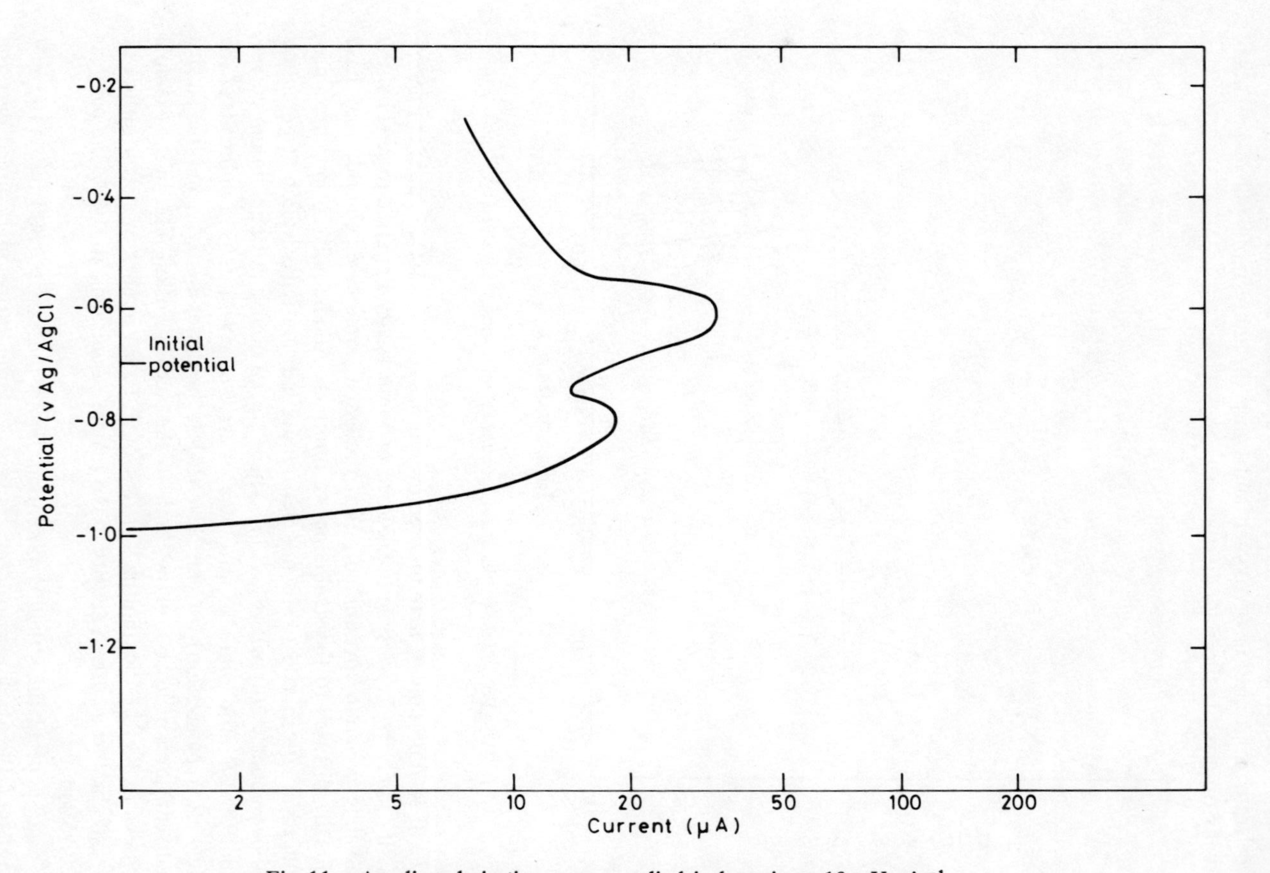

Fig. 11 – Anodic polarisation curve – cylindrical specimen 10 mV min^{-1}.

extremely small. However, the critical balance between corrosion and oxygen diffusion to the steel is clearly shown by our observation that the potential of an immersed specimen covered by silt, and therefore in a particularly low oxygen environment, increased from −900 mV to −200 mV Ag/AgCl on re-exposure to flowing sea water. In this case the reason for the potential change is clear, but it is not always easy to determine whether different potentials observed on specimens exposed to identical conditions of immersion are due to differences in the current required to maintain 'passivity', or to different rates of diffusion of oxygen to the steel surface.

Concrete with Access to Air

In concrete at or near the tidal zone, or possibly in concrete exposed to water on one side and air on the other, as in a hollow reinforced concrete tower in the sea [10], oxygen may be more readily available at the surface of the embedded steel, partly because of the higher concentration of oxygen at the free concrete surface, and partly because some of the capillary pores of the concrete may contain air rather than water. In these circumstances, potentials as high as +50 mV Ag/AgCl have been observed and potentials more negative than −300 mV would almost certainly indicate a breakdown in passivity and significant corrosion of the embedded steel. The 'hollow leg' effect is particularly difficult to assess since its cathodic behaviour will depend on a number of factors, including water penetration through the wall thickness, influenced in turn by sea water pressure (depth) and concrete porosity and/or permeability.

STEEL IN CRACKED CONCRETE

While great efforts are made to produce 'sound' concrete in reinforced concrete marine structures, defects will almost certainly exist, which might range from the so-called 'design cracks' which are limited in codes to various surface widths (usually ranging from 0.1 to 0.3 mm), to much wider cracks or spalled areas caused by mechanical damage. The latter will almost certainly result in early reinforcement corrosion in marine atmosphere and splash zone conditions. The question of 'design' crack width, linked with depth of cover and load on the reinforcement in atmospheric conditions, has been reviewed in detail by Beeby in a Concrete-in-the-Oceans report [11], and he pointed out that corrosion processes underwater might lead to quite different conclusions from those drawn from 'normal' situations. However, the effect on corrosion of cracks or damage underwater has been given relatively little attention until recently.

It is particularly important to recognise that for fully immersed conditions there exists the possibility that corrosion at a damaged area could be greatly accelerated by galvanic coupling to a 'cathode' of reinforcing steel in the remaining undamaged concrete. As shown previously, the sea water will provide a very low electrolyte resistance path compared with even 50–75 mm of concrete cover over the reinforcement (Fig. 4), and there is therefore the possibility of including very large areas of 'cathode' in such a galvanic cell.

The Concrete Cathode

An important deduction from the delicate balance between corrosion and oxygen availability in fully immersed specimens previously discussed, is that the cathodic (oxygen reduction) current density must be of the same order as the critical 'repassivation' current density for the steel. For our cylindrical specimens the repassivation current seems to be about 50–100 μA, and cathodic polarisation of a passive fully immersed specimen shows that a current of about this magnitude would be drawn if it was coupled to corroding steel in sea water. Therefore, even if negligible self-corrosion occurred on the embedded steel, fully immersed concrete would be an extremely poor cathode; capable of supplying, (assuming that the effective rebar surface area and the superficial concrete surface area are approximately the same), not more than ~ 5 μA cm^{-2} when coupled to steel corroding in sea water. At the other extreme, where self-corrosion accounts for all the oxygen available, the steel embedded in concrete would be anodic to bare steel and would tend to 'cathodically protect' it to the limit of the repassivation current.

The balance between these extremes of anodic and cathodic behaviour of embedded steel is well illustrated by an experiment in which a sound cylindrical concrete specimen (A) was coupled to a similar specimen (B) in which a 1 mm wide parallel side slot had been cut to expose a small area of steel. The free potential of specimen (B) (−640 mV) was dominated by the exposed steel, while specimen (A) eventually reached a more active potential (−740 mV). When the specimens were first coupled together (Fig. 12) specimen (A) cathdically protected the steel as expected, being itself anodically polarised at a current of ~ 10 μA; but after about 90 minutes specimen (A) passivated with a reversal of current (~ 80 μA, specimen (B) anodic) and with a corresponding shift of the couple potential towards a less negative value. Subsequently current reversals occurred at irregular intervals, accompanied by the potential changes shown in Fig. 12.

The difficulty of assessing the cathodic capacity of immersed concrete exposed in some way to air through either tidal or 'hollow leg' effects has already been discussed. However, another deduction from the balance between corrosion and oxygen availability in immersed concrete specimens is that it would probably require very little extra oxygen access to ensure that steel in *sound* concrete remained passive, and therefore to ensure that quite large areas of concrete cathode would be available to take part in a galvanic corrosion cell. Even if the cathodic capacity of this concrete is very small, a sufficiently large cathode polarising a small anodic area could then result in significant corrosion. It is worth noting that during its cathodic cycles, specimen (A) in the couple shown in Fig. 12 was polarising a very small area of steel (0.2 cm^2) at up to 80 μA, corresponding to a relatively large anodic current density of 400 μA cm^{-2}. This, if applied continuously, is equivalent to nearly 4 mm per year corrosion.

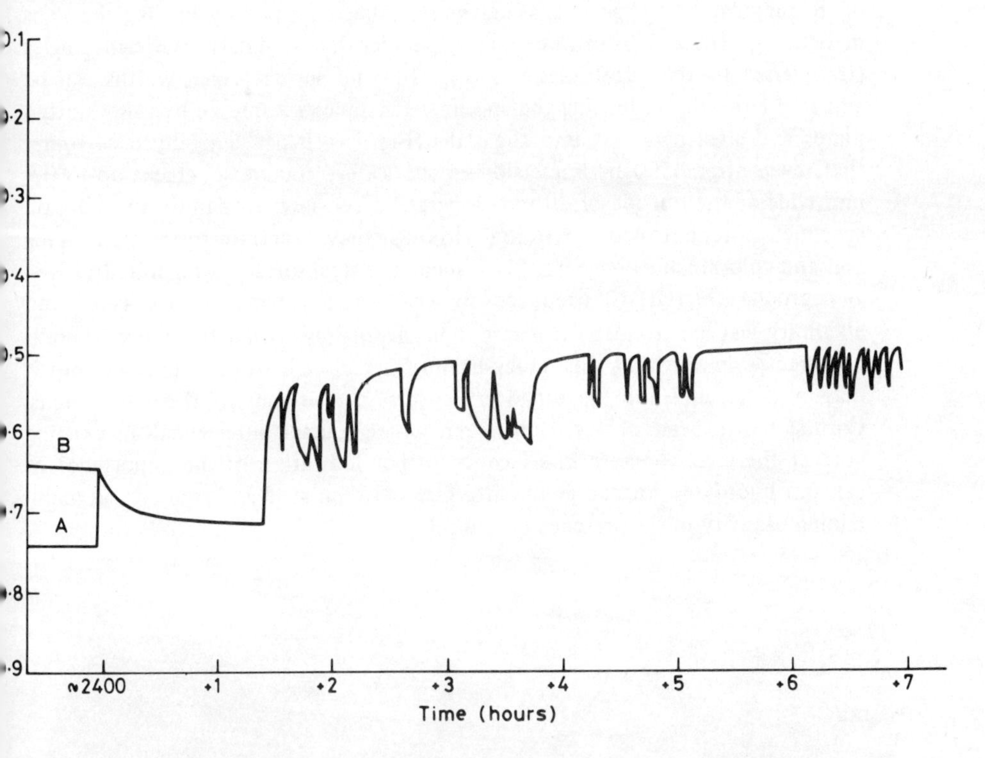

Fig. 12 – Potential of couple between sound specimen (A) and defected specimen (B).

The Cracked Concrete Anode

It is apparent from the experiment described above that with a 1 mm wide parallel-sided slot, the exposed steel at the tip of the 'crack' merely acted as a small area of bare steel in sea water. A real crack penetrating through concrete to the steel would normally be narrower than this at the surface and, in most cases, it would taper to become even narrower near the steel. Furthermore, such a crack would probably not initially disrupt the protective film at the steel/concrete interface.

Rather more realistic cracks were produced in our cylindrical specimens by cutting a circumferential slot in the middle of the cylinder and pulling in a tensile machine until the concrete cracked. The slot determined the position of the crack and served to support wedges which were inserted to maintain a predetermined external crack width (Fig. 13). These cracks still tend not to taper sufficiently and in fact surface crack widths greater than about 0.5 mm could only be achieved by slight yielding of the bar. In more recent experiments on rectangular 'slab' specimens, cracks have been formed by loading the slabs in bending. These slabs can be cracked parallel (longitudinal) or at right angles (transverse) to the reinforcing bar (Fig. 14) and surface crack widths can be adjusted either by re-loading the specimens in a metal frame, or by using slotted plates and steel bars cast into the slabs (Fig. 15). It has generally been found that, even after 4500 hr immersion in sea water, 'transverse' cracks up to 0.3 mm wide at the outside of 40 mm concrete cover have no significant effect on specimen potential or on corrosion, although they penetrate to the reinforcing bar, and chloride must have quickly reached the steel surface. Aragonite ($CaCO_3$) and brucite ($Mg(OH)_2$), produced by the reaction between sea water and alkalinity leached from the concrete, deposit, mainly within the crack. There is no direct evidence that this 'crack-healing' affects corrosion behaviour, but it may well account for the long-term corrosion resistance of these specimens. Corrosion has been observed, however, where a crack intersected an existing void at the steel/concrete interface, a further indication of the importance of calcium hydroxide formed at the interface between steel and concrete, in maintaining passivity in the presence of chloride.

Fig. 13 – Partially prepared specimen showing 1 mm wide crack.

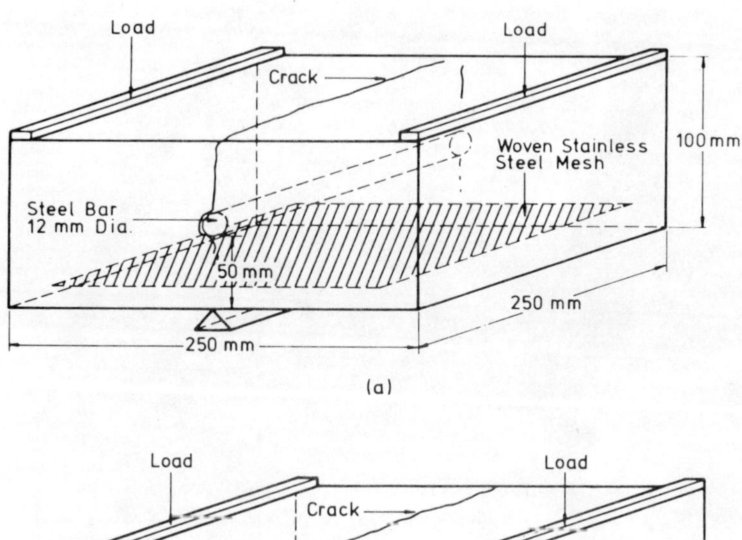

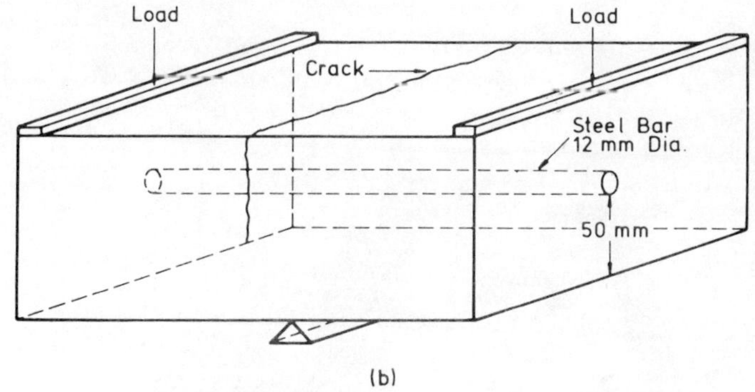

Fig. 14 – Slab specimens, (a) transverse crack, (b) longitudinal crack.

In contrast to the 0.3 mm cracks, cracks 0.6 mm and 1 mm wide on cylindrical specimens have generally been the site of active corrosion. This was particularly evident in an experiment in which specimens were polarised to a controlled potential of −200 mV Ag/AgCl to represent coupling to a large area of cathodic concrete. Quite significant currents were sustained (Fig. 16) representing considerable corrosion of the steel near the crack (Fig. 17), although in the absence of the 'large cathode' no significant corrosion occurred. The effect of crack geometry is shown more clearly by the 'slab' specimens, in which transverse and longitudinal cracks in a range of widths have been polarised to −200 mV Ag/AgCl. This work is still in progress, but results clearly show that corrosion occurs with quite narrow longitudinal cracks, whereas steel intersected by transverse cracks at least up to 1 mm wide, remain passive.

Fig. 15 – Slab specimens with 1 mm cracks, (a) transverse, (b) longitudinal.

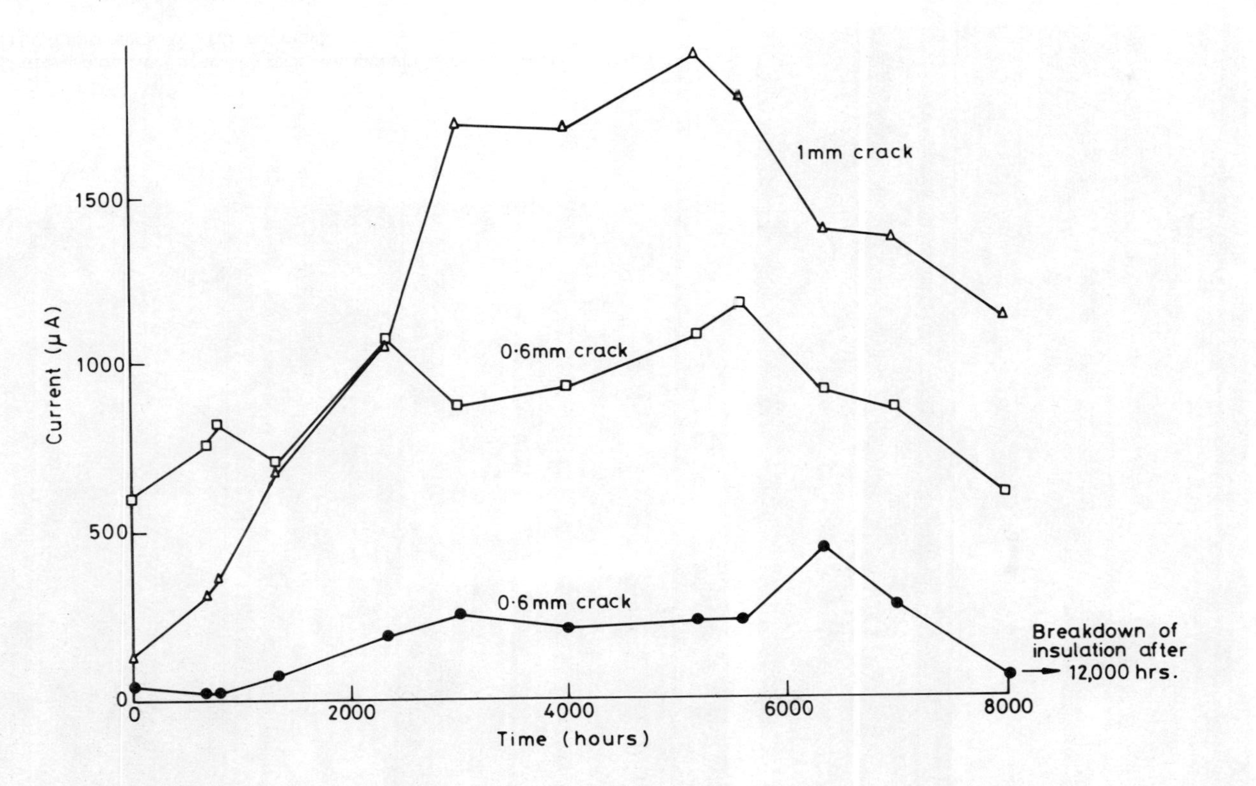

Fig. 16 – Variations of current with time for cracked specimens polarised at −200 mV Ag/AgCl.

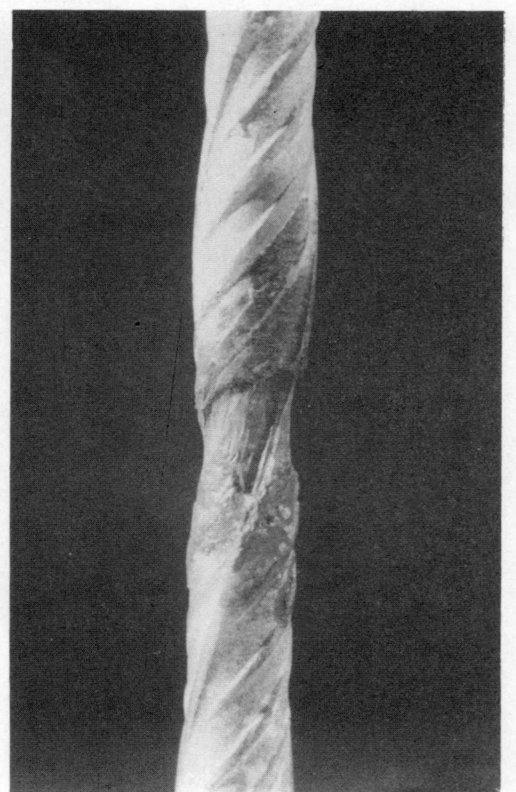

Fig. 17 – Corrosion on steel beneath cracks on specimens polarised at −200 mV Ag/AgCl, (a) 0.6 mm crack, (b) 1.0 mm crack.

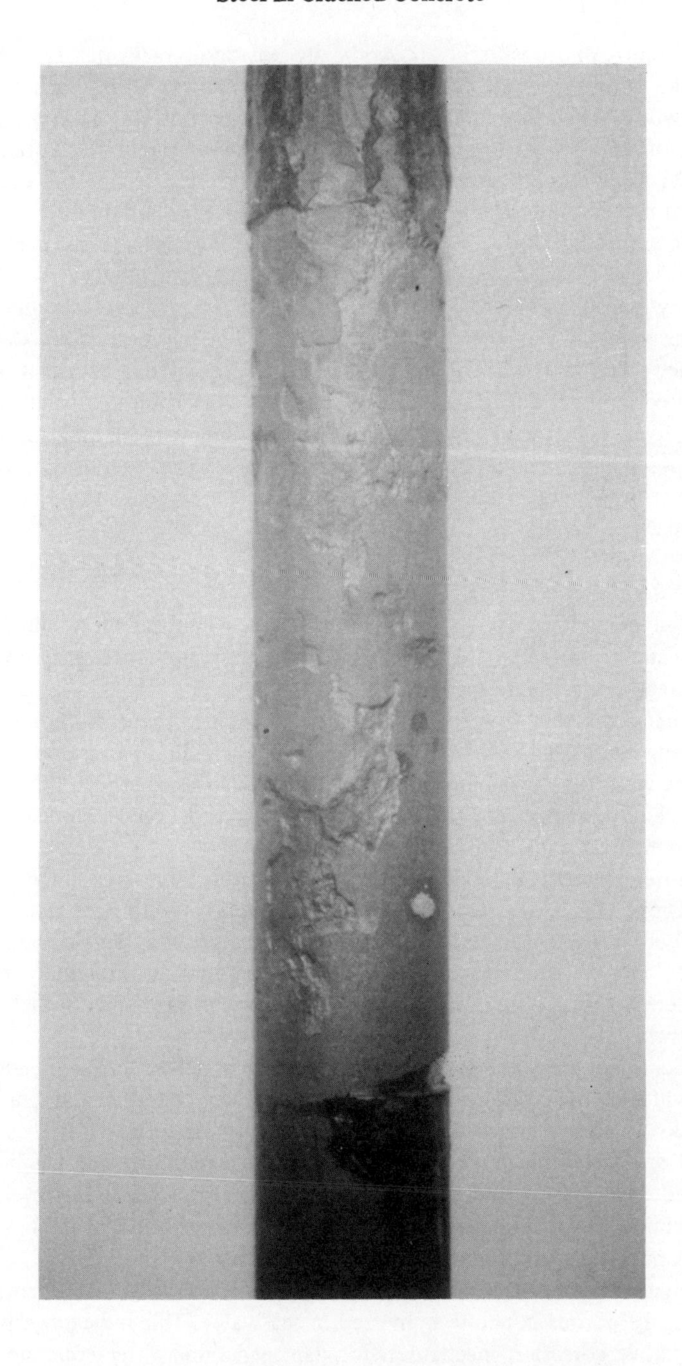

Fig. 18 – Localised corrosion of steel embedded in low-fines concrete.

When corrosion occurs at an anodically polarised crack, rust is sometimes seen at the mouth, but corrosion products within the crack are black or white/ greenish-white. The appearance of greenish-white ferrous corrosion products is evidence of galvanic corrosion when the active anodic area is essentially free from oxygen, and these products are quickly converted to the more familiar red-brown rust on exposure to air. The black product in the intermediate oxidation state approximating to Fe_3O_4, is more stable, but this also oxidises slowly in the presence of air and moisture to form hydrated ferric oxides.

Highly porous concrete specimens, produced from a low-fines mix, introduce large voids at the steel reinforcement with direct access of sea water. As with other voids intersected by cracks, galvanic corrosion is confined to areas not in direct contact with the concrete (Fig. 18). Anaerobic corrosion has also been observed on steel in low-fines concrete specimens after long immersion, and it is possible that similar conditions might arise in severely cracked or crushed concrete at areas of damage. However, corrosion rates under these conditions seem unlikely to be very high.

SUMMARY

(1) For all practical purposes steel embedded in sound concrete immersed in sea water is protected from corrosion, even though chloride may diffuse to the steel/concrete interface.

(2) Normally the very low rate of oxygen diffusion in immersed concrete is a limiting factor, and embedded steel becomes anodic to bare steel exposed to sea water. However, in some circumstances (particularly in tidal areas or in a 'hollow leg' where there is better access of oxygen), embedded steel can remain cathodic to bare steel.

(3) Because of the high conductivity of sea water, large areas of embedded steelwork can be included in galvanic interactions with bare steel or steel exposed at areas of damage to the concrete. However, a 'large cathode/small anode' couple, which can result in serious corrosion, is only likely to occur if there are small areas of cracking or damage in a large 'cathodic' area of concrete.

(4) Cracks in concrete are not necessarily the site of active corrosion, and there is evidence of crack-healing reactions between concrete and sea water. Cracks up to 0.6 mm wide (surface width) may be tolerated if they run at right angles to the reinforcing bar, although some corrosion has been observed at voids intersected by cracks only 0.3 mm wide. However, there are indications from other Concrete-in-the-Oceans work that different conclusions might apply to cracks of other geometry.

(5) Sulphate-reducing bacteria corrosion has been observed on steel embedded in highly porous concrete immersed in sea water. This indicates a possible alternative corrosion mechanism for damaged underwater concrete in the absence of oxygen.

REFERENCES

[1] Cornite Euro-International du Beton, *Bulletin d'Information No. 148,* January 1982.

[2] C. L. Page and K. W. J. Treadawy, *Nature,* **297**, 5862 pp. 109–115 (1982).

[3] N. J. M. Wilkins and P. F. Lawrence, *Concrete in the Oceans – Technical Report No. 6,* 1980. Concrete in the Oceans Phase II Project, Pla. Interim Reports Nos. 1 and 2.

[4] M. Pourbaix *et al., Atlas of Electrochemical Equilibria in Aqueous Solutions,* Pergamon Press, 1966.

[5] C. L. Page, *Nature,* **258,** 514 (1975).

[6] M. Pourbaix, *Corrosion Science,* **14,** 25 January (1974).

[7] D. A. Housmann, *Materials Protection,* **8,** 10, pp. 23–25 (1969).

[8] R. C. Robinson, *ACI Publication SP49-7,* 1975, pp. 83–93.

[9] R. D. Browne *et al., Concrete in the Oceans Technical Report No. 5.*

[10] H. Arup, *Steel in Concrete Newsletter No. 1,* Korrosioncentralen, Copenhagen, October 1978.

[11] A. W. Beeby, 'Cracking and Corrosion', *Concrete in the Oceans Technical Report No. 1,* 1978.

Chapter 9

The influence of chlorides and sulphates on durability

W. R. HOLDEN, C. L. PAGE and N. R. SHORT, University of Aston, Birmingham, UK

SUMMARY

The use of poorly prepared aggregates, contaminated with mixtures of salts containing chloride and sulphate ions, presents risks of corrosion of reinforcement in concrete. There is an urgent need to assess these risks, particularly with regard to the durability of reinforced concrete structures in the Middle Eastern Gulf regions where salt-contaminated aggregates proliferate.

In this paper, techniques of pore solution expression and analysis are used to quantify the extent to which chlorides and sulphates react with different cements to form insoluble products when various dosages of salts are included in the mix materials. The influence of cement composition is examined for a range of cements produced in Britain. Kinetics of chloride ion diffusion in different types of cement matrix are also considered.

The importance of the above factors in relation to the mechanism of chloride-induced pitting of reinforcing steel is examined and implications regarding corrosion risks are discussed.

INTRODUCTION

Embedded steel reinforcement in dense concrete normally shows good long-term durability. It is generally accepted that the steel is passivated in view of the high alkalinity of the associated pore solution, which in Portland cement concretes often has a pH value in excess of 13. Corrosion problems arising in this environment are a result of depassivation of the steel owing to the presence of aggressive agencies such as acidic gases or chloride ions, the latter being present as original constituents of the mix or as a result of penetration from the environment.

In the UK use of chloride-containing admixtures is now discouraged for concrete containing embedded metal [1] although penetration from the environment may still occur, particularly in the case of structures exposed to sea water or de-icing salts. However, in the Middle East the deterioration problems are much more severe on account of the particularly aggressive environment [2]. Adverse climatic conditions introduce problems in the manufacture, placing and curing of concrete that may result in cracking, reduced strength, increased permeability and reduced durability of the hardened product. Chlorides may be present as original constituents of the mix in the form of contaminated aggregates or mixing water or as wind-borne contamination of the reinforcement. Ingress of chlorides can also take place through cracks or by absorption from the surface. Furthermore, the presence of sulphates in combination with chlorides may accentuate corrosion problems for reasons that will be considered in this paper.

The mechanism of steel corrosion in concrete, with particular reference to the electrochemical behaviour of the steel and physical performance of the concrete cover, has been recently reviewed [3]. The presence of chloride ions can cause depassivation of the steel even when the associated pore solution has a high pH value and corrosion is often in the form of intense localised attack (pitting). Experience from corrosion problems in the field suggests that under UK conditions, the risk of corrosion in concrete made from OPC is small when chloride ion contents (by weight of cement) are $<0.4\%$ and high for contents $>1.0\%$ [4]. Although the mechanism of depassivation by chloride ions is not yet fully understood, it is known that the risk of corrosion increases as the ratio of the concentration of aggressive chloride ion to concentration of inhibitive hydroxyl ion in solution increases [5].

The levels of free chloride and hydroxyl ions in the pore solution are governed by several factors. The concentration of hydroxyl ions would be expected to increase with the alkalinity of the cement although several processes have been suggested which may tend to reduce particularly high alkalinities [3]. There is considerable evidence that the C_3A content of the cement is important since it is known that C_3A may form an insoluble complex, calcium chloroaluminate hydrate (Friedel's salt), thus reducing the concentration of chloride in the pore solution [6]. The presence of sulphate ions may result in liberation of chloride ions owing to the preferential formation of calcium sulpho-aluminate hydrates [7]. Other factors such as temperature, crystallographic form of C_3A and the decomposition of calcium chloroaluminate hydrate by carbonation may also be important in determining the extent of chloride complexation. Furthermore, it has been suggested that chloride present on mixing is complexed to a greater extent than chloride penetrating hardened concrete from an external environment [8].

A critical factor concerning sustainability of corrosion is the mobility of chloride and hydroxyl ions in the vicinity of the metal/concrete interface. For corrosion to proceed, the relative rates of transport of chloride and hydroxyl ions near anodic sites must be such as to give rise to locally elevated chloride concentrations and depressed hydroxyl concentration. It is therefore significant that the diffusivity and activation energy for chloride ion transport through different hardened cements have been found to vary widely, the mechanism of diffusion depending on the pore structure and mineralogy of the cement [9].

A further factor is the diffusion of oxygen to the steel surface but apart from the case of fully submerged concrete this is probably not as important as the relative diffusion rates of chloride and hydroxyl ions.

In view of questions raised by the above considerations an extensive programme of work on the effect of cement matrix variables on the corrosion behaviour of steel in concrete has been initiated. This programme involves:

(i) the expression and chemical analysis of pore solution from within a wide range of hardened UK and overseas cement pastes;
(ii) the determination of effective diffusivities and activation energies for chloride ion transport through a selection of the above range of hardened cement pastes;
(iii) electrochemical study of the corrosion of steel in a wide range of cement pastes and concretes using rest potential, and polarisation resistance [10] techniques;
(iv) field trials using concrete samples manufactured, cured and exposed under different environmental conditions, and incorporating cement matrix variables identified as being significant from section (i)–(iii).

Some initial results of sections (i) and (ii) of the programme are discussed in this paper.

EXPERIMENTAL

(a) Expression and Analysis of Pore Solution

The cements used were two varieties of Ordinary Portland Cement (OPC-A and OPC-B), two blended cements prepared from OPC-B and 30% pulverised fuel ash (PFA) or 65% ground granulated blast furnace slag (BFS) and a Sulphate Resisting Portland Cement (SRPC). The compositions of the various materials, expressed in percentages by weight of the constituent oxides, are shown in Table 1.

Table 1 – Chemical analyses of cements and blending agents

Material	CaO	SiO_2	Al_2O_3	Fe_2O_3	SO_3	MgO	Na_2O	K_2O	Ignition loss
OPC-A	62.8	20.8	5.1	3.4	2.9	1.3	–	0.96	1.6
OPC-B	63.0	20.3	7.1	2.7	3.3	1.3	0.47	0.60	0.8
SRPC	64.0	20.2	4.1	5.3	2.6	1.4	0.28	0.39	1.1
BFS	42.4	33.3	10.8	0.3	–	8.7	0.37	0.50	–
PFA	2.9	46.6	24.0	9.5	0.9	2.1	2.00	3.80	3.3

Samples of cement paste were made with distilled water to W/C 0.5 in three parallel series: (1) with no additions, (2) with 0.4% chloride (as Cl by weight of the cement), (3) with 0.4% chloride and 1.5% sulphate (as SO_3 by weight of the cement) added as AR anhydrous sodium chloride and anhydrous sodium sulphate dissolved in the mixing water. After being mixed thoroughly they were cast into cylindrical PVC moulds, 49 mm diameter × 75 mm length, and compacted by vibration. The completely filled moulds were tightly sealed, stored at 22°C and demoulded at 84 days. Immediately after demoulding the pore solution was expressed using a pressure vessel of similar design and construction to those used by other workers [11–13]. The maximum pressure applied was *ca* 350 MPa and expressed pore solution was collected in a plastic syringe from the fluid drain at the base of the apparatus. Samples were stored in polystyrene vials and sealed to avoid undue exposure to the atmosphere. The concentration of chloride in the expressed pore solution was determined using a standard spectrophotometric technique [14] and pH determined by titration of expressed pore solution against 10 mM nitric acid using phenolphthalein as indicator.

(b) Determination of Chloride Diffusivity

The range of cements used was as in (a). A convenient technique for the study of chloride diffusion kinetics in hardened cement pastes has been established and details of specimen preparation, the diffusion cell and calculation of effective diffusivity of chloride ions are reported elsewhere [9].

RESULTS AND DISCUSSION

(a) Detailed results of the analyses of pore solutions from these and other cements are to be published separately and only general trends will be considered here. A summary is presented in Table 2, the data in most cases being averages of results obtained for three specimens. Using this data it is possible to establish a number of important points. From a general observation of the results in Table 2, it is evident that the cements vary quite markedly in their alkalinity and ability to complex chloride ions.

Considering the results for the series (2) experiments, a ranking of the cements in order of free $[Cl^-]/[OH^-]$ may be expected to have some value as a means of assessing the degree of corrosion protection afforded to embedded steel. Within this ranking the major difference was between the low C_3A (1.9%) SRPC and the other cements, presumably because of the former's limited capacity to form Friedel's salt ($3CaO.Al_2O_3.CaCl_2.10H_2O$). However, there was also a substantial difference in the performance of the two OPC samples which again probably reflects their C_3A contents OPC-A (C_3A content 7.7%) having a lower chloride complexing ability than OPC-B (C_3A content 14.3%).

Table 2 – The concentration of chloride and hydroxyl ions (mmole 1^{-1}) in pore solution expressed from the hardened cement paste specimens.

Cement	Series (1) No additions			Series (2) + 0.4% Cl			Series (3) + 0.4% Cl, 1.5% SO_3		
	$[Cl^-]$	$[OH^-]$	$\left[\frac{Cl^-}{OH^-}\right]$	$[Cl^-]$	$[OH^-]$	$\left[\frac{Cl^-}{OH^-}\right]$	$[Cl^-]$	$[OH^-]$	$\left[\frac{Cl^-}{OH^-}\right]$
OPC-A	2	589	0.003	83	741	0.112	215	1318	0.163
OPC-B	3	479	0.006	41	661	0.062	153	1047	0.146
OPC-B/ 30% PFA	2	339	0.006	39	457	0.085	145	851	0.170
OPC-B 65% BFS	5	355	0.014	28	457	0.061	147	741	0.198
SRPC	2	347	0.006	110	501	0.220	257	1000	0.257

The influence of the blending agents (BFS and PFA) was to cause a modest reduction in alkalinity of the pore solution. This reduced alkalinity of the blended cements has led to the suggestion that these materials will offer reduced corrosion protection to embedded steel [15], but long-term experience of concrete manufactured from blended cements has indicated that they may, in fact, be superior to OPC in chloride-containing environments [16, 17]. This may be partly explained by the fact that there is no reduction in the chloride binding capacity. Indeed, the chloride binding capacity of the slag cement was significantly higher than that of the corresponding OPC. Thus the ranking of blended cements in terms of $[Cl^-]/[OH^-]$ compared fairly well with OPC. However, it should be emphasised that these findings may not be generally applicable to cements containing slags or pozzolanas from other sources, see, for example, reference [13].

Consideration of the results for series (3) experiments shows that the combined additions of sodium chloride and sodium sulphate, in all the cements studied, resulted in a substantial decrease in chloride binding capacity. This presumably reflects the tendency of sulphate ions to react preferentially with the C_3A phase thus inhibiting the formation of Friedel's salt. An increase in pore solution pH was also observed as hydroxyl ions entered solution to balance the anions removed in the form of insoluble complex salts. However, the net effect of 1.5% SO_3/0.4% Cl additions was to yield a higher free $[Cl^-]/[OH^-]$ than was the case for 0.4% Cl in the absence of sulphate. This supports the suggestion that corrosion risks are likely to be significantly increased in circumstances where concrete materials are subjected to contamination with both chlorides and sulphates, as is commonly the case in the Middle Eastern Gulf regions. A further point, which although not of direct concern to the subject of this discussion is nevertheless significant, is the increased risk of alkali-aggregate reaction that would be expected to arise from the high pH values recorded for samples containing sodium chloride and sodium sulphate.

(b) Values of effective diffusivity of chloride ions in hardened cement pastes at 25°C are given in Table 3. From these results, it can be seen that cement composition has a strong influence on the diffusion of chloride ions in hardened cement pastes. Although values for the two different OPC pastes are similar, values for the two blended cements are significantly lower, and for the SRPC significantly higher, than the OPC. As a second means of assessing likely corrosion protection it is thus possible to rank the cements in terms of their capacity for limiting chloride ion supply and thus discouraging pitting of embedded steel. Possible reasons for the large differences of diffusivity, based on complexing ability and pore structure have been considered elsewhere [10]. In this context, the fact that the C_3A contents of the OPC-A and OPC-B samples were significantly different whilst their values of chloride diffusivity were fairly close suggests that C_3A content is of minor importance in relation to the kinetics of chloride ion transport.

Table 3 – Effective diffusivity of chloride ions at 25°C in various cement pastes of W/C 0.5.

Type of cement	Diffusivity ($\times 10^8$) ($cm^2 s^{-1}$)
OPC-A	3.14
OPC-B	4.47
OPC-B/30% PFA	1.47
OPC-B/65% BFS	0.41
SRPC	10.00

CONCLUSIONS

The technique of pore solution expression and analysis has shown that the cements studied vary considerably in their alkalinity and ability to complex chloride ions. A rank order of cements has been established in terms of $[Cl^-]/[OH^-]$ in the expressed pore solutions. This ranking shows that there was a major difference between the SRPC and the OPC samples and a significant difference between the two types of OPC. These differences probably reflect the difference in C_3A contents of the cements. The addition of PFA or BFS to OPC resulted in a modest decrease in the alkalinity of the cement but gave superior chloride complexing ability, with the overall result that the ranking of the blended cements compared well with the corresponding OPC.

The combined presence of chloride and sulphate ions resulted in a substantial decrease in the chloride binding capacity of the cements. Thus the corrosion risk to embedded steel is likely to increase in these circumstances. Furthermore there is an increase in the alkalinity of the pore solution, which enhances the risk of alkali–aggregate reaction.

Diffusivities of chloride ions in hardened cement pastes have been used to establish a further rank order of the cements in terms of their ability for limiting chloride ion supply to anodic corrosion sites. There is little difference in ranking between the different OPC, but the blended cements are significantly better and SRPC significantly worse than the corresponding OPC.

Electrochemical investigations are being carried out to test the validity of these rankings and it is hoped that results will be available for oral presentation.

ACKNOWLEDGEMENTS

The authors are grateful to the SERC and to the Department of the Environment, Building Research Establishment for financial support of this research. Sincere thanks are also due to Mr K. W. J. Treadaway for his advice and encouragement.

REFERENCES

[1] British Standard Code of Practice, CP110 Part 1:1972 (amended May 1977).
[2] E. A. Kay, P. G. Fookes and D. J. Pollock, *Concrete*, **15**, 22 (1981).
[3] C. L. Page and K. W. J. Treadaway, *Nature*, **297**, 109 (1982).
[4] L. M. Everett and K. W. J. Treadaway, *BRE Information Paper IP 12/80*, 1980.
[5] V. K. Gouda, *Brit. Corros. J.*, **5**, 198 (1970).
[6] F. M. Lea, *The Chemistry of Cement and Concrete*, 3rd edn., p. 232. Edward Arnold, London, (1970).
[7] J. Calleja, *7th Int. Congress on the Chem. of Cement, Paris, 1980*, Vol. 1, VII-2/1, Paris, France, 1980.
[8] R. D. Browne and M. P. Geoghegan, *Proc. Symp. Corrosion Steel Reinforcements in Concrete Construction*, p. 79. Society of Chemical Industry, London, 1979.
[9] C. L. Page, N. R. Short and A. El Tarras, *Cement and Concrete Research*, **11**, 395 (1981).
[10] C. Andrade and J. A. González, *Werkstoffe und Korrosion*, **29**, 515 (1978).
[11] P. Longuet, L. Burglen and A. Zelwer, *Rev. de Mat. de Const. et de Trav. Pub.*, **676**, 35 (1973).

[12] R. S. Barneyback Jr. and S. Diamond, *Cement and Concrete Research,* **11**, 229 (1981).

[13] C. L. Page and O. Vennesland, *Materials and Structures,* in press.

[14] A. I. Vogel, *A Textbook of Quantitative Inorganic Analysis,* 4th edn., revised by J. Bassett, p. 754. Longman, London, 1978.

[15] V. K. Gouda and W. Y. Halaka, *Brit. Corros. J.,* **5**, 204 (1970).

[16] F. M. Lea and C. M. Watkins, *National Building Studies Research Paper No. 30.* HMSO, London, 1960.

[17] M. Regourd, H. Hornain and B. Mortureux, *Silic. Ind.,* **42**, 19 (1977).

Chapter 10

The mechanisms of the protection of steel by concrete

H. ARUP, Korrosionscentralen, Glostrup, Denmark

INTRODUCTION

It is not altogether uncommon to hear, or to find the opinion expressed, that concrete protects steel by preventing ingress of water and oxygen. However, most of the ordinary and even higher quality, constructional concretes allow ample access of both water and oxygen. It is, thus, now generally accepted knowledge that steel in concrete is protected by the passivity induced by the highly alkaline nature of the porewater, and that this passivity persists not in spite of the water and oxygen present, but as a result of presence of water and oxygen.

It is the intention of the present paper to review this statement carefully and to consider other parameters, external and internal, which may become important if and when the initial passive state is no longer tenable.

For the purpose of this discussion it may be helpful to define four states (of protection or corrosion), in which steel in concrete may find itself.

THE FOUR STATES OF CORROSION OF STEEL IN CONCRETE

1. The passive state

Passivity is the normal, protected state of steel in concrete, brought about by the high pH and the availability of oxygen. In this state, the corrosion rate is always insignificantly low (0.1 m/year). In the absence of chlorides the passive potential range is very wide, from +200 mV to −700 mV SCE at pH 13 but, in aerated concrete, steel normally exhibits a potential in the range +100 to −200 mV SCE. The 'quality' of the concrete, including such parameters as w/c ratio, pozzolanic additions, permeability and electrical resistance determine the ability of the concrete to resist depassivating forces (to be discussed later) and influence the characteristic potential, but the passive steel is still essentially non-corroding.

2. The pitting corrosion

Pitting is typically brought about by the presence or ingress of chloride ions. The state of pitting corrosion is characterized by the galvanic action between large areas of passive steel acting as cathode and small anodic areas, where the local environment has a high chloride concentration and a low pH.

The average potential is between that of the passive steel and that of the local anode, typically −200 to −500 mV. Electrochemical potential mapping will reveal high potential gradients near the anodic areas.

3. General corrosion

General corrosion is the result of general loss of passivity, due either to carbonation or to the presence of excessive amounts of chloride.

The electrochemical potential is similar to that of corroding steel in other environments, typically −450 mV to −600 mV SCE, and the potential gradients are not very great.

4. Active, low-potential corrosion

In environments, where the access of oxygen is so limited that the passive film cannot be maintained, embedded steel may become active in the still highly alkaline concrete. The equilibrium potential of iron in a solution of pH 13 is around −1000 mV, and the iron dissolves as the complex ion FeO_2. The corrosion rate, however, is very low – as low or lower than in the passive state – and, if shifting conditions make oxygen more available, the steel will repassivate easily.

In this condition, pitting is impossible, because the potential is lower than the potential of the pitting anode, and if a small area of the embedded steel should become exposed to the external environment at the bottom of a crack, this part of the steel will become cathodically protected.

FACTORS DETERMINING CORROSION RATES

The passive state

Figure 1 reviews the factors responsible for the maintenance of the passive state. The figure emphasizes the correlation between chloride content (Cl/OH ratio) and potential and is a reminder that the 'chloride threshold' is a function of potential (among other things) and that a 'pitting potential' is a function of chloride content. Such relations, which are universally accepted in corrosion research on stainless steels and other passive metals, have only occassionally been considered in studies of steel in concrete.

It has already been mentioned, that embedded steel in this state is well protected, but a small passive current is required to maintain the protective oxide film. The mechanisms involved in the cathodic reduction of oxygen on the passive surface of steel in concrete are not at all clear. Moreover, the reaction rate and its dependence on factors such as cover and potential cannot be adequately interpreted as diffusion controlled and/or activation controlled [1]. Those factors, which seem to have an influence on the magnitude of the passive current density are discussed below.

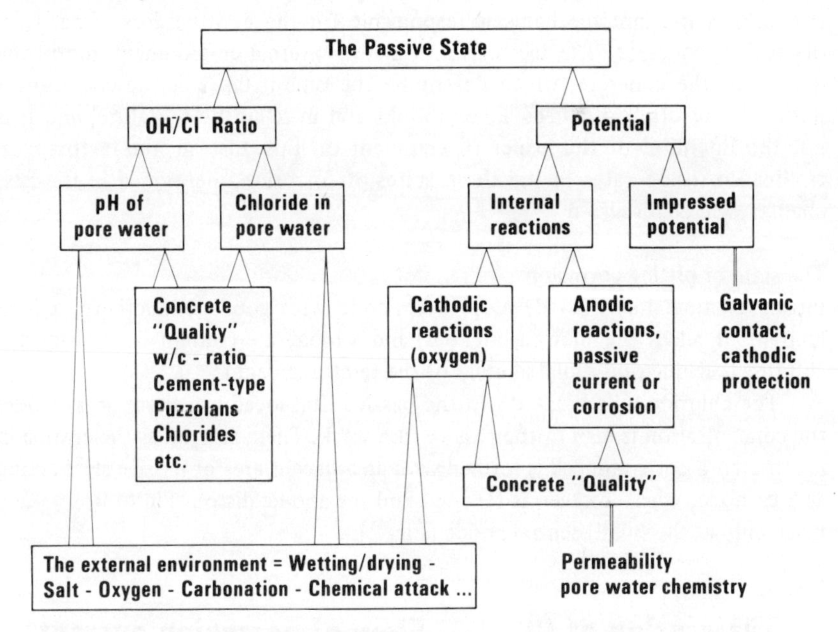

Fig. 1 – An overview of the factors involved in the maintenance of the passive state of steel in concrete.

The classical way to study passive current densities in liquids is to measure the current as a function of potential (or the opposite) in a de-aerated solution to eliminate the cathodic reduction of oxygen. This approach is also possible for steel in concrete [2, 3], but the de-aeration process is more difficult and may introduce chemical changes at the steel/concrete interface.

The passive current densities by this technique are approximately three orders of magnitude lower than for steel in alkaline solutions with a composition similar to that of the pore solution. This is probably due to the high diffusion resistance in the concrete and in the precipitated layer of portlandite near the steel [4], which restricts the movement of dissolved iron species away from the oxide/concrete interface. In comparison, the differences in passive current denisties for steel in different types of concrete are not so great [3], and the significance of these differences in terms of concrete 'quality' has not been determined.

As a point of discussion, it might be argued that it may be an advantage if the passive current density is not too low, as this – other factors being equal –

will help to maintain a lower (passive) potential in situations with a limited oxygen supply and this may keep the steel below the critical potential for chloride initiated attack.

It is well known that carbonation and penetration of chloride are the two most important mechanisms responsible for the eventual loss of passivity of steel in concrete. The factors, both in the external environment and in the quality of the concrete, which determine the time it takes before corrosion is initiated, are often discussed in textbooks and in scientific literature, and it is not the intention of this paper to comment on this. Instead, the factors controlling corrosion rates in the three states of corrosion, mentioned in the first chapter, will be discussed.

The state of pitting corrosion

Pitting is most likely to develop in concrete with good conductivity, a high content of alkali (i.e. non-carbonated) and a moderate content of chloride (or chloride reaching only isolated areas of the reinforcement).

The chloride ions break down the passive film locally in those areas where the concentration is high or the passive film weak. Then, as soon as the corrosion is initiated a corrosion cell is formed with an adjacent area of passive steel acting as a cathode, where oxygen is reduced and the anodic dissolution of iron taking place only at the small, central anode (Fig. 2).

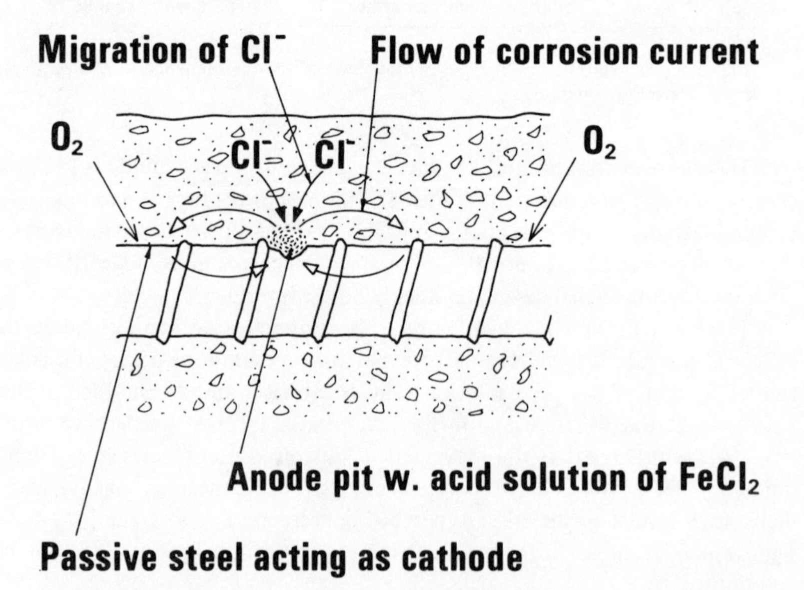

Fig. 2 – Electrochemical corrosion processes around a corrosion pit on steel in concrete.

Several factors then work together to maintain or aggravate the development of the existing pit rather than to spread the corrosion or nucleate new pits. Acid is produced at the anode (the pit) and alkali at the cathode thereby causing the pH to shift in opposite directions. The pH-shift (to lower values) at the anode is very pronounced, both because of the high current density and because of the hydrolysis connected with the further oxidation of Fe^{2+} to Fe^{3+} at the edge of the rust nodule. The migration of Cl towards the anode (and away from the cathode) is another contributing factor, but the pronounced drop of the average corrosion potential, brought about by the polarization of the cathode, is perhaps the strongest factor in preventing formation of new pits in the neighbourhood, at least until much higher chloride concentrations are established elsewhere.

On the other hand, the potential shift may also start a development towards a stifling of the attack. The current density at the anode will decrease, both because of the lower driving force and because the area of the anode increases as the low pH liquid at the anode neutralizes the concrete at the rim of the anode. Consequently, the electrochemically produced concentration gradients will be flattened by the diffusion of ions in the opposite direction and this could result in the repassivation of the pit.

To complete the picture, one further characteristic of the pitting corrosion should be mentioned. The corrosion products formed are soluble at the conditions of low pH near the anode, therefore, considerable amounts of corrosion can occur without spalling of the concrete.

The development of pitting corrosion will be influenced by several properties of the concrete:

(i) A low electrical resistivity of the concrete favours the development of the pit. The effective cathode/anode area ratio may become very large, and the migration of ions is facilitated.

(ii) A high alkali content (in this case both soluble alkalis and precipitated calcium hydroxide) will hamper the lateral growth of the pit and favour its deepening.

(iii) The access of oxygen is also important and is a factor in determining the total amount of corrosion. However, with a large cathode/anode area ratio intense pitting can result even with a limited oxygen supply. The oxygen access will be more plentiful if the concrete becomes partially dry, but the stimulating effect of this on pit development will be more than nullified by the higher resistivity and the drying out of the liquid near the anode.

The state of general corrosion

As mentioned earlier, general corrosion may be caused by carbonation or by the presence or ingress of chloride ions in excessive amounts, so a large number of closely situated pits are formed. Both anodic and cathodic processes take place everywhere on the surface, and the pH-shifts normally associated with each of these processes cancel each other. This means that the anodic dissolution takes place in a near neutral or alkaline environment, where oxygen has access. The corrosion product, in this case, is solid rust, which causes spalling at a relatively early stage.

In practice, general corrosion caused by carbonation or by chlorides have different characteristics. In cases, where carbonation has penetrated deeply, the concrete is likely to be rather permeable and semi-dry, and the rate of corrosion, once it starts, is probably controlled by the relatively high resistivity and lack of water rather than by diffusion of oxygen. The 'time-of-wetness', known to be an important factor in atmospheric corrosion, is an appropriate measure of corrosion rate under these conditions, but it applies of course to the conditions at the steel/concrete interface and there is probably a gradual increase in corrosion rate with increasing humidity. K. Tuutti [5] has confirmed this correlation, using an electrochemical corrosion cell to monitor corrosion rate in carbonated concrete, and he observed, that there is also a sudden drop in corrosion rate if the humidity becomes so high that the capillary pores become waterfilled, restricting oxygen diffusion.

Conversely, corrosion in chloride-rich concrete may more often be found in concrete in sea water or in bridges with snow and salts, where water is abundant, resistivity is low, but where diffusion of oxygen through the water filled pores is likely to be the rate-determining factor.

The state of active, low potential corrosion

The fact that depassivation of steel can occur under conditions of very restricted oxygen supply, means that the current density of the cathodic reduction of oxygen must have been less than the passive current density. This corresponds to a very, very low corrosion rate, of the order of 0.1 m/year. Once the active state is established, repassivation will only occur if the cathodic current density is higher than the passivation peak. The corrosion rate can, therefore, theoretically become somewhat higher when the steel has become active. Experimentally, however, the passivation peak has been found not to be very pronounced for steel in concrete.

The active state can also be the result of a period of cathodic protection in low-oxygen environments because cathodic protection not only removes the oxygen near the steel surface but also increases the pH value, which increases the solubility of the ferroate ions.

In the low potential state the potential is such that hydrogen evolution probably contributes significantly to the cathodic reaction, but still at a very low current density. A further depression of the potential by excessive cathodic protection will promote a steep rise in the current consumption and the evolution of hydrogen†.

Very little is known at this moment about the effect of cement type or concrete composition on the development of the active state. The pH of the porewater is probably an important factor, but more research is needed.

REFERENCES

[1] N. Wilkins, 'A "State of the Art"' Summary from the 1st International Seminar on Electrochemistry and Corrosion of Steel in Concrete'. *Steel in Concrete Newsletter,* Jan. 1980. Korrosionscentralen, Denmark.

[2] F. O. Grønvold, C. M. Preece and H. Arup, 'Corrosion protection of steel by concrete, in particular by low porosity cement mortars'. *8th International Congress in Metallic Corrosion, Mainz, 1981.*

[3] C. M. Preece *et al.,* 'The influence of cement type on the electrochemical behaviour of steel in concrete'. *Conference on Corrosion of Reinforcement in Concrete Construction, London, June 1983.*

[4] C. L. Page, *Nature,* **258**, 514–515 (1975).

[5] K. Tuutti, *Corrosion of Steel in Concrete.* Swedish Cement and Concrete Research Institute. ISSN 0-346-6906.

[6] H. Arup, *Steel in Concrete Newsletter,* Jan. 1980. Korrosionscentralen, Denmark.

†It is interesting to note, the prestressing tendons in grouted steel ducts, which may be assumed to be in the low-potential, active state [6] do not exhibit any tendency to fail by hydrogen embrittlement, but hydrogen embrittlement has been observed in cases where the tendons were subject to galvanic contact with corroding aluminium or galvanized steel.

Chapter 11

Corrosion rate of reinforcements during accelerated carbonation of mortar made with different types of cement

J. A. GONZALEZ, C. ALONSO, and C. ANDRADE, Instituto Eduardo Torroja de la Construccion y del Cemento, Madrid, Spain

INTRODUCTION

The corrosion of steel reinforcements in carbonated concrete is attributed to the pH lowering of the pore concrete solution due to precipitation of $Ca(OH)_2$ by atmospheric CO_2 to $CaCO_3$. Most authors agree that the final pH value is about 9.

We have already reported [1] some measurements by linear polarization resistance (*Rp*) technique of steel corrosion, during accelerated carbonation of the concrete, and the changes induced by different relative humidities of conservation of the specimens, on the reinforcement corrosion rates. The corrosion rate was practically negligible at a relative humidity (R.H.) of 50% in the absence of chlorides. To obtain high corrosion rates either chlorides or very high R.H.s were required.

The above studies are here extended to the carbonation of mortars made with four different cement types. We also measured the corrosion during the carbonation of $Ca(OH)_2$-saturated solutions, with and without NaOH and KOH additions. The results disprove some of the assumptions in the field and give new insights into the phenomenon.

EXPERIMENTAL METHOD

Materials

Corrugated steel bars, 6 mm diameter and 8 cm length were used. The air mortar/steel interphase and the lower end of the steel were covered with adhesive tape. The surface exposed to attack was one of 5 cm^2.

Tests in mortar

Four different types of cements were used: (1) Portland cement with low C_3A content; (2) cement with 35% of slag; (3) cement with 35% of natural pozzolanas; and (4) cement with 30% of fly-ash. The respective chemical analyses are shown in Table 1.

The mortar followed the Spanish Standard RC-75 [2] with a 1/2 water/cement ratio and a 1/3 cement/sand ratio. The mortar was made without additives and with 2% anhydrous $CaCl_2$ in relation to cement weight, as a corrosive agent.

Table 1 – Chemical analysis of the cements.

Cements	Portland	With slag	With pozzolana	With fly-ash
L.I.	2.6	–	4.2	1.9
I.R.	1.4	0.3	3.1	16.2
SiO_2	20.8	25.1	23.6	21.8
Al_2O_3	4.1	8.3	6.3	6.0
Fe_2O_3	4.4	2.3	4.0	3.2
CaO	61.7	53.9	50.6	44.8
MgO	2.0	5.3	3.8	2.1
SO_3	2.2	2.9	3.1	3.0
Na_2O	0.24	0.7	0.41	0.10
K_2O	0.54	0.82	0.91	0.68
free 60	0.6	1.8	0.8	0.4
S^{2-}	–	0.4	–	–
MnO	–	0.4	–	–
CO_2	–	0.1	–	–
Comb. H_2O	–	0.4	–	–

Tests in dissolution

Two types of solutions were prepared: (a) a saturated solution (8 g/l) of $Ca(OH)_2$ (pH = 12.6); and (b) a saturated solution of $Ca(OH)_2$ also 0.1 M in NaOH and 0.1 M in KOH (pH = 13.2).

The type of corrosion cell has been described in previous reports [3,4].

Experiments

Tests in mortar

The test specimens were 2 × 5.5 × 8 cm and contained two embedded identical steel bars. The counter electrode was a central embedded carbon bar [1].

They were cured in an environment with over 90% R.H. at 20 ± 2°C for the first 24 h. They were then cured for 6 days under partial immersion in distilled water, then dried for 24 h at 50% R.H. and finally held for 7 days in the carbonation chamber.

Carbonation was affected at 20 ± 2°C and 50% R.H. The chamber was flushed with CO_2 in two different ways that we define as 'continuous' and 'intermittent'.

In the 'continuous' carbonation, the chamber was continuously flushed with CO_2 during the first 30 minutes, after which CO_2 was added from time to time. As the reference electrode was previously introduced in the carbonation chamber, we continuously measured the corrosion potential, E_{corr}, and the Rp only when the E_{corr} was almost stabilized.

In the 'intermittent' carbonation the CO_2-flushing was stopped to measure outside the chamber the E_{corr} and the Rp of the two embedded identical steel bars after 1, 5, 15, 30 min, 2–3 hours and once daily for 7 days.

After 7 days accelerated carbonation (14 days after mixing) the specimens were kept in the open in the laboratory at R.H. ≈ 50% for a total of 21–25 days of life.

Tests in solution

The carbonation was produced by slowly bubbling CO_2 through the solutions (as bubbling at a high rate can increase the CO_2-pressure, decreasing the pH beyond the range of interest) [5].

The pH solution was continuously recorded as well as the E_{corr} of the steel bars. The Rp was measured before bubbling CO_2 and after completion of the test, as it changes very much during carbonation.

The carbonated solutions were kept well stoppered for 5 further days.

Procedure

The pH was measured with a AMEL pH-meter, with an ORION combined electrode of pH range 0–14. It was calibrated with two buffers of pH 7, 12 and 13.

The Rp measurements were taken with an AMEL-552 potentiostat, with positive feedback that compensates instrumentally the IR-drop between the working and reference electrodes (always saturated calomel electrode). To ensure a good electrical contact between the reference and the mortar a wet filter paper was placed between them.

The calculation of corrosion intensity, i_{corr} from Stern's formula:

$$I_{corr} = \frac{B \times \Delta I}{E} = \frac{B}{Rp}$$

has been already described [3, 6]. B values of 26 and 52 mV were used for steel in the active and passive states, respectively.

RESULTS

We give in Fig. 1 the evolution with time of i_{corr} and E_{corr} of steel bars embedded in mortar made up with Portland cement (low in C_3A). 'Continuous' and 'intermittent' carbonation produce quite different results which led us to search for the origin of this difference.

All the results reported up to now by us [6–8] correspond to 'continuous' accelerated carbonation. However, most of the results presented here were obtained, from 'intermittent' carbonation as we feel that it better represents real-life conditions.

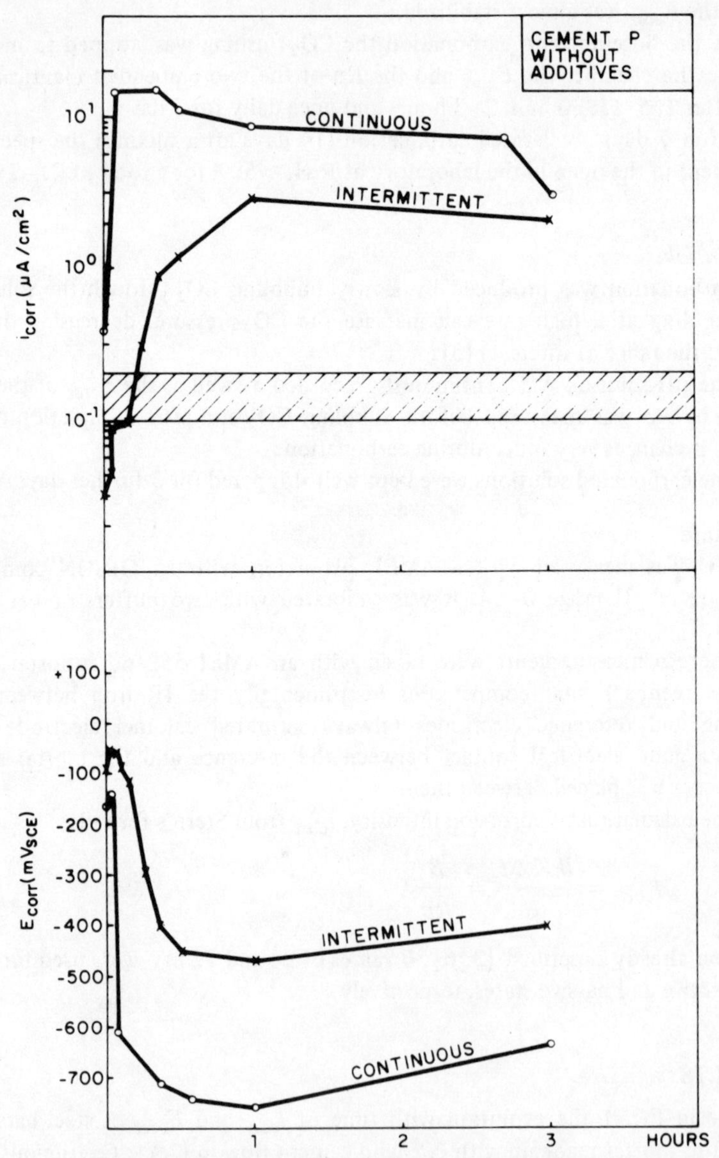

Fig. 1 – Effect of the mode of carbonation on the development of i_{corr} and E_{corr} during accelerated carbonation for steel bars embedded in mortar made with Portland cement.

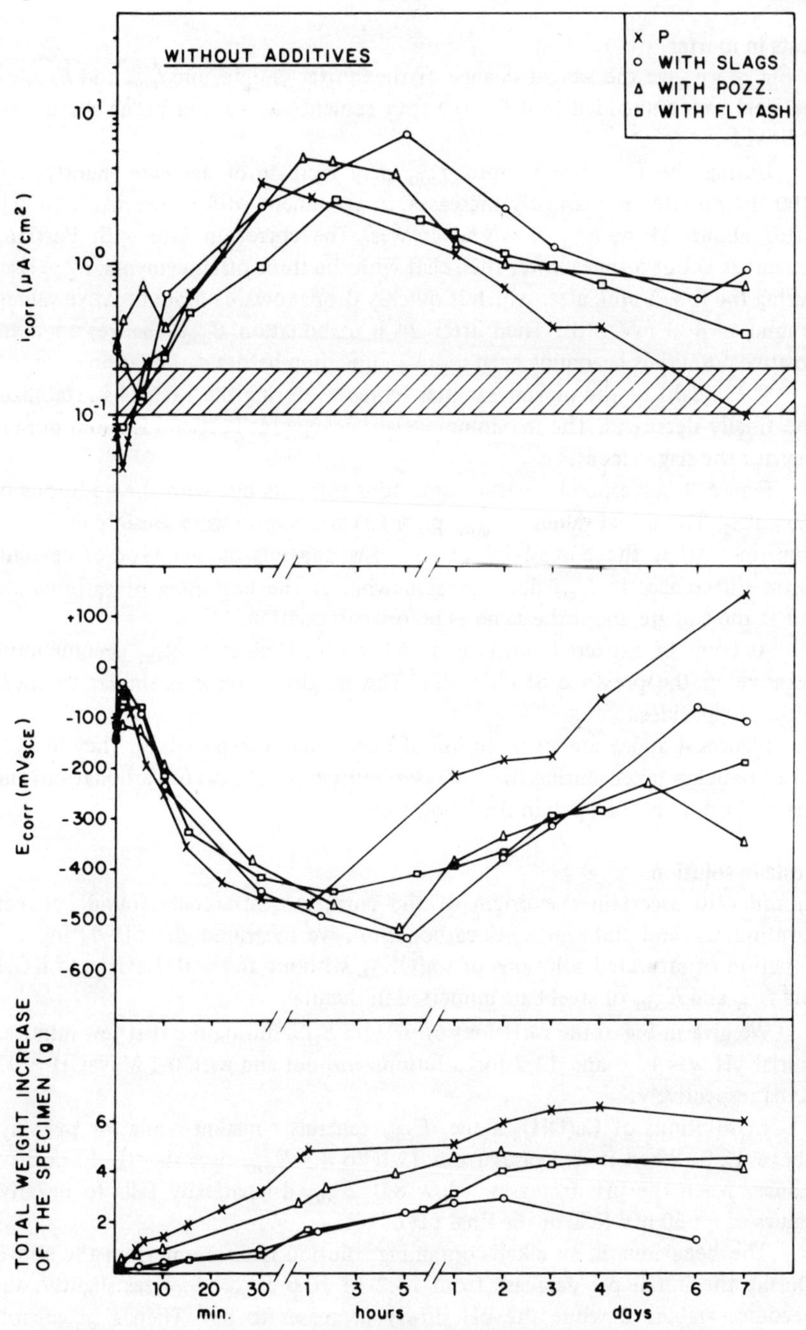

Fig. 2 – Non-additive-containing mortars: weight increase of the specimens and evolution of i_{corr} and E_{corr} of steel bars while the specimens are in the carbonation chamber.

Tests in mortar

In Fig. 2 we give the weight change of the mortar sample, and i_{corr} and E_{corr} of the steel bars embedded in it for the four cements tested and in the absence of additives.

During the first five minutes i_{corr} may increase or decrease slightly, but after 10 minutes it markedly increases, in agreement with earlier work [6, 7]. After about 24 h, i_{corr} slowly decreases. The corrosion rate with Portland cement is about 5 times lower than that with the three other cements. E_{corr} rises during the first 5 min, after which it quickly drops towards more negative values, around – 400 mV SCE. Then after 24 h carbonation E_{corr} changes to more positive potentials becoming even more anodic than before carbonation.

The weight of the mortar samples increases during the first 24 h, stabilizes and finally decreases. The maximum weight change for Portland is about double that for the slag cement.

Figure 3 corresponds to the same four cements but with the additions of 2% $CaCl_2$. The initial values of i_{corr}, prior to carbonation were sensibly different, confirming that the corrosivity of Cl^- ions depends on the type of cement. These differences in I_{corr} decrease somewhat at the beginning of carbonation but at the end are about the same as before carbonation.

As could be expected, both the initial and final values of E_{corr} become more negative in the presence of chlorides. The weight increase is similar to those without chlorides.

Figures 4 and 5 are an extension of Figs. 2 and 3 respectively. They include measurements taken during the 7 days of curing, the 7 days of carbonation, and the 7–11 days in the open in the laboratory.

Tests in solution

In order to ascertain the origin of the corrosion differences found between 'continuous' and 'intermittent' carbonation, we measured the pH during carbonation of saturated solutions of $Ca(OH)_2$, without and with NaOH and KOH, and i_{corr} and E_{corr} of steel bars immersed in them.

We give in Fig. 6 the variations of pH and E_{corr} during the first few minutes. Initial pH was 12.6 and 13.2 for solutions without and with 0.1 M NaOH + 0.1 KOH respectively.

In solutions of $Ca(OH)_2$ alone, E_{corr} remains constant while the pH stays above 12.0. When it decreases from 12.0 to 8.5 E_{corr} rises slightly to positive values. When the pH decreases below 8.0, E_{corr} dramatically falls to negative values of −660 mV SCE at the final pH of 6.1.

The behaviour in an alkali-containing solution is different from the above. During the initial pH decrease from 13.2 to 10.6 E_{corr} increases slightly, and becomes stabilized while the pH slowly decreases to 8.0. Then E_{corr} steeply decreases down to −680 mV, while pH decreases down to a stable value of 7.0.

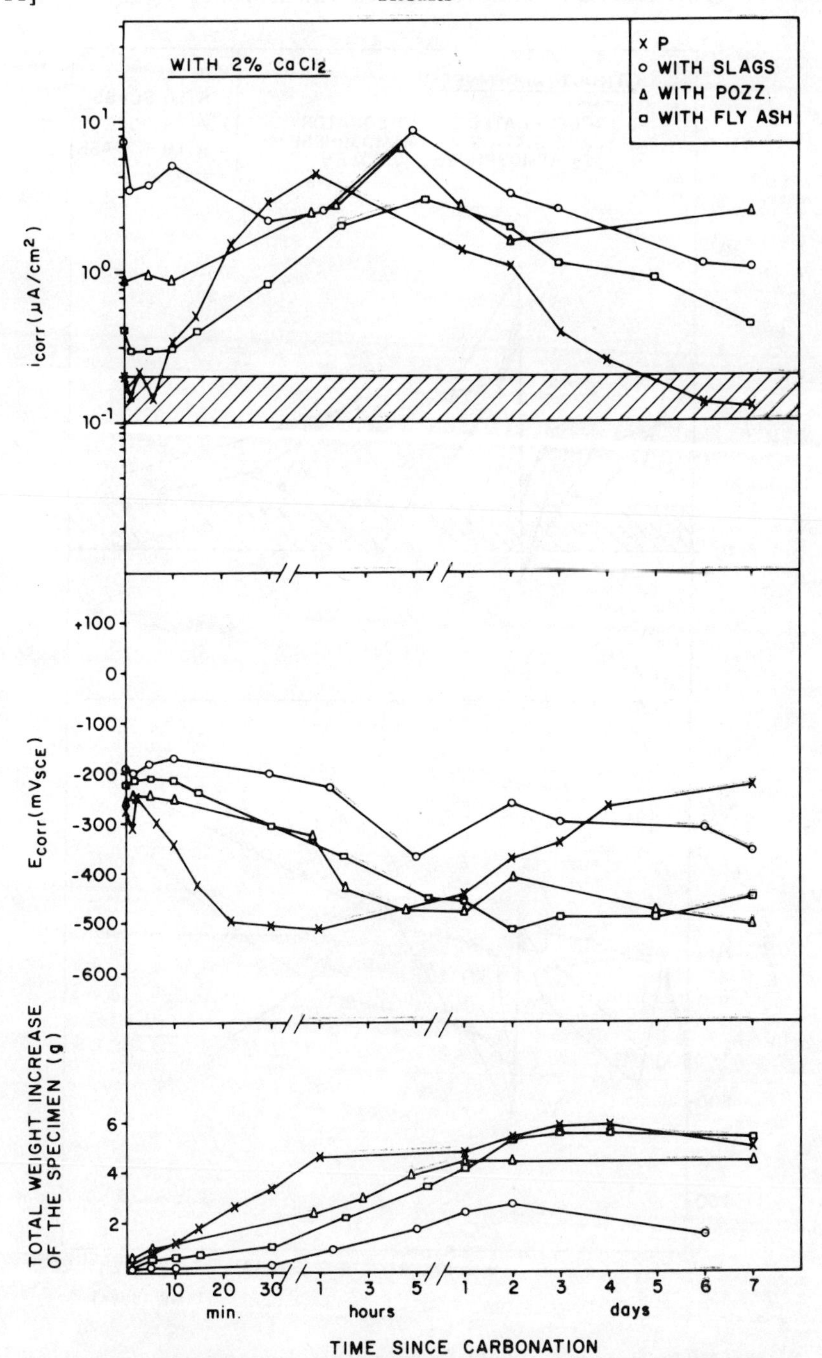

Fig. 3 – 2% $CaCl_2$-containing mortars: weight increase of the specimens and evolution of i_{corr} and E_{corr} of steel bars while the specimens are in the carbonation chamber.

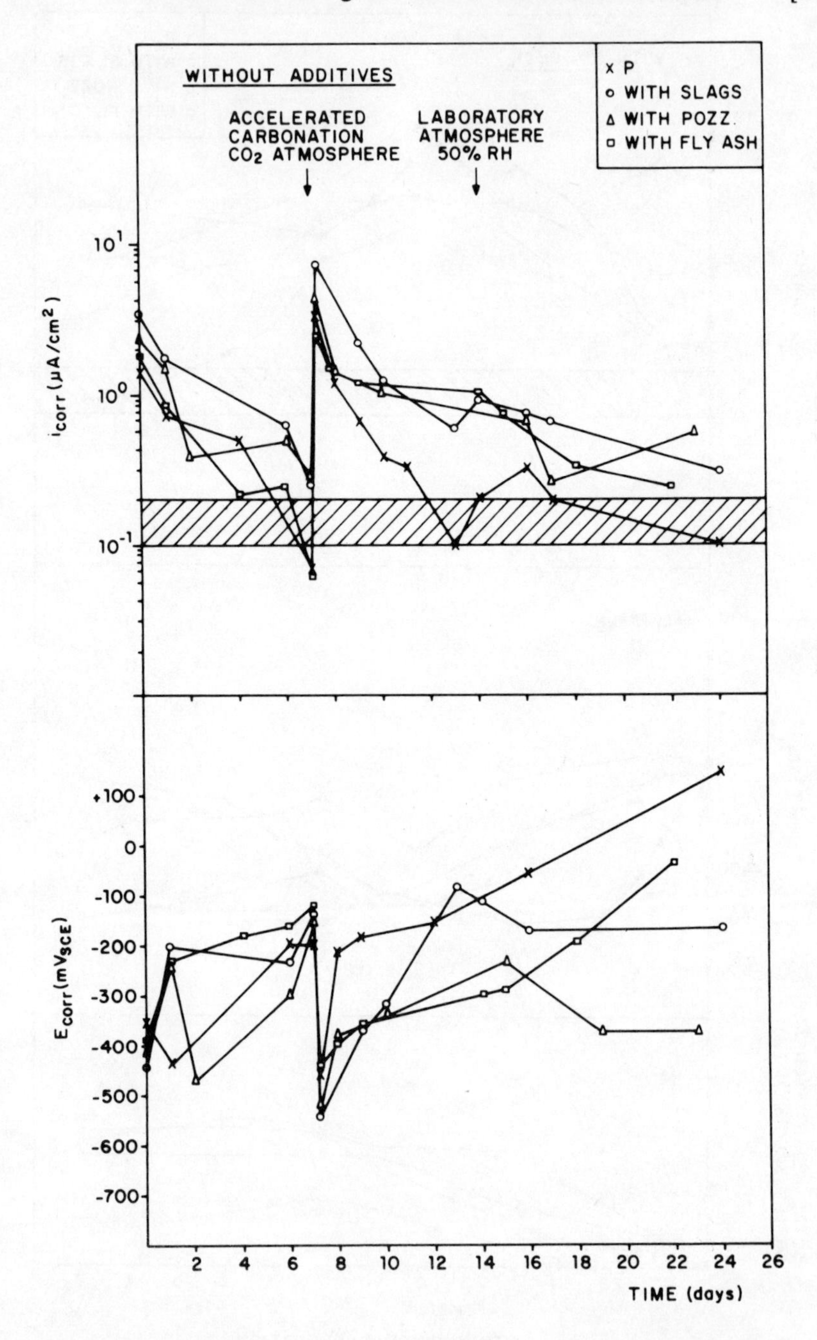

Fig. 4 – Evolution of i_{corr} and E_{corr} for steel bars embedded in the non-additive-containing mortar.

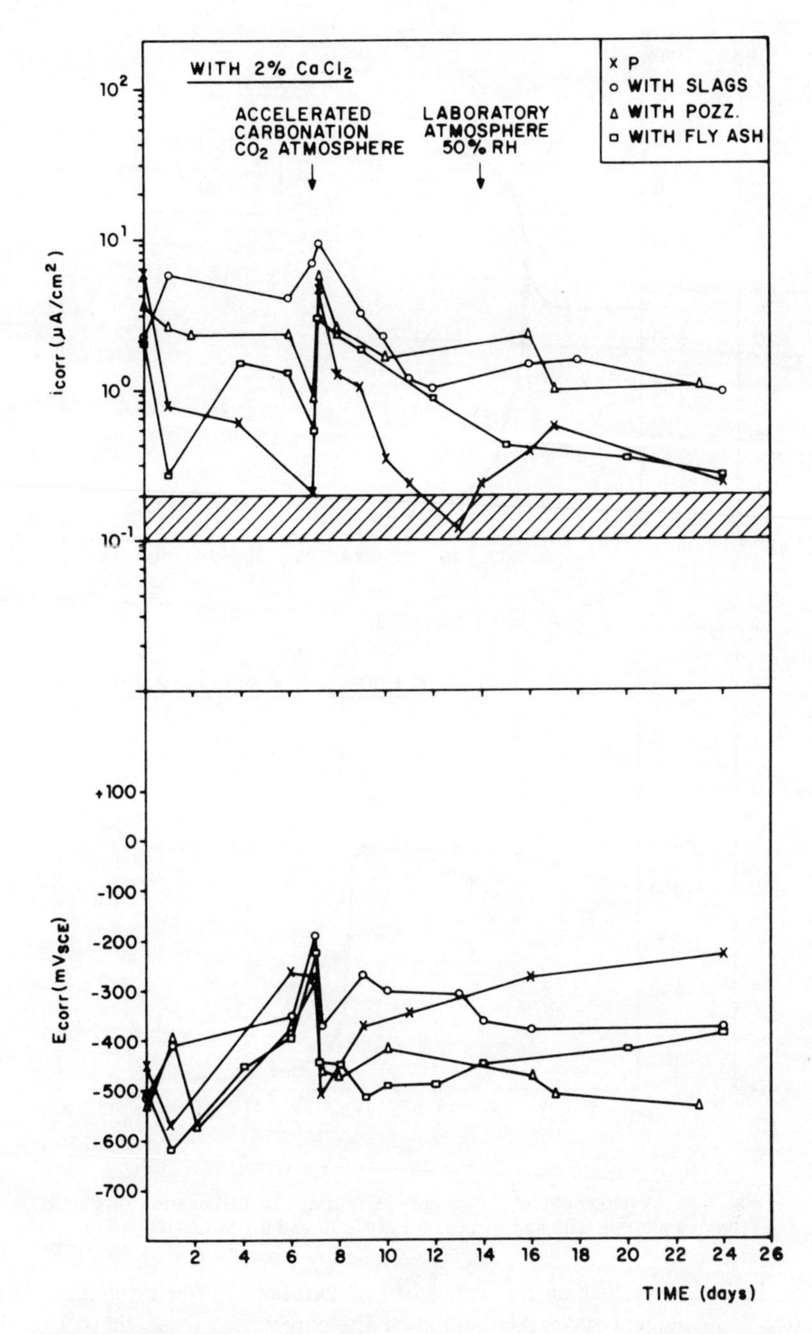

Fig. 5 – Evolution of i_{corr} and E_{corr} for steel bars embedded in the 2% $CaCl_2$-containing mortar.

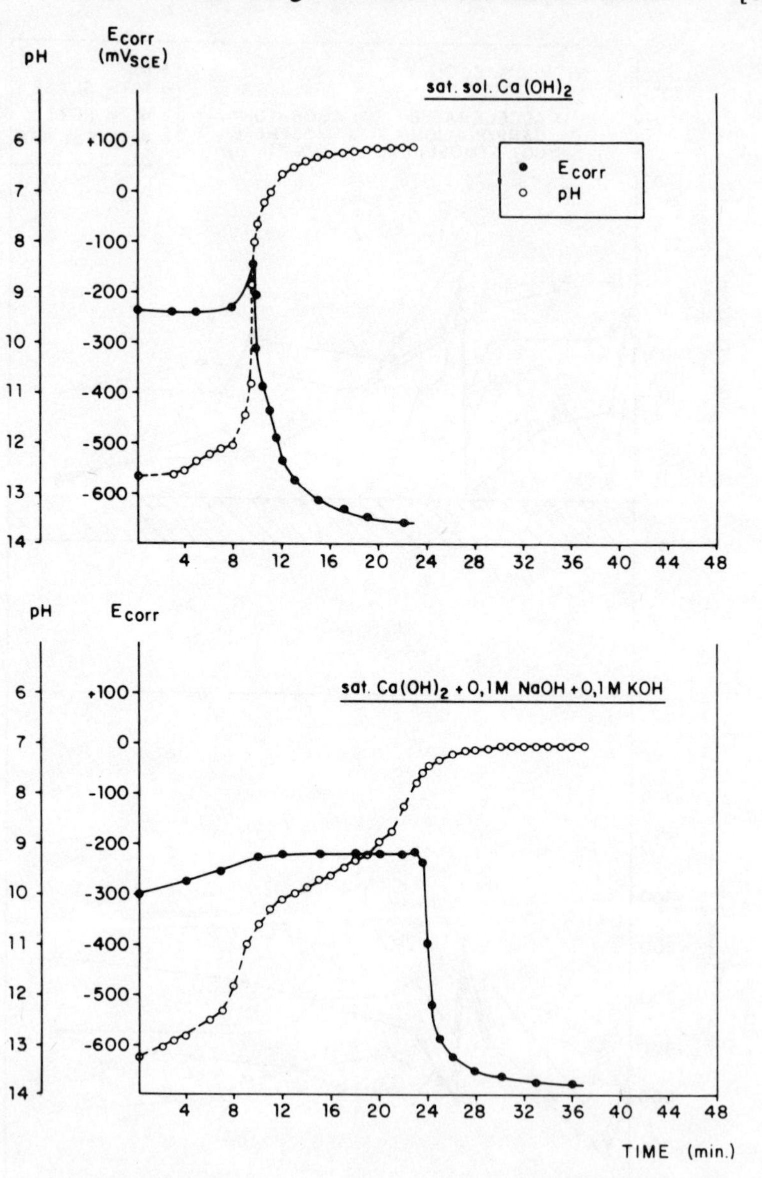

Fig. 6 – Development of E_{corr} and pH during the carbonation of $Ca(OH)_2$-saturated solution with and without 0.1 M NaOH and 0.1 M KOH.

In Fig. 7 we include the values of i_{corr} before and after carbonation, and after 5 subsequent days. At the end of the experiment while the pH of the alkali-containing solution increases from 7.0 to about 10.5 that of the $Ca(OH)_2$-saturated solution increases only from 6.1 to 7.0.

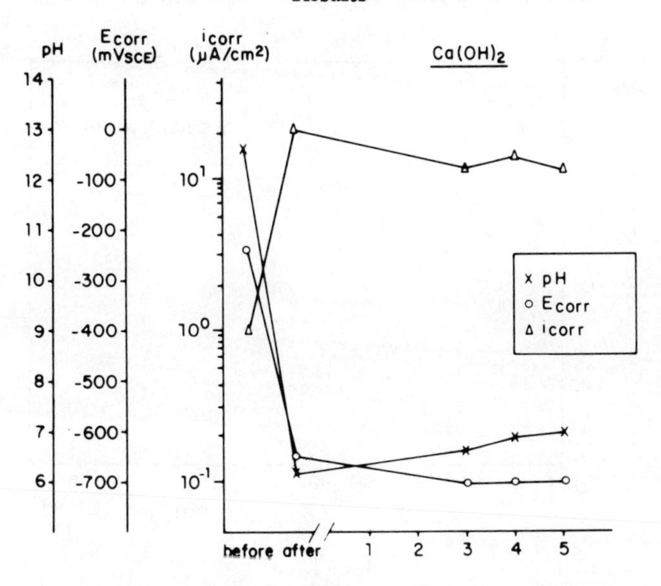

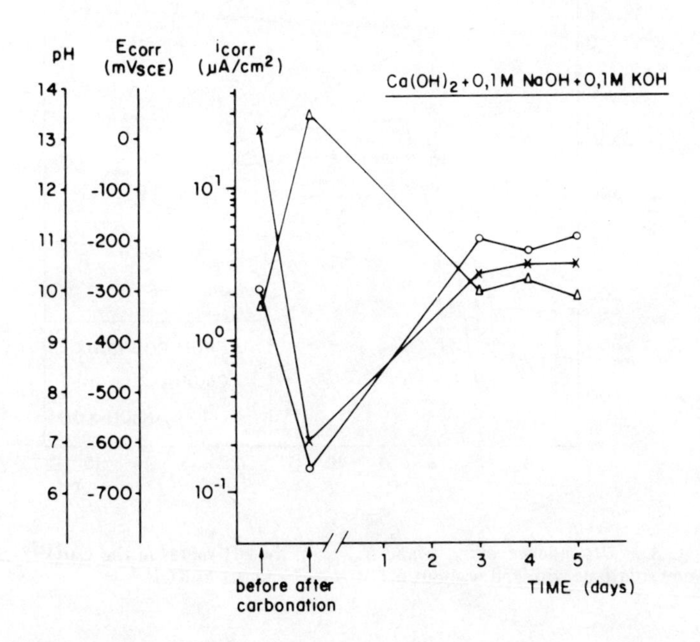

Fig. 7 – Values of pH E_{corr} and i_{corr} until 5 days after carbonation of the $Ca(OH)_2$-saturated solutions with and without 0.1 M NaOH and 0.1 M KOH.

Finally, we give in Fig. 8 the dependence of E_{corr} and i_{corr} on the pH solution.

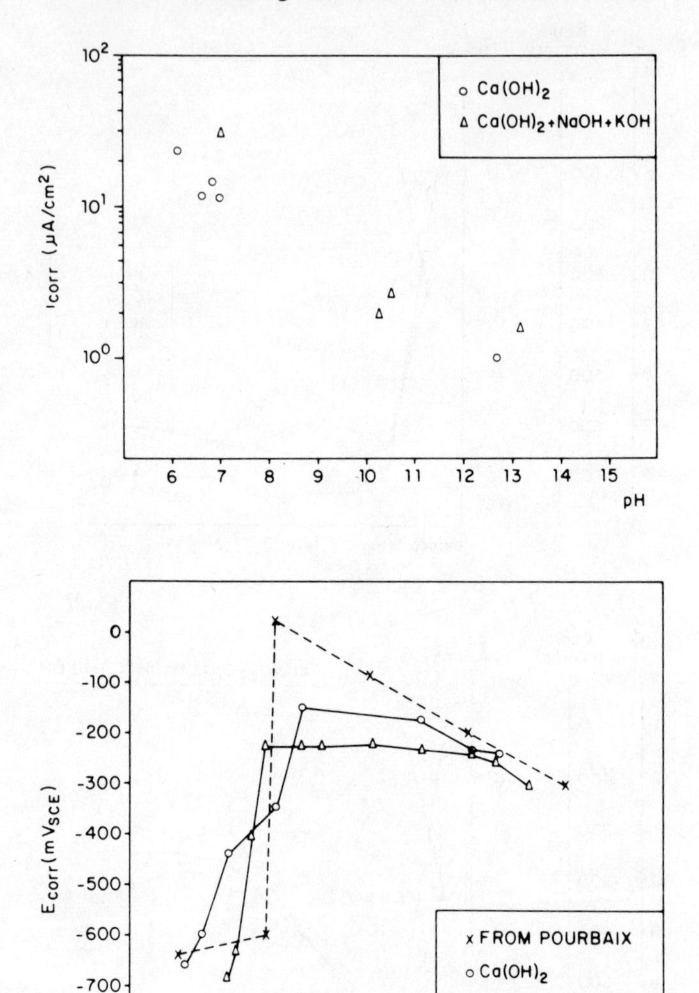

Fig. 8 – Dependence of i_{corr} and E_{corr} on the pH values in the $Ca(OH)_2$-saturated solutions with and without 0.1 M NaOH and 0.1 M KOH.

DISCUSSION

As the potential, according to Nernt's equation, increases by 59 mV for every pH unit decrease, we have attributed [6] the initial increase of E_{corr} of steel bars upon carbonation to a pH lowering in the pore mortar solution. When the pH decreases below a threshold value, the passive layer is quickly destroyed, this is easily detected by the steep decrease and increase of E_{corr} and i_{corr}, respectively. In this process an important role was assigned to the water liberated during carbonation, which should decrease the resistivity of the mortar, thereby facilitating corrosion.

Tests in solution

Slow bubbling of CO_2 through a $Ca(OH)_2$-saturated solution precipitates $CaCO_3$ with a fast pH decrease:

$$Ca(OH)_2 + H_2O + CO_2 \rightarrow CaCO_3 \downarrow + H_2O \quad (1)$$

If the CO_2 bubbling continues in presence of water, CO_3H^- are formed from the CO_3^{2-} according to the known reversible reaction:

$$CaCO_3 + H_2O + CO_2 \rightleftarrows Ca(HCO_3)_2 \quad (2)$$

Thus, in an accelerated carbonation if there is enough CO_2 and water the precipitated $CaCO_3$ from $Ca(OH)_2$ is redissolved to give $Ca(CO_3H)_2$ which will acidify the solution to produce more acid values (between 6 and 7) than that given for the presence only of CO_3^{2-}.

When the amount of dissolved or 'aggressive' CO_2 is less than a critical value the CO_3H^- will precipitate as CO_3Ca and the pH becomes nearly neutral.

The addition of NaOH and KOH as can be seen in Fig. 6 changes the process because Na_2CO_3 and K_2CO_3 are much more soluble than $CaCO_3$. During carbonation they remain in solution, stabilizing the alkalinity until they are completely transformed into $NaCO_3H$ and KCO_3H. Morover, the simultaneous presence of $NaCO_3H$ and Na_2CO_3 gives a buffer of pH = 10.4 that is the final spontaneous pH value in our test in the $Ca(OH)_2$-saturated solution with NaOH and KOH.

The steep decrease of E_{corr} takes place at pH $\approx$ 8.0 in agreement with the value by Pourbaix [5] for dissolution of the passivating layer on steel in an aerated solution. Stronger oxidants than oxygen are needed for the preservation of passivity. We show in Fig. 8 the dependence of E_{corr} on pH for the two solutions tested (full line) and that reported by Pourbaix in other solutions (dashed line). The disagreement in the potentials at pH between 12 and 8 is due to the fact that we cannot wait for the stabilization of every potential, because it changes continuously.

Accelerated carbonation of the mortar

In the light of the reported results it is possible to improve the interpretation of the changes observed in some variables during the accelerated carbonation of the mortar.

In the first stages of accelerated carbonation the precipitation of calcium carbonate produces a pH decrease whose magnitude depends on alkaline reserve and alkali content of the cement, while E_{corr} becomes more positive (more noble) and i_{corr} stabilizes. Further action of CO_2 and of the H_2O molecules, liberated by the carbonation and not used up in hydration of the paste, change CO_3^{2-} into (HCO_3^-) according to reaction (2). The consequent pH decrease may destroy the passivating layer in steel if pH = 8 is reached, unless oxidants such as nitrites [5] are present, as already reported [6]. The depassivation is easily seen by a dramatic increase of i_{corr}.

During 'intermittent' or slower carbonation the liberated water is both lost by evaporation and used up by hydration of the cement paste. Consequently the changes in pH E_{corr} and i_{corr} are less marked than during 'continuous' or fast carbonation (Fig. 1). It is remarkable that in tests in solution at pH = 10.4 it is $i_{corr} \sim 2\ \mu A/cm^2$ (Figs. 7 and 8) of the same order found in 'intermittent' carbonation (Figs. 1 to 3), while 'continuous' carbonation is accompanied by $i_{corr} > 10\ \mu A/cm^2$ which according to Fig. 8 corresponds to pH = 6–7.

Once carbonation has been completed, keeping the specimen in the open at 50% R.H. dries it progressively, which both changes the bicarbonates back to carbonates and increases ohmic resistance; this produces a decrease of i_{corr} and an increase of E_{corr} towards more noble values (Figs. 4 and 5).

On the other hand, if a completely carbonated mortar is kept at a high R.H. or partially immersed, two phenomena can occur: (a) firstly water will favour bicarbonate formation according to reaction (2) provided there is enough dissolved CO_2, which will depend on the air content of the surrounding atmosphere; and (b) water will tend to progressively hydrate the silicates, liberating $Ca(OH)_2$ which will tend to raise the pH, both because of the liberated hydroxyls, and because of neutralization of bicarbonate according to:

$$Ca(OH)_2 + Ca(HCO_3)_2 \rightarrow 2\ CaCO_3\downarrow + 2\ H_2O \tag{3}$$

As a result of the above reactions the aqueous phase will contain the ions CO_3^{2-}, HCO_3^-, OH^-, Ca^{2+}, Na^+ and K^+ (besides SO_4^{2-}), giving a pH value which will finally depend on:

(1) alkali content of the cement;
(2) proportion and hydration degree of the cement silicates;
(3) relative ambient humidity;
(4) partial pressure of CO_2 in the air or in the water.

From Fig. 7 it seems logical that the mortar pH (together with ohmic resistivity) controls the corrosion behaviour of the steel bars.

Atmospheric carbonation

Carbonation under atmospheric conditions is slow (CO_2 partial pressure in the air is 0.3%) and should not produce bicarbonates; which explains why they are usually not mentioned in the relevant literature. Only in two cases are bicarbonates commented upon [9, 10] in accelerated carbonation or upon action of very pure water on carbonated concrete.

In concrete that has spontaneously (i.e.slowly) carbonated down to the reinforcements the corrosion rate of steel will remain low providing R.H. is at 50–60%, both because the pH will be fairly alkaline and because of the high resistivity of dry concrete.

However, if carbonated concrete comes into contact with aerated water the situation can become similar to that obtained under accelerated carbonation: the pH can decrease and subsequently catastrophically increase the corrosion rate.

Comparison of the four cements tested

In agreement with other workers [1] we have found that pure Portland cement is able to fix more CO_2 (i.e. takes longer to reach complete carbonation) than those of lower clinker contents. The range of corrosion rates during carbonation was 1:5, corresponding to the extremes of pure Portland and slag cement respectively.

In order to correlate in some way the above differences in i_{corr} with pH of the aqueous phase we have measured the pH of water which has been in contact with the carbonated mortars for 3 days. The pH values are given in Table 2.

Table 2 – pH values of the water of conservation of the carbonated mortars after 3 days of contact.

Cements	Portland		With slag		With pozzolana		With fly-ash	
Additives	Without additives	With 2% $CaCl_2$	Without additives	With 2% $CaCl_2$	Without additives	With 2% $CaCl_2$	Without additives	With 2% $CaCl_2$
pH	9.98	9.47	8.34	8.23	8.23	8.30	8.66	8.22

The lower pH produced by the addition of $CaCl_2$ is due to the common-ion effect of Ca^{2+}. Portland cement produces a higher pH than the other three which have about the same pH value. The higher corrosion rate in slag cement as compared with the two other blended cements could be due to its sulphide content.

The hatched area in Figs. 1–4 shows the frontier between tolerable and dangerous corrosion. It includes results of previous work [6, 7] and could perhaps be somewhat subjective. In the case of steel 0.1–0.2 $\mu A/cm^2$ corresponds to a penetration depth of about 0.11 mm after 50–100 years in use, providing the attack is uniform. Also, we estimated 1–2 $\mu A/cm^2$ as the maximum permissible value of the i_{corr} of reinforcements embedded in concrete, as the service life of concrete structures should be some 50–100 years.

CONCLUSIONS

(1) Accelerated carbonation of the concrete in CO_2-rich atmospheres generates the appearance of bicarbonates in the pore solution and a consequent pH decrease down to pH ≈ 6–7 at which steel corrosion could be catastrophic.
(2) The passive layer on steel seems to disappear at pH ≈ 8, in agreement with Pourbaix's results.
(3) The pH reached in the pore solution in carbonated concrete depends on:

- alkali content of the cement,
- proportion and hydration degree of the cement silicates,
- relative ambient humidity, and
- CO_2 partial pressure in the air or in the water.

The pH value and the concrete resistivity seem to control the corrosion rate in carbonated concrete.

REFERENCES

[1] M. Venuat, *Materiaux et Constructions* (1978), **11**, 142.
[2] RC-75 'Pliego de Prescripciones Técnicas Generales para la Recepción de Cemento', 1975, Madrid, Ministerio de Obras Públicas y Urbanismo.
[3] C. Andrade and J. A. Gonzalez, *Werkstoffe un Korrosion* (1978), **29**, 515.
[4] C. Andrade and A. Macias, 13th Internat. Galvanizing Conference, London, May 1982, 9/1.
[5] M. Pourbaix, *Lectures on Electrochemical Corrosion.* Plenum Press, New York and London, 1973, pp. 67 and 217.
[6] J. A. Gonzalez, S. Algaba and C. Andrade, *British Corrosion Journal* (1981), **15**, *135.*
[7] J. A. Gonzalez and C. Andrade, *Corrosion y Protección* (1980), Enero–Febrero, 15.
[8] J. A. Gonzalez and C. Andrade, *British Corrosion J.* (1982), **17**, 1, 21.
[9] I. Bizok, *La corrosión del homigón en masa y armado.* Urmo, p. 220.
[10] N. Aschan, *Nordisk Betong* (1963), **3**, 275.
[11] J. Keucher, J. Olden, H. Polster, *Bauplanung-Bautechnik* (1971), **25**, 8, 378.

Part 3
DIAGNOSIS, MONITORING AND REPAIR

Chapter 12
Corrosion monitoring of steel in concrete

J. L. DAWSON, UMIST, Manchester, UK

ABSTRACT

The paper outlines methods available to assess and monitor the corrosion of steel in concrete. In particular, emphasis is given to the use of electrochemical noise and impedance measurements since these offer advantages both for the inspection of existing structures and as an aid to engineering materials research.

A.c. impedance is a laboratory-based technique which can quantify the rate process, provide information on interfacial properties and changes in resistivity of the porous matrix. Electrochemical noise analysis is a new development in the study of concrete, the UMIST technique uses the measurement of the low frequency, small amplitude fluctuations of the corrosion potential. These noise signals are spontaneously generated by the naturally occurring corrosion process and can be used as a means of monitoring the corrosion penetration rate and identifying when, for example, chloride ions are causing film breakdown and pitting.

The use of these modern *in situ* electrochemical measurements will mean less reliance on destructive testing. As research methods they are more effective and provide a greater understanding of fundamental mechanisms; ultimately the quantification of data should lead to recommendations for the use of various reinforced materials. All the techniques therefore have a role to play in fundamental studies using exposure of prism samples.

Inspection of existing structures will continue to rely on the traditional methods which can assess the condition of the concrete, particularly those areas requiring obvious remedial attention. Electrochemical noise can be used on-site to complement the corrosion potential mapping techniques as it provides a rate measurement at selected areas without having to remove the concrete cover.

INTRODUCTION

The use of mild steel as a reinforcement for concrete structures is a well established practicable and economic solution for many civil engineering constructions. Excellent service spanning a number of decades has already been achieved by many reinforced structures and adequate performance can be anticipated provided the initial construction has followed established codes of practice. However, over the past fifteen years there has been an increasing number of reports on corrosion failures; particularly those problems associated with de-icing salts, marine environments, calcium chloride additives and contaminated aggregates.

The corrosion protection mechanism essentially relies on the maintenance of alkaline conditions developed during the hydration of the cement and pH values above 13 are frequently observed for Portland cement pastes. The ingress and reaction of atmospheric carbon dioxide and sulphur dioxide will decrease the alkalinity to below a pH of 9 and exacerbate the effect of chloride ions, the latter being the major cause of depassivation of the steel. The prime feature of corrosion protection is the long-term ratio of chloride to hydroxyl, but this is moderated by the influence of the calcium salts, tricalcium silicate, dicalcium silicate, tricalcium aluminate and tetracalcium alumino-ferrite. The pore water chemistry and the contribution from the surface chemistry involving micro and macro pore walls coated with hydrated gels is a subject area of continuing research and it is important that we understand the process since they control the electrochemistry of the corrosion reactions.

These are complex matters further complicated by macro-galvanic corrosion cells developed within large structures owing to differential effects, for example of oxygen diffusion. The influence of these factors has been covered by the many previous workers in the field but attention is drawn to the comprehensive publications in 1982 on the 'Durability of concrete structures' [1] and the review article on 'Aspects of the electrochemistry of steel in concrete' by Page and Treadaway [2]. In the latter paper one of the conclusions was that there was a need for improved electrochemical monitoring techniques for both laboratory experimentation and *in situ* measurements of corrosion rates in structures.

Research at UMIST over the past few years by the Electrochemistry Group in the Corrosion and Protection Centre has been directed towards increasing the understanding and limitations of standard electrochemical monitoring methods [3, 4] and also at developing various techniques [5–7] for a range of practical situations. Our more recent work [8–11] in the use of these modern techniques for studies in concrete is therefore the subject of the present paper. However, not only are these developments of scientific interest to the corrosionist but they also have implications for both on-site inspection of structures and fundamental studies related to the use of various concretes. In this context it is the non-destructive aspects of the measurements and the quantification of the data which could provide future guidelines for the industry.

CORROSION MONITORING AND INSPECTION TECHNIQUES

A cursory assessment of the various techniques used to study reinforced concrete may suggest that the requirements of the engineer concerned with a practical on-site examination are somewhat different from the aims of a materials scientist interested in laboratory based fundamental studies. However, in both cases essentially the same techniques are used, but in an investigation prior to repair or replacement or an investigation to assess the effectiveness of a new concrete formulation. The examinations would be first used to quantify the extent of the corrosion damage and this would be followed by a requirement to predict future performance; a major difficulty is that no one method provides sufficient information hence there is an element of judgement required in assessing the corrosion hazard. Since corrosion is an electrochemical process the traditional electrochemical methods have been surprisingly inept at assisting in the on-site monitoring of corrosion in concrete, the reason for this lack of appreciation will become obvious later in the paper. First it is useful to list the various techniques available, the non-electrochemical indicate the composition and environmental factors which contribute to corrosion whilst the electrochemical measure the corrosion state and *in situ* kinetics.

(1) Visual inspection

This often provides the first indication of a corrosion problem but is unable to provide sufficient information for a comprehensive survey. Initially hairline cracks may be observed at right angles to the surface and these typically follow the line of reinforcement. They result from the internal tensile forces caused by the build-up of corrosion product around the reinforcement. Fracture of the concrete, delamination of the outer layers of cover and rust staining are the obvious symptoms of the underlying problem.

Once the cover has been removed and the concrete has been cut back to expose the bar then the type of corrosion, the degree of pitting and nature of the corrosion products may all be important indicators of the cause of corrosion.

(2) Mechanical and ultrasonic tests

Delamination can be detected by the hollow sound produced on hitting the surface with a hammer. A comparative test, involving semi-quantification, uses the rebound or Schmidt hammer; here a spring-loaded device directs a known mass onto a plunger, which is in direct contact with the concrete, and the return distance expressed as a percentage of the initial spring tension is called the rebound number. This technique effectively defines the area of delamination and may provide a comparative indication of concrete quality provided account of voids and the size and type of aggregate is known. These site values will not be as reliable as compressive strength measurements obtained in the laboratory.

Ultrasonic pulse velocity measurements between two transducers can be used to assess the elastic properties and density of the concrete, the 'Pundit' is a typical site instrument which can also detect the presence of internal cracks.

(3) Core sampling, chemical and physical tests

Core samples from selected areas are the normal method of assessing the condition of the concrete and the reinforcement. The depth of the reinforcement and its position may be located by an electromagnetic covermeter, the cores can therefore be obtained with and without samples of the steel.

The depth of carbonation is obtained by the application of phenolphthalein indicator, the purple stain indentifying those areas which remain in an alkaline condition. Permeability, density and water absorption indicate the ease by which external contaminants penetrate the concrete, more sophisticated tests may include the measurement of the diffusion parameters for oxygen and chloride ions. Chemical analysis for chloride and sulphate distribution is often required. Samples for pore water analysis, pore size and surface area determination can be obtained by coring or by simple drilling to various depths and then retaining the dust samples for the laboratory examinations.

Electrical resistivity measurements are another indicator of the concrete composition and performance, a high resistance would suggest a decreased corrosion from, for example, galvanic effects. Typical guidelines have been proposed by a number of workers and in particular where corrosion potential measurements indicate possible corrosion, the values given in Table 1 are from the review and practical field study by Cavalier and Vassie [12].

Table 1

Concrete resistivity (Ω cm)	Corrosion
> 12,000	Usually not sign
5,000 – 12,000	Probable
< 5,000	Almost certain

(4) Potential mapping

Electrochemical measurements

Corrosion potential measurements using a reference electrode placed on the concrete surface which is connected via a high input impedance voltmeter ($> 10^9$ Ω) to the reinforcement cage or test prism can provide information on the state of passivity of the reinforcing steel. The technique is well known and is widely advocated for the inspection of bridge decks [12], the method being described in the American National Standard ANSI/ASTM C876-77.

The method has also been successfully used by Hans Arup and Frits Gronvold of the Danish Corrosion Centre in an automated instrumental version for the examination of over 5,000 concrete balconies as well as swimming pools and storage tanks. Their experience suggests that the copper/copper sulphate reference electrode should not be used for concrete [15] and the calomel electrode is to be preferred, also the surface should be mechanically abraded prior to the reference electrode contact point. They consider that an interpretation based on a simple criterion as suggested by Table 2 often leads to a wrong diagnosis; equally important from their experience, are the gradients of the iso-corrosion potential contours and where the contours bunch together this is often an indication of a corroding or corrosion risk area.

Table 2 – An oversimplified interpretation of corrosion potential measurements [14].

E_{corr} (vs $Cu/CuSO_4$)	Probability of corrosion
< -0.35	$>95\%$
> -0.20	$<5\%$
$-0.20 - 0.35$	approx. 50%

Potentiodynamic sweep or potential step measurements

This standard laboratory method is useful for polarization studies of corrosion in liquid electrolytes either for the identification of specific reactions or establishing the regions of passivity in the particular environment by means of polarization curves. However, the direct adoption for studies in concrete is open to criticism unless allowance is made both for the high resistivity of the environment, i.e. by IR drop compensation, and the very slow kinetics of the electrochemical corrosion reactions on reinforcement steel. In this context it should be appreciated that a concrete system may take hours or even days to achieve a new equilibrium condition after a d.c. perturbation. This is evidenced by the large time constants which are a measure of the kinetic parameters controlling the reinforcement corrosion [10]. Arup has already commented that for concrete studies the potential sweep method has to allow for the very restricted diffusion process [15]; reproducibility is only achieved by making the potentiostatic sweep over a period of weeks [15]. The conventional potentiodynamic method would therefore appear to have only a limited application, for example, in preliminary investigations, it is unlikely to yield kinetic information or data which can be quantified.

Linear polarization resistance measurement (LPRM)

LRPM is the traditional d.c. technique of measuring corrosion rates in aqueous systems and is based on the method of Stern and Gearey [18]:

$$\text{Corrosion rate} = \frac{\text{Const.}}{R_p}, \text{ where the polarization}$$

$$\text{resistance, } R_p. = \frac{\Delta E}{\Delta i}$$

is given by the slope of the linear portion of the polarization curve close to the corrosion potential, i.e. within ±10 to 20 mV. For i in $\mu A/cm^2$ and a corrosion rate in $\mu m/y$ then the constant is typically 11.5 × 52 for 'passive' steel and 11.5 × 26 for 'active' steel, the differences reflecting the change in Tafel slope or kinetics for the two systems.

The method assumes that steady state conditions apply and hence that a new dynamic equilibrium state has been achieved following the applied voltage or current perturbation [3, 4]. Some workers plot the decay of the transient versus time over two or three days in order to ensure that equilibrium is reached [19], also an appropriate mathematical interpretation can yield kinetic information. A number of workers have reported good correlations between weight loss and LRPM [20, 21, 22]. The technique is essentially laboratory-based, unless test probes were originally buried in the structure during construction or repair. Care must be taken to allow for the concrete (or pore solution) resitance, $R\Omega$, on ohmic resistance. An ohmic, or IR-drop, can be compensated for both electronically or numerically, also improved electrode design can assist the experimentation.

A.c. impedance

Corrosion at the steel/concrete interface is a complex process, the most appropriate laboratory technique is therefore the impedance method which provides information on the kinetic control mechanisms, either activation or diffusion, and on the film dielectric and electrochemical capacitive elements. The technique is able to measure the pore solution or concrete resistance, $R_{(\Omega)}$, the dielectric film which develops on the steel, $R_{(f)}$, as well as the electrochemical parameters such as the double layer capacitance, $C_{(dl)}$, charge transfer resistance, $R_{(ct)}$, and Warburg diffusion, W. These parameters define the corrosion processes and since each responds to the applied voltage or current perturbation within specific frequency ranges it is possible to quantify each component. Typically these are represented by an equivalent circuit and impedance is shown as a Nyquist or Complex Plane plot.

Impedance measurements can be made in the frequency domain by applying a small amplitude perturbation ± 10 to 20 mV as a sine wave of known frequency and observing the phase shift and modulus of impedance, the measurement is then repeated at different frequencies over the range of interest, 100 kHz to 1 mHz in the case of concrete; typical instrumentation is the Frequency Response Analyser. An alternative method is to use time domain measurements where a varying signal, white noise or pseudo-random, is applied and the resultant response from the corrosion cell is compared to the initial perturbation by means of a spectrum analyser. The use of a Fast Fourier Transform will then convert the time domain data into the more conventional frequency domain. In principle it should also be possible to use a single perturbation, as in the d.c.

method, and then use the Laplace transform to obtain the frequency data but unfortunately the errors encountered in the analysis of the low frequency data, obtained from the longer time transients, mitigate against this technique.

For concrete studies there is a requirement to obtain high and low frequency data and here digital techniques can be used to advantage, both for the accuracy of measurement and ease of interpretation. This can be achieved by a suitable computer program to control the instrumentation, and provide adequate replots of the data from which the required parameters can be assessed (see Fig. 1). We use a simulation based on an aggregate equivalent circuit model, this first uses the experimental values to give a plot which is compared to the original experimental data. The values used in the model can be modified in an interactive manner to achieve the best possible fit to the experimental plots. Time constant dispersion, which causes a depression of the complex plane plot, can be allowed for whilst diffusion impedence can be estimated using the Warburg coefficient, σ, and a diffusion factor, K, [23]. We have combined these control, interpretation and modelling programs together to provide an easily operated software package [24]. The use of impedance in the study of electrochemical corrosion of steel in concrete has been reported previously [9, 10, 11, 25, 26], further examples are given below and presented in the paper on corrosion repair assessment at this conference [27], these will indicate the application of impedance to practical research.

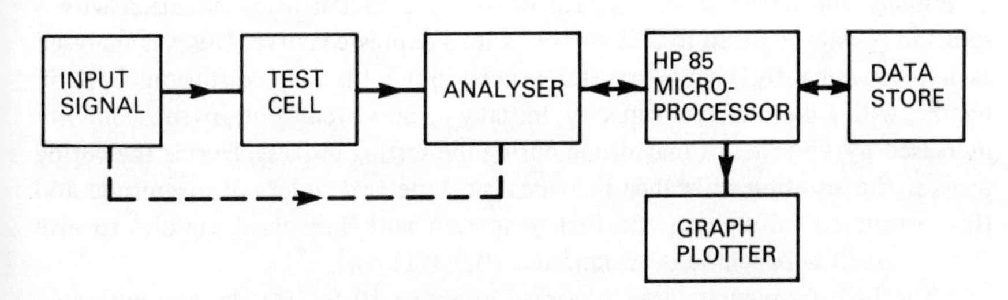

Fig. 1 – Block diagram of impedance instrumentation.

Electrochemical noise

Electrochemical noise is observed as spontaneous fluctuations of potential and current; the noise may result from thermal agitation of charge carriers [28], drift from changes in environmental factors and stochastic variations in the nucleation of the anodic reaction [29] or film formation process. The research at UMIST has been concerned with the measurement and analysis of the corrosion potential noise [7, 8, 11] since this offers advantages as a novel method for monitoring corrosion both in the laboratory and on-site.

The noise signal, or time record, was obtained between two identical electrodes, or between the reinforcement and a referenced electrode, using a sensitive digital voltmeter. The data was stored on a magnetic tape and interpreted by means of standard spectral analysis techniques, e.g. the Fast Fourier Transform (FFT) [7] or the Maximum Entropy Method (MEM) [8], the latter provides a smoother plot. The corrosion noise is typical of low frequency $1/f$ noise, it is non-gaussian and thus increasing the measurement time with the aim of improving the data collection by means of averaging techniques does not apply. The length of the time record is determined by the frequency range of interest, typically this was 10 Hz to 1 mHz, which required quarter of an hour to two hours sample time. Under most conditions a sample period of between twenty minutes and forty minutes is adequate. The standard deviation, or r.m.s., of the noise signal appears to be proportional to the corrosion penetration rate [7] and this is therefore of interest for site measurements since the rate data will complement the iso-corrosion potential mapping techniques. The latter method can therefore be used to first identify the likely corrosion hazard regions prior to the rate determination.

RESULTS

Impedance results obtained on concrete prisms during setting are shown in Fig. 2. Initially the response was typical of an aqueous corrosion situation with a solution resistance of up to 5 Ω cm^2 but with a depressed curve. This was analysed as a predominantly charge transfer process but with some diffusion, both in parallel with a double layer capacity, initially of 50 μF/cm^2 but this subsequently increased by an order of magnitude during the setting process. During the curing process the solution resistance increases by some two orders of magnitude and this continued throughout the first year even with immersed samples to give R(s) values of between 2,000 Ω cm^2 and 10,000 Ω cm^2.

The high frequency curve, observed at above 10 to 100 Hz, was not apparent during the initial setting period, this 'spur' appears as the curing proceeds and continues to develop over the exposure period. This high frequency effect can be modelled by a parallel R.C. combination whose resistance increased to a constant value over the first 20 to 50 weeks, Fig. 4. This high frequency response is not associated with the electrochemical double layer capacitance but appears to be related to the dielectric properties of the steel/concrete interface. It is tentatively suggested that this may be a feature of the portlandite or lime-rich film which grows on the steel during this period. The rate of development of this film resistance and its ultimate value depends on the type of pre-exposure, the environmental conditions and the degree of concrete cover. Further research is required before a full description of the processes involved in the growth of this high frequency dielectric phenomena can be presented.

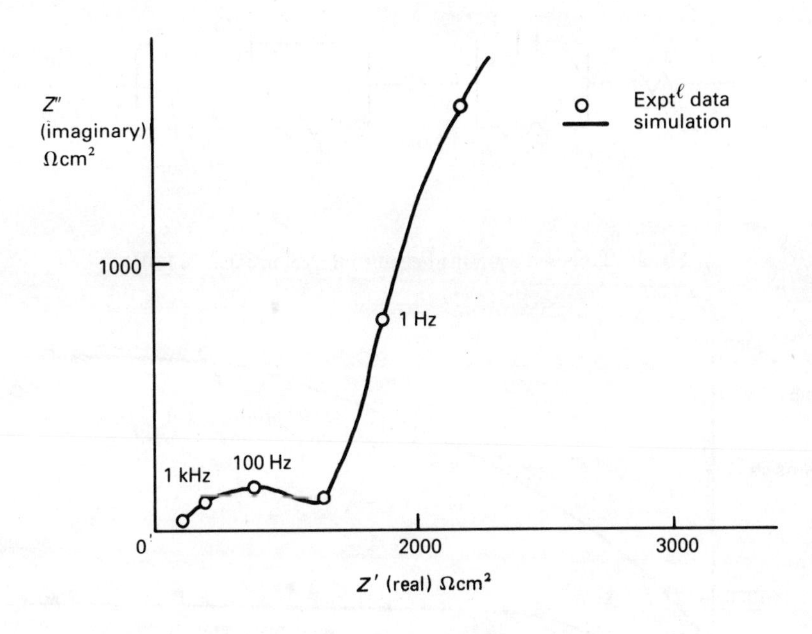

Fig. 2 – Impedance plot from steel in concrete.

The lower frequency impedance response, below 100 mHz, has a capacitance value of between 50 $\mu F/cm^2$ to 100 $\mu F/cm^2$ which is typical of an electrochemical double layer. The time constants of those electrochemical processes were between 1 s and 10 ks, indicative of a slow reaction, and is not surprising since the diffusion rates associated with cathodic reactions in concrete are restricted. Detailed analysis of the complex plane impedance by means of the 'IMPED' program [24] showed that the electrochemical response could be modelled by a charge transfer resistance in series with a Warburg diffusion term and both these are in parallel with the double layer capacitance, Fig. 3. The use of the interactive simulation portion of the computer program enables the model data to coincide with the experimentally obtained plots. Allowance was made for the depression of both the high and low frequency portions of the plots, this depression is the result of a dispersion of the time constants which is indicative of a non-uniform distribution of the electrochemical processes. Corrosion rate measurements were also obtained from the low frequency data; typically the values for charge transfer resistance, R_{ct}, were substituted for the polarization resistance, R_p, in the Stern–Geary equation and a value of 50 mV for the conversion constant.

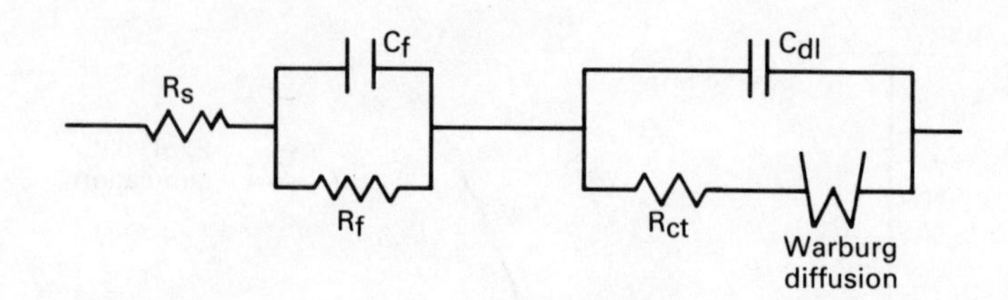

Fig. 3 – Proposed equivalent circuit for steel in concrete [10].

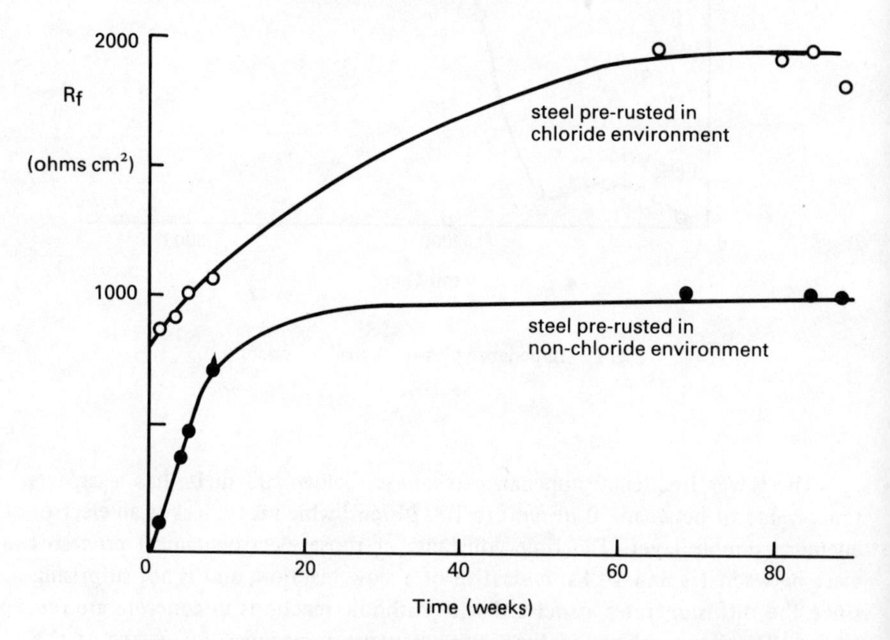

Fig. 4 – Increase in interfacial resistance during exposure in sea water [36].

It should be noted that these charge transfer resistances are the smallest during pouring and setting (Fig. 2), i.e. in an aqueous environment (Fig. 5). As the curing period commences the corrosion rate decreases and the solution resistance increases, the charge transfer resistances are of the order of 10 kΩ cm^2 during setting and increases, to between 50 kΩ cm^2 and 100 kΩ cm^2 during exposure, hence in general the observed corrosion rates in concrete are of the order of a few micrometres per year or less. When the corrosion rate falls to very low values the impedance technique is unable to provide sensible corrosion rate data; however, it should be noted that the actual rates are so small as to be practically zero.

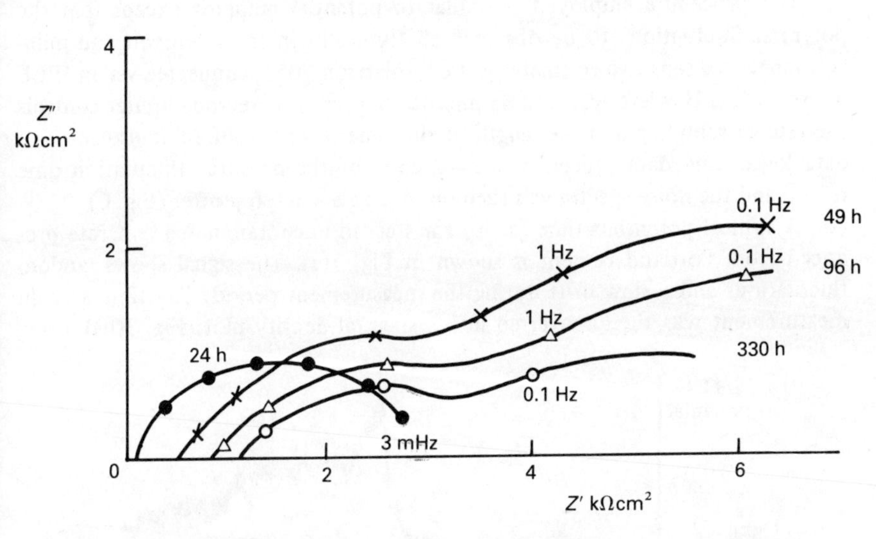

Fig. 5 – Impedance plots from concrete mix containing 4% $CaCl_2$ showing change from initial aqueous environment response and development of interfacial response [9].

Electrochemical noise results have been obtained on steel bars embedded in concrete prisms, on various test beams exposed at marine and industrial sites and on reinforcement steel within reinforced concrete structures. For the laboratory-based studies using concrete cubes or test prisms then measurements can be made either between two lengths of bar of electrodes cast into the samples or between the steel and an external reference electrode. Larger structures require an electrical connection to be made to the reinforcement and a reference electrode must be employed. We have found that the calomel electrode generates less noise than both the copper/copper sulphate electrode and silver/silver chloride electrode, neither is it subject to photovoltaic effects.

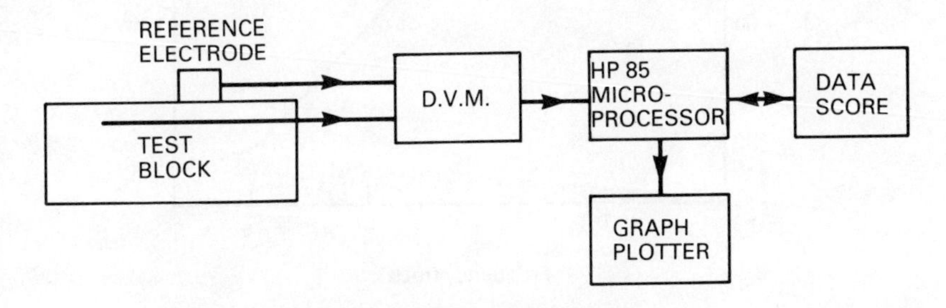

Fig. 6 – Block diagram of noise measuring instrumentation.

The procedure employed is similar to potential mapping except that the potential fluctuations to be observed are typically in the microvolt and millivolt range so a sensitive voltmeter is used, solartron 7055, connected via an IEEE interface to a Hewlett–Packard 85 microcomputer. The microcomputer controls the rate of sampling and the length of the time record required; it also acts as a data logger and data processor; a hard copy of the potential fluctuation time record and the noise spectra was then obtained via a graph plotter (Fig. 6).

A typical potential–time record for steel in uncontaminated concrete produced from Portland cement is shown in Fig. 7(a). The signal shows random fluctuations and a slow drift during the measurement period. This time domain measurement was then converted into a spectral density plot (Fig. 7(b)), using

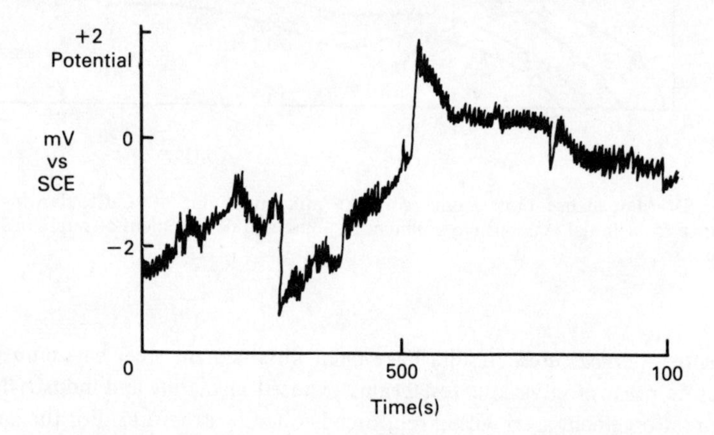

Fig. 7(a) – Potential/time record from uncontaminated concrete.

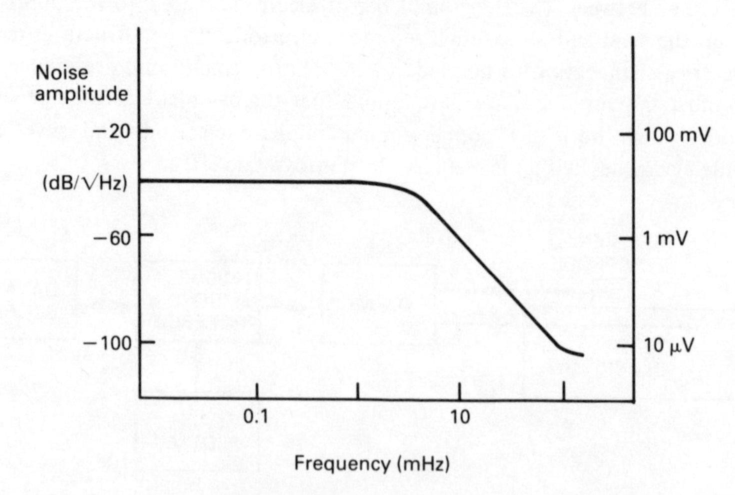

Fig. 7(b) – Noise spectrum from uncontaminated concrete.

the Maximum Entropy Method of spectral analysis [30]. The plot is a conventional presentation showing noise amplitude (dBV/√Hz) or logarithm of the voltage (V/√Hz) vs logarithm of frequency (Hz). At high frequencies the noise level is constant at a low level and is characteristic of a white noise with a gaussian distribution resulting from thermal vibrations and in many cases this could also indicate the noise floor of the measuring equipment. The noise amplitude increased with decreasing frequency, a feature of $1/f$ or pink noise [31], defined as a stochastic or non-stationary random process. At the low frequency the noise amplitude becomes constant, for some electrochemical systems it has been proposed that the spectra is the result of electro-crystallization processes [32, 33].

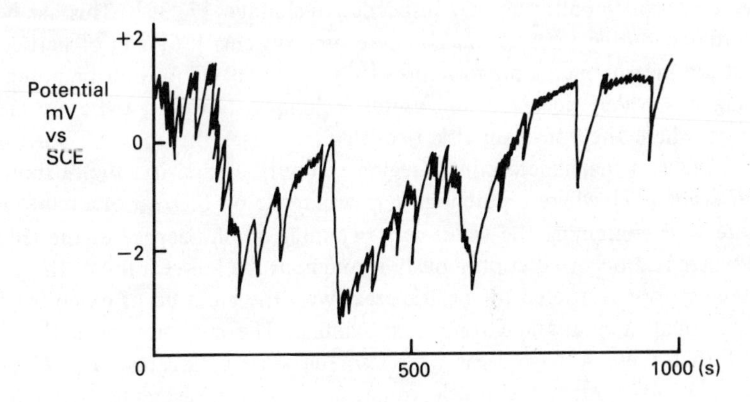

Fig. 8(a) – Potential/time record from chloride contaminated concrete.

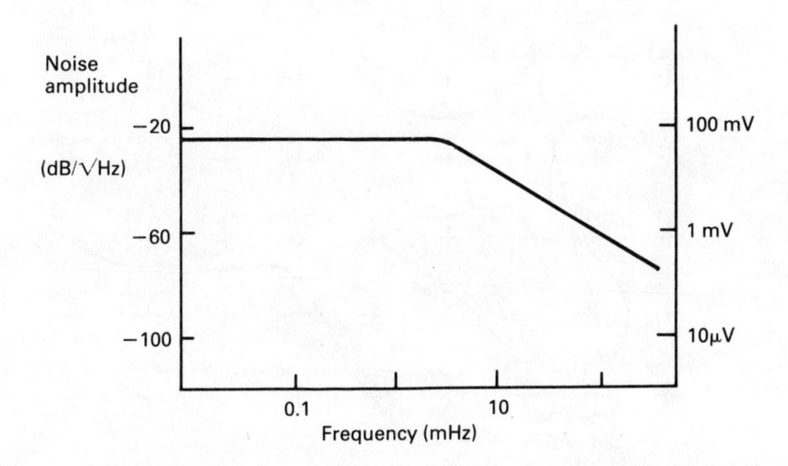

Fig. 8(b) – Noise spectrum from chloride contaminated concrete.

Noise measurements made on steels embedded in concrete which was contaminated with chloride showed a different type of potential fluctuation. The time record contained a series of 'glitches', recorded as a virtually instantaneous potential drop of up to a few tens of millivolts followed by a relatively slow exponential recovery. This appears to be a characteristic of corroding systems undergoing pitting [34, 35]. The noise spectrum is similar to the uncontaminated concrete but the roll-off or slope of the $1/f$ electrochemical noise plot decreased to values approaching −10 dB per decade, again a feature of pitting corrosion (Figs. 8(a) and 8(b)).

It has been our observation that the standard deviation or r.m.s. value of the noise signal is proportional to the corrosion rate and this is the basis of an on-site corrosion monitoring and inspection technique [7, 35]. This has been substantiated during research on concrete type systems [36] and of particular interest are measurements on structures [37]. As indicated previously, potential mapping is a widely accepted and useful technique; however, there are many instances where the corrosion risk is difficult to assess because the corrosion potential falls in that indeterminate region below true passivity. Fig. 9 shows a case in point [37] where a potential map was made on the rear of a reinforced concrete wall containing the water of a swimming pool. Seepage of the chlorinated water had occurred continuously throughout the ten-year life of the pool. This seepage was restricted to specific areas with the build up of extensive calcium carbonate deposits and some rust staining. The corrosion potential map showed that these seepage areas were also the areas of greatest risk, E_{corr} = −0.45 V vs calomel; however, electrochemical noise measurements indicated that although there was evidence of pitting on the reinforcement (from the presence of potential 'glitches' on the time record) the noise level as measured

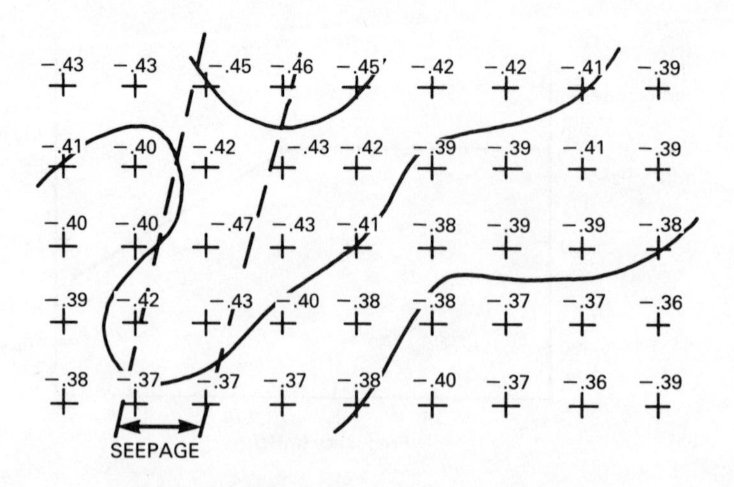

Fig. 9 – Potential map of pool wall (mV vs SCE).

by the standard deviation was relatively low (equivalent to 2 mpy). These data suggested that repairs or remedial work were unnecessary and indeed a subsequent destructive examination confirmed that the steel was in good condition apart from some minor shallow pitting.

DISCUSSION AND CONCLUSIONS

Electrochemical measurements are shown to be a valuable aid for monitoring the corrosion of steel reinforcement in concrete. Traditionally the inspection of structures has relied on destructive tests to provide the engineering data required to assess the present condition and predict the longer-term performance. The exception has perhaps been the use of corrosion potential maps used for specific applications, but although the mapping process is relatively quick the problem has been the inability to obtain a direct corrosion rate readily. The work at UMIST indicates that electrochemical noise measurements could be a valuable aid to the engineer since these provide quantification of the corrosion process and an indication of pitting corrosion. Further investigations and field studies will obviously be required before the technique is widely accepted but it is our opinion that noise will find many applications in corrosion monitoring.

Electrochemical noise has been studied by electrochemists and corrosion scientists for the past decade, their research interests being directed to understanding this stochastic phenomenon at a fundamental level [32] or as an aid to investigations of passive films [38]. These investigators have tended to carry out their experiments with potentiostatic control of the test electrode, but this can introduce further noise from the electronic circuit [39] and, as Barker [40] indicated in 1969, deterministic process are probably better studied by more conventional techniques. The source of the noise is obviously a subject for continuing research, however, at UMIST we have been interested in developing these techniques as an aid in corrosion rate measurements and in their application for corrosion monitoring, hence our interest in the r.m.s. or standard deviation of the noise amplitude. Also we consider it preferable to use spectral density plots [7, 8, 38, 39]. To present the noise as an amplitude is preferred since the alternative plot with the noise as an autocorrelation function [32, 33] does not make for ease of comparison between different sets of data. These studies suggest that the electrochemical noise method can be a useful tool for the investigation of corrosion on reinforcement steel in both the laboratory and on-site.

Electrochemical impedance is essentially a laboratory-based technique which may be used for studies on relatively small electrodes or sections of reinforcement bar. Information is obtained on changes of concrete/pore resistance, i.e. the observed 'solution' resistance, R_s; corrosion rate data and an indication of the relative magnitudes of the charge transfer and diffusion processes can often be observed, also the development of interfacial dielectric film can be monitored. Electrochemical noise on the other hand provides no information on the film resistance or film properties.

The importance of these *in situ* measurements is in their ability to quantify the corrosion and related processes and as such they are a valuable aid to inspection and research studies. The techniques should therefore greatly assist the materials scientist, the corrosionist and the concrete technologist in their development and research work particularly as there will no longer be the need for extensive destructive testing.

REFERENCES

[1] *Durability of Concrete,* **pub.** ILEM (1982).

[2] Page, C. L. and Treadaway, K. W. J., *Nature,* **297**, 109 (1982).

[3] Callow, L. M., Richardson, J. A. and Dawson, J. L., *Br. Corr. J.,* **11**, 123 (1976).

[4] Callow, L. M., Richardson, J. A. and Dawson, J. L., *Br. Corr. J.,* **11**, 131 (1976).

[5] Callow, L. M., Hladky, K. and Dawson, J. L., *Br. Corr. J.,* **15**, 20 (1980).

[6] Hladky, K. and Dawson, J. L., *Corr. Sci.,* **21**, 317 (1981).

[7] Hladky, K. and Dawson, J. L., *Corr. Sci.,* **22**, 231 (1982).

[8] Hladky, K. and Dawson, J. L., *Corr. Sci.,* in press (Paper presented at the Int. Conf. on Corrosion Testing & Research, UMIST, 1982.

[9] Dawson, J. L., Callow, L. M., Hladky, K. and Richardson, J. A., *Proc. Corrosion/79, NACE, Houston, 1978,* Paper 125.

[10] John, D. S., Searson, P. C. and Dawson, J. L., *Br. Corr. J.,* **16**, 103 (1981).

[11] Searson, P. C., Dawson, J. L. and John, D. G., *Corr. Sci.,* in press (Paper presented at the Int. Conf. on Corrosion Testing & Research, UMIST, 1982.

[12] Cavalier, P. G. and Vassie, P. R., *Proc. Instn. Civ. Engrs.,* **70**, 461 (1981).

[13] Arup, H. and Gronvold, *Sreel in Concrete Newsletter,* **7**, 8, Jan. (1981), Korrosion Scentralen, Denmark.

[14] Van Deveer, J. R., *J. Am. Concr. Inst.,* **12**, 597 (1975).

[15] Arup, H., *Steel in Concrete Newsletter,* Korrosion Scentralen.

[16] Preece, C., Seminar on Electrochemistry and Corrosion of Steel in Concrete, May 1982, Copenhagen.

[17] Kaesche, H., *Zem-Kalk.-Gips,* **12**, 289 (1959).

[18] Stern, M. and Geary, A. L., *J. Electrochem. Soc.,* **104,** 56 (1957).

[19] Maahn, E., Seminar on Electrochemistry and Corrosion of Steel in Concrete. May 1982, Copenhagen.

[20] Andrade, C. and Gonzalez, J. A., *Werkstaffe Korros.,* **29**, 515 (1978).

[21] Gonzalez, J. A., Algaba, S. and Andrade, C., *Br. Corros. J.,* **15**, 135 (1980).

[22] Moore, W. B. R. and Hughes, D. T., *Proc. 8th Int. Congress on Metallic Corrosion,* Mainz, 1981, pp. 1820–1825.
[23] Dawson, J. L. and John, D. G., *J. Electroanal. Chem.,* **110,** 37 (1980).
[24] IMPED Series of Programs, Corrosion and Protection Centre Industrial Services, UMIST.
[25] Lemoine, L. and Tache, G., *Proc. Int. Symp. Corrosion & Protection Offshore/79, France, 1979.*
[26] Wenger, F., Galland, J. and Lemoine, L., *Proc. Int. Symp. on Behaviour of Offshore Concrete Structures, France, 1980,* Paper no. 11.
[27] John, D. G., Coote, A. T., Treadaway, K. W. J. and Dawson, J. L., Chapter 17 herein.
[28] Hooge, F. N., *Physica,* **60,** 130 (1972).
[29] Bindra, P., Fleischman, M., Oldfield, T. W. and Singleton, D., *Disc. Farad. Soc.,* **56,** 331 (1973).
[30] Burg, J. P., *Modern Spectrum Analysis,* 42–48 (IEEE, New York, 1978).
[31] Halford, D. *Proc. I.E.E.E.,* **56,** 251 (1968).
[32] Blanc, G., Epelboin, I., Gabrielli, C. and Keddam, M., *J. Electroanal. Chem.,* **62,** 59 (1975).
[33] Blanc, G., Epelboin, I., Gabrielli, C. and Keddam, M., *J. Electroanal. Chem.,* **75,** 97 (1977).
[34] Hagyard, T. and Williams, J. R., *Trans. Farad. Soc.,* **57,** 2288 (1961).
[35] UK Patent Application 8200196 (1982).
[36] Searson, P. C., Ph.D. Thesis, Manchester (1982).
[37] Dawson, J. L., Al-Zank, I. A., Lomas, J., Mossavi, A. N. and Searson, P. C., Paper submitted to *Corrosion 1983.*
[38] Tachibana, K. and Okamoto, G., *Reviews on Coatings and Corrosion,* **IV,** No. 3, 229 (1981).
[39] Bertocci, U., *J. Electrochem. Soc.,* **128,** 520 (1981).
[40] Barker, G. C., *J. Electroanal. Chem.,* **21,** 127 (1969).

Chapter 13

Analysis of structural condition from durability results

R. D. BROWNE, M. P. GEOGHEGAN and **A. F. BAKER,** Taywood Engineering Ltd, Southall, Middx., UK

INTRODUCTION

Concrete structures have been promoted to society over the years as having an indefinitely long life requiring negligible maintenance. Protection to the embedded steel against the external environment was believed to be adequately provided for by the cover thickness and the quality and type of the concrete in the cover zone, obtained primarily from relevant codes of practice. Although many concrete structures have survived 50 years or more, it was apparent to us during our investigations in the early 1970s to support the design of North Sea concrete platforms, that many marine structures had been damaged by the environment owing to reinforcement corrosion after some 10–15 years from construction.

Subsequent investigations for our own company and overseas clients revealed that even faster environmental penetration was possible with major visible distress to the structure [1, 2, 3]. Further, damage in the UK from de-icing salts on bridges and the use of $CaCl_2$ in construction, as well as, CO_2 penetration from the atmosphere, particularly to cladding for buildings, illustrated the misunderstanding by Engineers as to the permanency of their structure in different environments and for different concretes. Moreover, it also became clear that:

(a) *The method of undertaking inspections of concrete structures was too limited* – relying on a visual examination only, with few tools or techniques available to identify the extent of environmental penetration and level of corrosion damage to the steel and the concrete.
(b) *Techniques of repair in many cases were ineffective* – similar damage with time re-occurred in repaired areas and undamaged areas degraded further.

Resulting from our studies to support our own Company's needs related to offshore design, we pioneered in the UK the use of a package of testing techniques to obtain a comprehensive picture of the state of a structure which gave a stronger basis for defining remedial measures. Existing and modified standard concrete testing methods were used in combination with physical, electrical, chemical and electrochemical techniques in conjunction with a range of measurements on the surface of the structure and on drilled cores and dust samples. Individual techniques and their combined usage have been developed, further aided by computer processing and computer graphics, speed of data processing and its presentation, to greatly aid the assessment of the state of the structure.

Our paper attempts to describe the approach we now adopt to inspection whether to meet statutory requirements for the North Sea, or for undertaking investigations on other structures where damage has been or may be occurring.

We have obtained a large amount of data from structural inspections in three continents and, bearing in mind the cost of repairs, we are endeavouring to use the know-how acquired to improve current practice with a view to improving the achievement rate of design life in the future for concrete structures for different environments.

We believe that routine, selective and more detailed inspection of a structure will become more important, as the identification of the lowest degree of environmental penetration becomes possible. This will allow earlier preventative measures to be taken, before the processes of damage become serious. Remedial techniques may then be used before expensive restoration or even replacement is necessary.

THE SURVEY

The objectives of a structural survey are obviously dictated by the clients' needs. While these may seem to differ superficially, the underlying requirement is for the following to be identified:

(a) What is the present state of environmental penetration and deterioration?
(b) What will be the future rates of penetration and deterioration?
(c) Are repairs required now or in the future? If so, which type and how much repair should be undertaken.

In order to achieve these objectives, our approach to structural surveys is designed and based on four main items.

The Structure's History and Environment

Details are required of the structure's environment. General climate information and atmosphere/ground chemical contamination levels, both general and local to the structure, are collated from records or if not available, obtained from measurements during the survey. Site survey and test records such as aggregate, cement and concrete cube test certificates and photographic records are obtained, if possible. In addition, we need to know the life which the client requires of his structure – it may now be different from the original design life.

The Field Specification

The Field Specification is a planning document, advising both the client and inspectors of the survey objectives, methods and scope. It contains all the client's available information pertinent to the survey, e.g. previous survey results, maintenance and repair records. It also contains pro-forma sheets used to record the mass of information on site in a logical manner. These contain a plan of each element to be covered by the surveyor, a drawing to locate the element in the overall structure, and a reference grid system of the element to relate all the test data.

The Field Specification provides the inspectors with a list of each task they are required to perform, together with a work schedule. This allows work to proceed in a planned sequence such that non-productive survey tasks, e.g. surface cleaning or erecting scaffolding, can be fitted into the main survey work to reduce downtime to a minimum.

The Global Survey

This provides information for the structure as a whole and would normally consist of a visual and photographic survey combined with limited non-destructive test (NDT) measurements. The global survey is designed to provide a fast scan of the structure such that areas for more comprehensive investigation may be selected. The basis of selection will normally be provided by NDT results indicating actively corroding areas, or by the structure's past history, for example, an area which has been repaired and requires checking. Other areas representing typical and sound areas of concrete will also be selected as a control. Pro-formas for the areas chosen from the global survey will then be prepared and added to the field specification.

In order to speed the survey process and to obtain conformity between different inspectors' interpretation of visual defects, a classification system is used. Table 1 shows a simplified visual defect classification scheme which gives guidelines to interpret the differences between say, pop-out, spalling and delamination, together with instructions as to what comments or measurements are required.

The Detailed Survey

The detailed survey employs a comprehensive range of NDT techniques to establish the quality of concrete cover and the corrosion state of the embedded steel. Concrete samples are also obtained. These will normally be in the form of drilled dust samples through the thickness of the structure and, where possible, 100-mm diameter cores. Such samples are analysed by a wide range of techniques available in the laboratories.

This stage of the survey provides the central, most important information required for the durability interpretation which is aided by computer plotting and contouring of NDT results and sample data set out in standardised formats.

Table 1 – Visual inspection. Simplified defect classification (T.E. system).

Code	Feature	Description	Cause	Details to be given on inspection	
A1	Cracking (general)	Jagged separations of concrete from no gap upwards	Overload, corrosion, shrinkage	Direction, width, depth	Cracking
A2	Pattern cracking	As cracking but formed as pattern	Differential volume change between internal and external concrete.	Surface area, width	
B1	Exudation	Viscous gel-like material exuding through a pore	Alkali aggregate reaction	Severity	Surface deposits
B2	Incrustation	A crust (white) on the concrete surface	Leaching of lime from cement	Severity/dampness	
B3	Rust stains	Brown stains	Corrosion of rebar, tying wire or surface steelwork	Severity	
B4	Dampness	The extent of water on the surface should be stated	Leakage, rundown	Severity	
C1	Pop-out	Shallow, conical depression	Development of local internal pressure, i.e. expansion of aggregate particle	Surface area, depth	Concrete loss
C2	Spall	A fragment detached from a larger mass	Exertion of internal pressure, i.e. by rebar corrosion or exertion of external force	Area, depth	
C3	Delamination	A sheet spall	Exertion of internal pressure over a large area	Area, depth	
C4	Weathering	Loss of the concrete surface	Environmental action wears away the laitance and paste	Area, depth	
D1	Tearing	Similar to cracks	Adhesion to slipform shuttering	Width, depth	Construction defects
D2	Honeycombing	Voidage between coarse aggregates	Lack of vibration	Area, depth, severity	
E1	Construction joint	Line on concrete surface, maybe feather-edged or porous-looking	Joint between two pours	Any associated deterioration	Construction features

However, the final interpretation must be made in conjunction with results from the other survey information. The engineers can then identify potentially damaging processes that are under way and grade the structure in terms of its present and future durability and repair.

It should be stressed that the detailed survey is made as basic or comprehensive as the client's requirements dictate.

TESTING TECHNIQUES AND DATA PLOTTING

When interpreting survey results it is necessary not only to know the probable accuracy of NDT and sampling techniques but also the pitfalls inherent in the techniques. All NDT techniques used in isolation require very careful interpretation. Used in combination, discrepancies in results and misleading results can often be accounted for; the combination of techniques is more powerful than all of the individual measurements made.

Testing Techniques

Our technique is to use a number of tests to explore the interaction between the level of chloride or carbon dioxide at the steel surface, and its 'activeness to corrosion', the conductivity/resitivity of the concrete and the ability of oxygen to penetrate the cover zone to permit corrosion of the steel to occur. This approach, and descriptions of the techniques used have been published previously [4]. However, Table 2 summarises the concrete properties which we investigate and the techniques used.

Table 2 – Techniques employed in durability surveys

	PROPERTIES EXAMINED	TESTING TECHNIQUES	OTHER CONSIDERATIONS
ON SITE	*MECHANICAL*		
ON SITE	Concrete surface quality	Schmidt hammer rebound number–type N equipment	Errors due to surface softening, cracks or laminations etc.
LAB	Compressive/tensile strength	Test on cores	To BS 6075
	CHEMICAL		
ON SITE	Carbonation	Depth measurement using phenolpthalein spray	
ON SITE	Chloride, sulphate and moisture contents	Drillings to collect dust samples in 10 to 25 mm bands through the concrete using percussive drill	Analysis in accordance with BS 4550. Results are curve fitted to produce penetration rate coefficients.

Continued next page

Table 2 – *Continued*

	PROPERTIES EXAMINED	TESTING TECHNIQUES	OTHER CONSIDERATIONS
	CHEMICAL		
LABORATORY	Concrete mix proportions	To BS 1881, Pt. 6, Analysis of hardened concrete	
	Cement type	Aluminium : iron ratio	OPC = Al:Fe > 1.5 SRPC = Al:Fe < 0.9
	Original water/cement ratio		
	Minerological composition	X-ray diffraction using diffractometer (copper $k\ \alpha$ radiation)	
	PHYSICAL		
ON SITE	Condition of embedded steel	Electrochemical potentials	Careful interpretation using ASTM C876-80 interrelated with experience and results from other NDT techniques.
	,,	Electrical resistivity measurement	Based on soil resistance measurement.
	,,	Ultrasonic pulse velocity using Pundit with exponential heads (indirect method)	Variable head spacing with best fit line spacing versus time used to calculate velocity.
	,,	Endoscope – visual inspection of embedded steel	
	Condition of cover	Fe depth metre – cover, reinforcement size/spacing	Problems where cover over 100 mm, where mesh is used etc.
LABORATORY	Permeability of cover	Water permeability – oxygen diffusion coefficient	Very susceptible to concrete preconditioning, i.e. laboratory conditions of storage
	Coefficient of thermal expansion		
	Concrete density/porosity		Usually determined during permaeability testing
	General (steel and concrete)	Visual and microscopic examination using optical and polarised light microscopes	Identification of aggregate source, presence of slag, pfa etc.

Number of Readings and their Accuracy

An important consideration in undertaking surveys concerns the number of readings to take in one area. For example, how many Schmidt hammer measurements should be taken in a survey area? With only a few readings you cannot be confident of their accuracy or detect which readings are significant, but taking too many would not be cost-effective. We have therefore employed statistical techniques to optimise the number of readings.

Figure 1 shows the result of applying Tukey's test [5] to Schmidt hammer results taken in a survey area divided into 10 squares. Obviously, by increasing the number of measurements per square, the ability to detect significant results increases but Fig. 1 shows the basis for defining a reasonable sampling frequency and also suggests the optimum number of readings. In this case, four readings per square shows a marked improvement over taking two per square. Five or six per square also improves the reliability of results still further, but taking more than six readings per square will involve much more work than the benefit justifies, indicating that five or six readings per square (i.e. 50 or 60 readings per area) would be adequate.

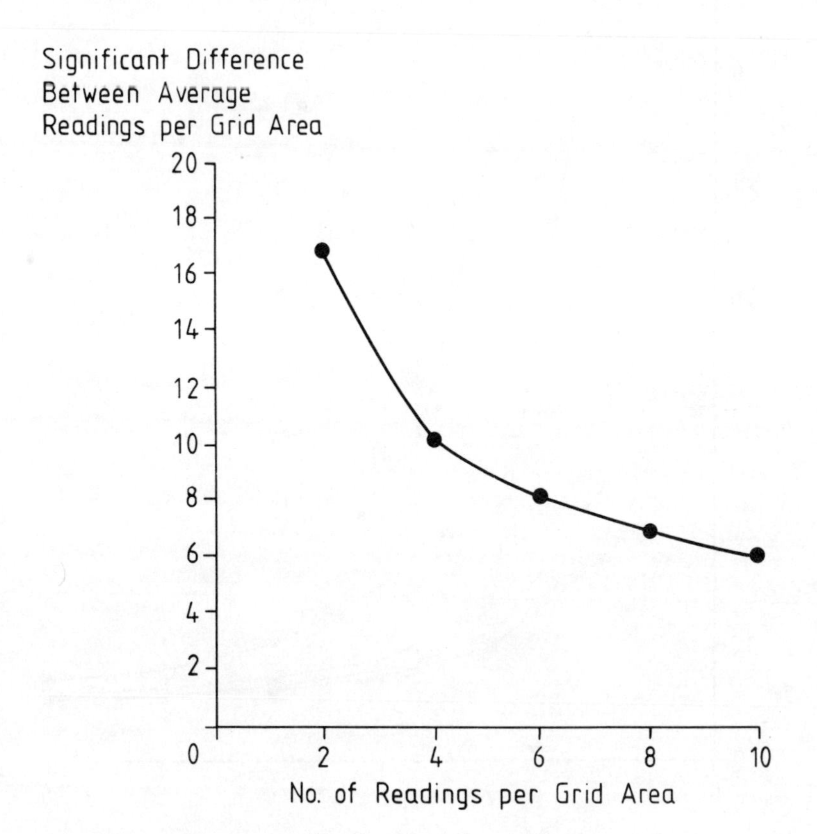

Fig. 1 – Optimising the number of Schmidt hammer readings to obtain reliable results from cost-effective surveys.

Figure 2 performs the same kind of analysis on cores. It shows that by taking three or four cores from an area and doing two or three strength tests from each of these cores, a greater confidence is obtained such that the actual strength will be within ± 5 N/mm² of your results.

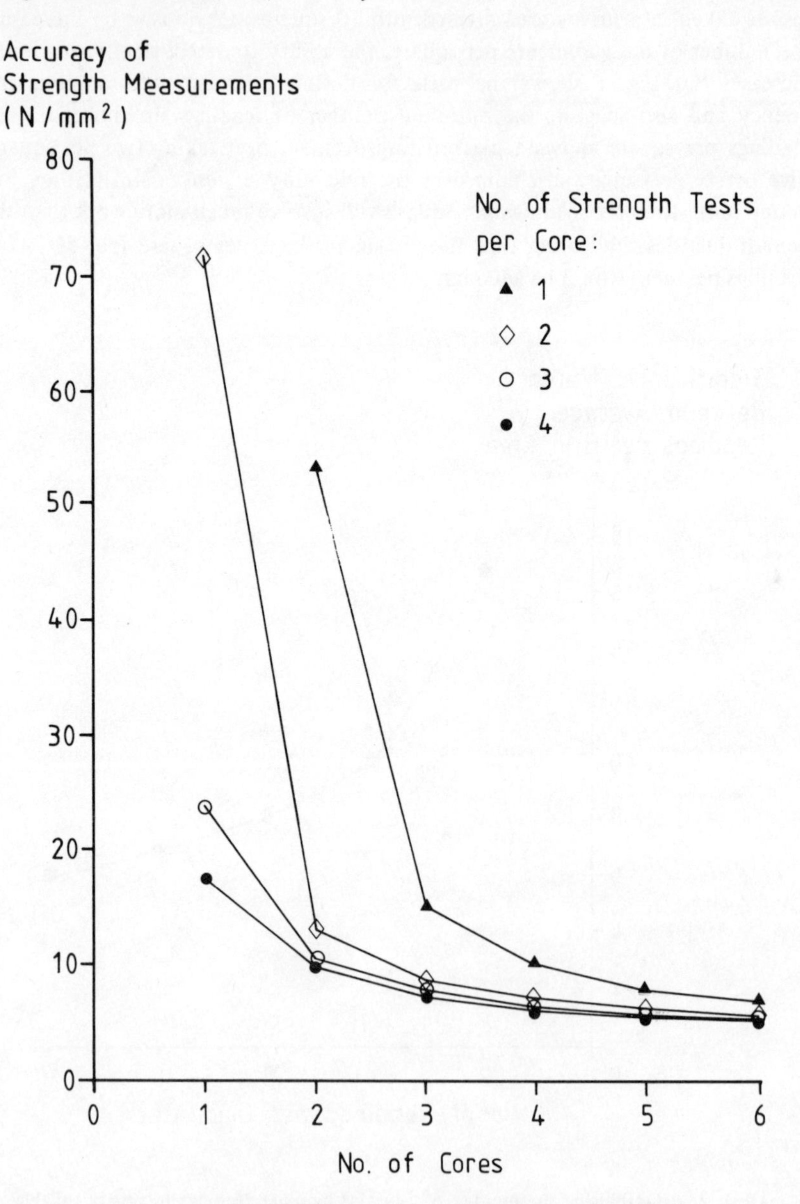

Fig. 2 – Optimising the number of core strength determinations.

Use of the Computer

On a major survey, the amount of raw data from potential resistivity and other NDT results can be overwhelming. We have therefore developed computer graphic techniques to produce print-outs of contour maps and 3D surfaces directly from the raw data. At the same time the data can be scanned to highlight critical areas in the contour maps and low results can be highlighted from statistical analysis. Figure 3 illustrates a typical computer presentation of data and shows how all the information from Schmidt hammer, resistivity and potential results ((a), (b) and (c) in Fig. 3) can be quickly scanned for areas of possible deterioration.

Figure 3 also illustrates the value of graphic techniques in interpretation for a structural component. In this case, for a wall, it can be seen from the vertical profiles that variation in results up a wall could be identified. This was associated with the differences in moisture content with height, frequently found in walls.

The computer techniques have minimised the tedious and time-consuming process of using draftsmen to produce visualisation of the data, thereby speeding up the time in preparing survey results and bringing forward the time when result interpretation can begin, the most difficult and demanding part of the process.

FURTHER PROBLEMS IN RESULT INTERPRETATION

The interpretation of the range of NDT results from the survey in terms of present state and likely future performance, is undertaken by using general guidelines outlined in Table 3 for both reinforcement corrosion and quality of the concrete cover. Apart from electrochemical potential levels, the interpretation guidelines have generally been assessed empirically from our past surveys and research work. The potential level guidelines were described by Van Daveer [6] after an extensive study of unpaved concrete bridge decks in the United States. Our surveys have generally tended to confirm these guidelines for other structures but apparent exceptions have been identified.

Below are summarised some of the factors which can effect NDT measurements to give apparently false results unless their magnitude can be assessed and taken into account.

Electrochemical Potential Measurements

Three particular factors which can have quite a significant effect on potential measurements are as follows:

1. *Junction Potentials*

(a) Carbonation

Carbonation effects on measured potentials have not been systematically assessed, but from our investigations an effect of producing potential readings *less negative* than expected, is suggested (Fig. 4(a)).

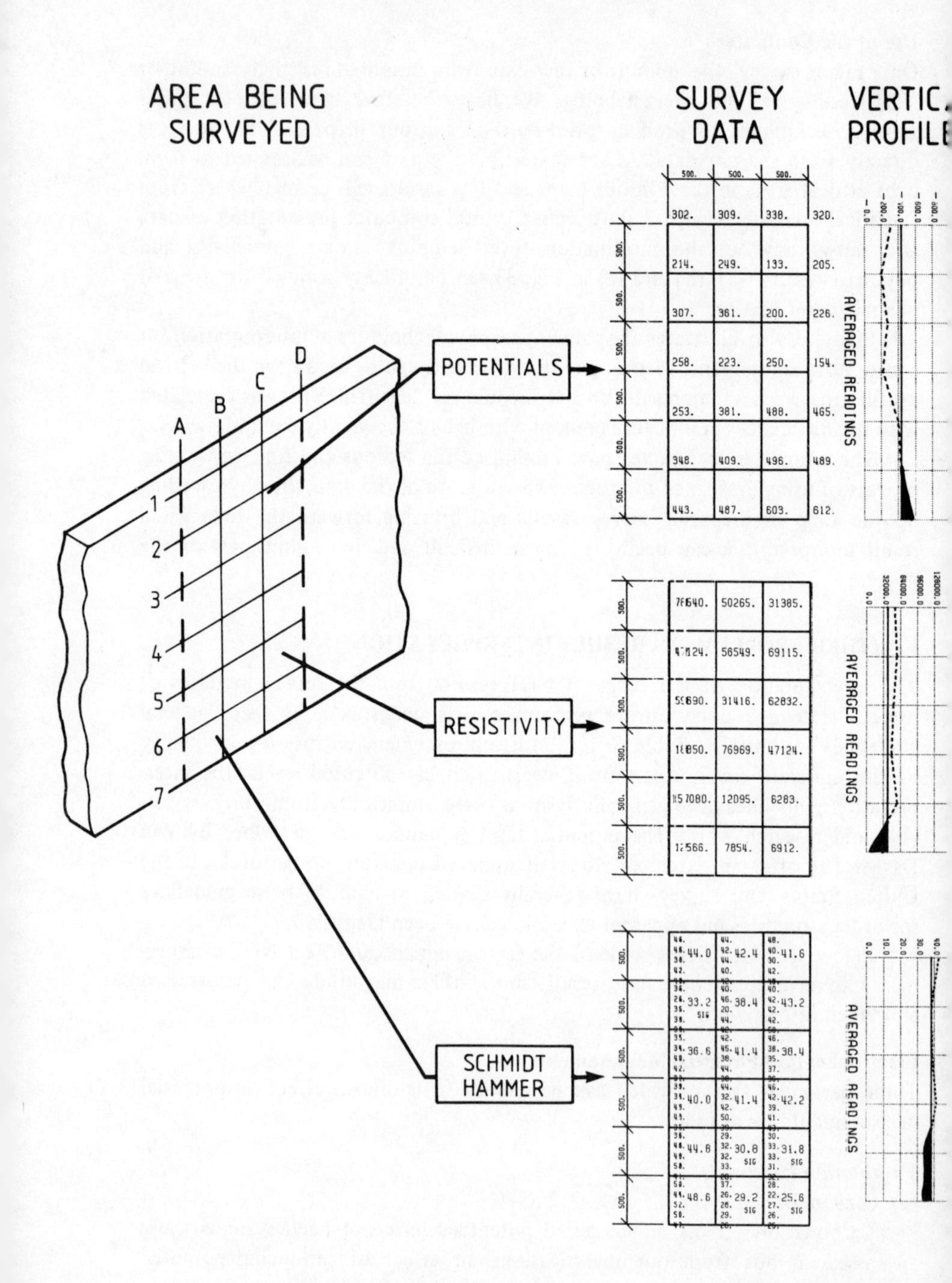

Fig. 3 – Visual plotting of non-destructive test results on a concrete wall.

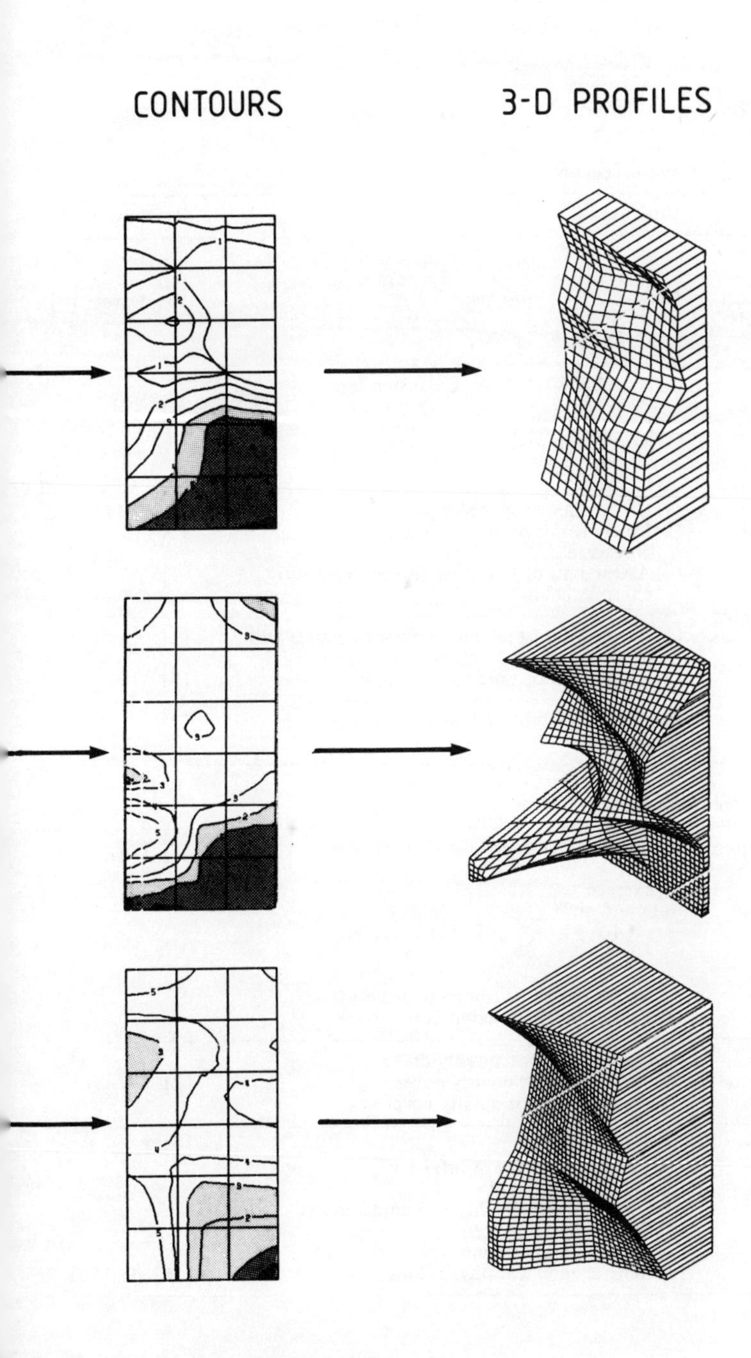

ables the rapid identification of critical areas.

Table 3 – General Interpretation of Inspection Survey Results

1. Corrosion of the reinforcement

Technique	Interpretation		References
Electrochemical potentials (CSE)	mV	Chance of Corrosion	Van Daveer [7]
	less −ve than −200	5%	
	−200 to 350	50%	
	more −ve than −350	95%	
Concrete resistivity	Ω cm	Corrosion rate	T.W. [2]
	<5,000	Very high	
	5,000 – 10,000	High	
	10,000 – 20,000	Low	
	> 20,000	Negligible	
Cover to reinforcement	(1) Observance of specification (2) Likely forms of deterioration due to rebar corrosion (3) Assessment of time to activation (see also chloride below)		CP110 [15] T.W. [3]
Chloride drillings	(1) Activation of the reinforcement possible with levels >0.35–0.4 wt% of the cement.		[2]
	(2) Assessment of time to activation		[3]
Core samples through rebar	Assessment of type and amount of corrosion		[2]

2. Quality of the concrete cover

Technique	Interpretation		References
Schmidt N-type hammer	Comparative hardness of the cover zone.		[2] T.W.
	Average rebound no.	Quality	
	> 40	Good, hard layer	
	30–40	Fair	
	20–30	Poor	
	< 20	Concrete delaminated near surface	
Pundit (indirect method)	>4 km/sec – good quality cover 3–4 km/sec – fair quality cover <3 km/sec – poor quality cover		[2] T.W.
Concrete cores	Assessment for: (1) Visual assessment of integrity (2) Concrete strength (3) Concrete permeability to liquid and gas (4) Diffusivity to oxygen (5) Depth of carbonation		[2] T.W.
	(6) General chemical/physical tests		[16] BS 1881

One possible reason for this occurring is the generation of a liquid junction potential at the carbonated/uncarbonated concrete layer interface, due to the difference in hydroxy ion concentration (via the dissolved $Ca(OH)_2$ in the pore water), between the almost neutral carbonated layer and highly alkaline concrete.

The equation controlling the magnitude of the junction potential [7] is:

$$E = (t_- - t_+) \frac{RT}{F} \ln \frac{a_2}{a_1}$$

Where t_- and t_+ are the transport numbers of anion and cation, respectively, and a is the activity of the ions of the two solutions.

The maximum magnitude of this effect is around 100 mV but this is enough to lead to misinterpretation of surface potential measurements.

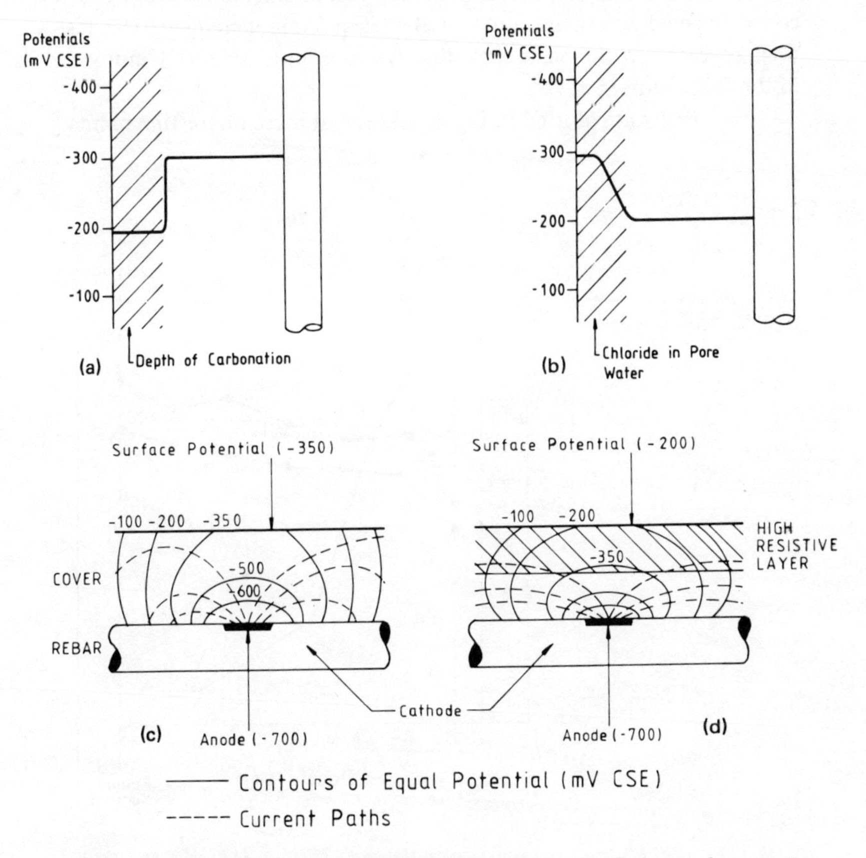

Fig. 4 – Factors affecting potential measurements. (a) Carbonated concrete, (b) chloride-contaminated concrete, (c) normal concrete, and (d) resistive concrete.

(b) *Chloride penetration*

In situations where chloride has ingressed into the concrete, but has not yet reached the reinforcement, a junction potential effect could occur owing to the varying concentration (activity) of chloride in the pore water. However, in this case, as the more concentrated solution is nearer the surface, we would expect from the above formula that potentials *more negative* than the true reading would tend to be measured (Fig. 4(b)). Again theory would suggest that the maximum junction effect would be 100 mV.

This led us to investigate potentials on a chloride-contaminated block of concrete in the laboratory. We were particularly interested in the layer closest to the concrete surface, as this was where the high chloride concentration was. Also, we expected to find much more negative potentials at the corroded rebar.

We took the first set of potential measurements by drilling into the concrete cover and taking a potential at each 5 mm increment. This test was then repeated in the same way that we took potentials at 1 mm intervals in the first 5 mm of cover.

The results are plotted in Fig. 5, and reveal three interesting points.

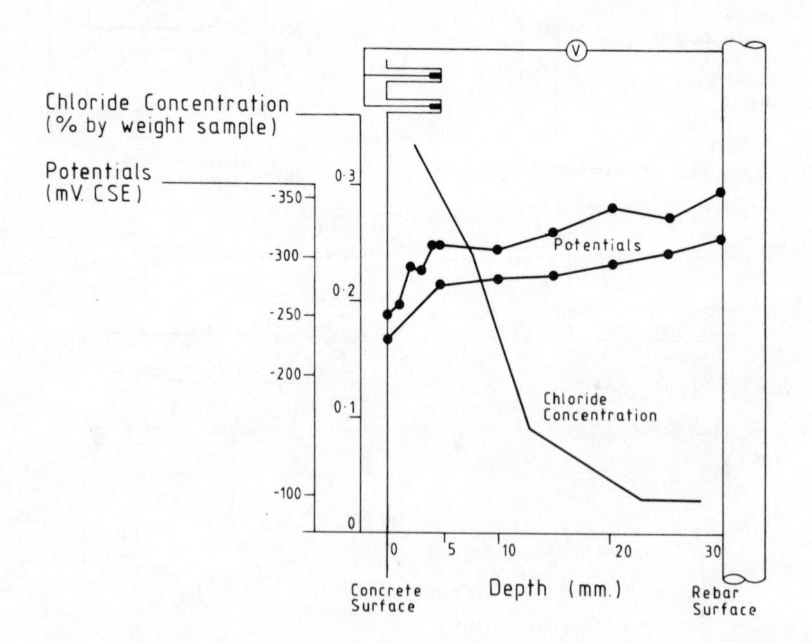

Fig. 5 – Measured potentials through the cover zone of a chloride-contaminated concrete specimen.

Firstly, in the first 5 mm, where the highest chloride level had been recorded, the results became *more positive* by 50 to 60 mV (i.e. an increase in potentials of 10 mV/mm). This directly conflicted with what we expected to happen from theory, which suggested that more negative potentials should be found.

Secondly, the readings obtained between 5 and 30 mm depth only became more negative by 50 to 60 mV (i.e. 2 mV/mm). These readings did not display a marked, more negative, jump in potentials at the reinforcement which Fig. 4(c) would suggest.

Lastly, apart from one reading at the reinforcement, the change in potentials through the cover did not stray outside the 50% corrosion risk interpretation of Van Daveer (−200 to −350 mV) [6].

Obviously, systematic study is required to resolve this difference between theory and experiment, and to assess if the junction effect is significant.

2. *Highly Resistive Concrete Surface Layers*

Concrete resistivity is particularly sensitive to moisture content [3]; and a highly resistive, dried out, surface layer could affect potential measurements. A significant depth of carbonation, which also tends to create a highly resistive concrete, could also produce the same effect. When surface potentials are taken they are essentially remote from the reinforcement due to the cover. The potentials thereby taken are, in fact, 'mixed' potentials, i.e. potentials taken at the anode areas are affected by the cathode areas and are, therefore, less negative than the actual potential immediately adjacent to the corroding reinforcement (Fig. 4(c)).

The effect of a high-resistance surface layer could be to enhance this effect as the macro-cell current paths would tend to avoid the highly resistant concrete (see Fig. 4(d)). The effect being more noticeable at the concentrated anode areas rather than the highly polarizable cathode areas. The overall effect being to record a less negative potential over the anode areas than expected.

The magnitude of this effect on potentials would depend on the thickness of the highly resistive layer but could range from negligible to a difference of up to 200–300 mV in theory.

3. *Polarization Fields*

In extreme cases, the macro-cell effect described above could lead to serious misinterpretation of potential results. The anode areas could polarize the adjacent cathode areas that the latter may be interpreted as areas of activation from the Van Daveer criteria. Two particular examples of this are related below.

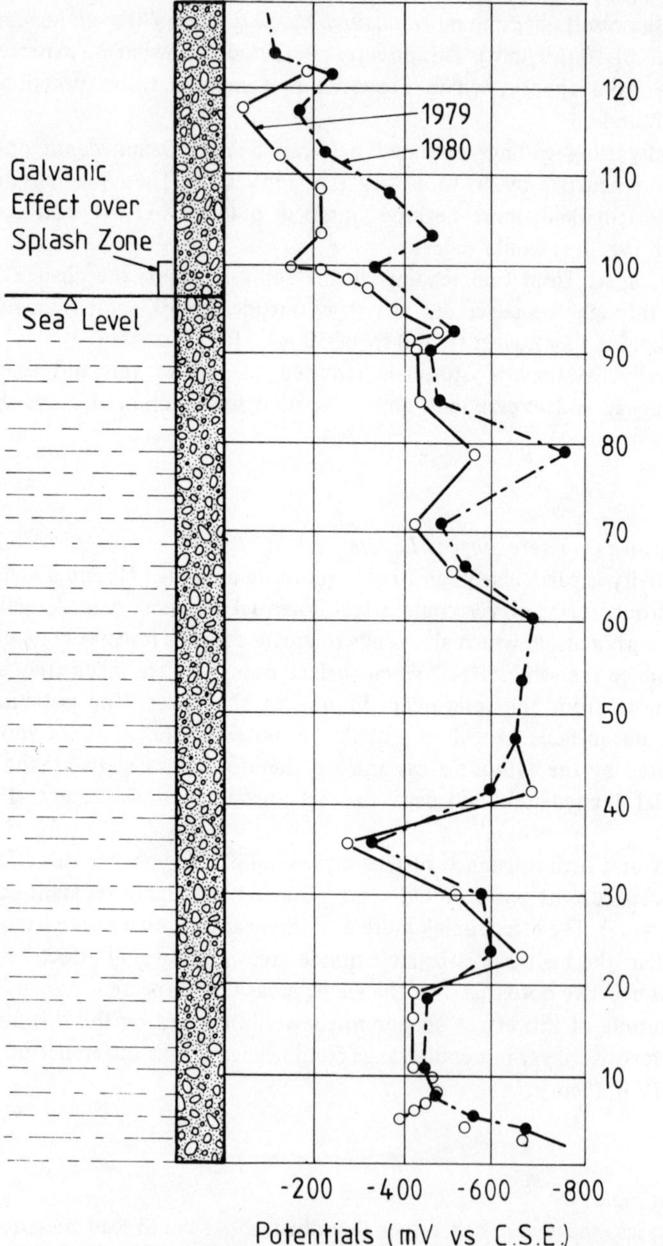

Fig. 6 – Reinforcement potentials taken down the inside of the central shaft of an offshore platform. This illustrates the way in which the submerged concrete has a polarising effect on the splash zone concrete [8].

(a) *The Splash–Underwater Zone Connection Effect*

Figure 6 shows a set of potential readings taken up the inside surface of the hollow central core of an offshore structure [8]. Below sea level the potential levels range, in general, from −400 to −700 mV CSE (except for one point for which there is no apparent explanation). However, at sea level the potentials dramatically became less negative to between 0 and −200 mV indicating passive reinforcement.

Of principal interest here is the area of rapid change between ± 5 m of the sea level (tidal range being a nominal 1 m). The general potential of the rebar on the outside of the leg taken by lowering a reference half cell into the water was found to be equivalent to −1050 mV vs CSE; thus the concrete immersed in sea water was polarizing that above the sea water. It is interesting that this galvanic effect extended to as much as 5 metres above sea level. Unless this effect was noted by the inspector, potentials taken blindly in a splash zone could be interpreted as active rebar where it was really just a polarization effect.

We have noted this interconnected zone mechanism on a number of marine structures in the splash zone. Sometimes, this effect has made it very difficult to interpret potential results to find the corroding areas.

(b) *The Hollow Leg Effect*

As stated above, the potential on the outside of the leg was around −1050 mV CSE and yet on the inside below sea level potentials ranging from −400 to −700 where recorded. This could be due to the hollow leg effect depicted in Fig. 7.

As the reinforcing cage is electrically interconnected, the potentials taken on the inside face reflect not only the nearest bar but also the potential contour field between the inner and outer layers of bar. If the outer layer of rebar becomes depassivated owing either to chloride ingress from the sea water or to inhibition of oxygen access it will take up a very negative potential. This will be reflected in the potential readings taken on the inside face as indicated by Fig. 7.

The actual potential field between inner and outer will depend not only on the resistivity of the concrete but also on the polarizability of the inner bars due to oxygen access. But the principal point is that unless this effect were considered, 'blind' interpretation of potentials taken on the inside surface would suggest that the underlying layer of rebar was active and corroding.

Each of the three effects described above (i.e. junction potentials, resistive concrete and polarization fields) could account for high active areas being interpreted as being passive, and vice versa, if the Van Daveer criteria were rigidly adhered to and no account was taken of the effects described above. In extreme cases the interaction of these factors may be so great that the structure may have to be 'internally calibrated' to devise a criteria of potential levels specific to that structure.

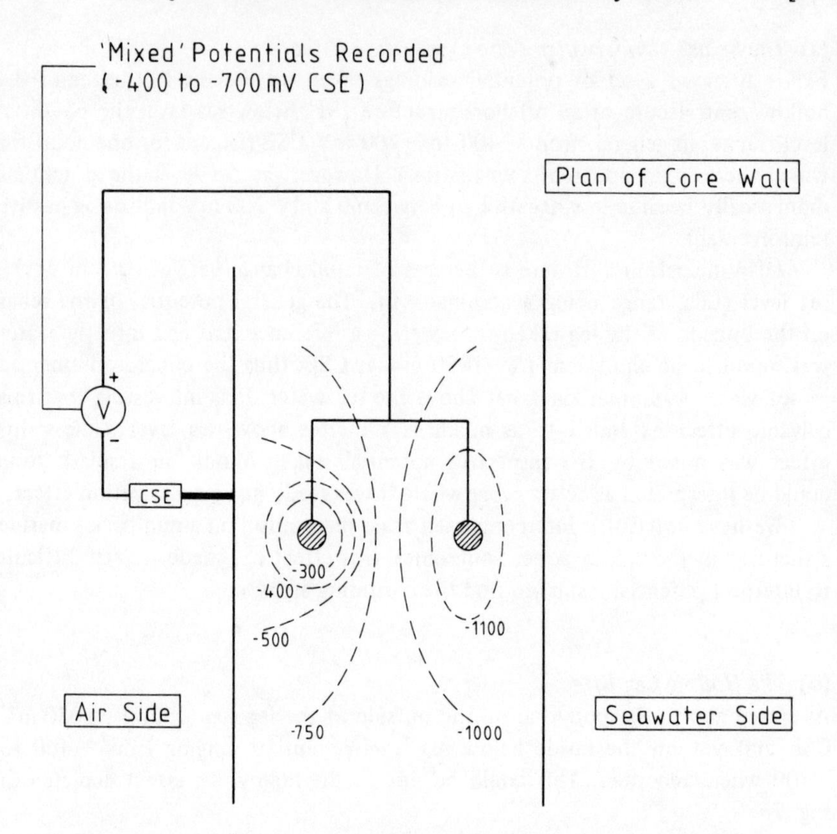

Fig. 7 – Potential field set up by the difference in potential of electrically connected internal and external reinforcement in the wall of a hollow leg offshore structure.

Resistivity Measurements

Unlike potential measurements, resistivity measurements tend not to be affected by extraneous factors and even the problem of a highly resistive surface skin can be fairly easily overcome. However, the resistivity measurement is only valid for the time of measurement and unless repeated at intervals, any effect of diurnal cycle needs to be taken into account. This is particularly prevalent when undertaking surveys in hot climates, e.g. the Middle East, where a resistivity measurement made during the dry season could indicate a value in excess of 100,000 Ω cm (and hence no corrosion) yet after the once-yearly downpour, it could reduce to 5,000 Ω cm [3].

If this were not taken into account very anomalous survey results could be produced. Resistivity measurements of cover samples from cores in the 'as-received' and saturated state can aid interpretation here.

Covermeter Readings

Certain problems have been encountered with the interpretation of covermeter readings. It has proved very difficult to measure cover to groups of reinforcing bars or mesh, where the spacing of steel in the mesh is smaller than the covermeter, even where a moderate cover exists. The magnetic covermeter has, in such cases, been found to be more successful at locating bars than assessing the depth of cover. In these cases, a positive determination of cover depth by drilling has been used as an on-site calibration of the instrument.

Ultrasonic Pulse Velocity Measurements

The empirical evaluation of 'indirect' measurements given in Table 3 is in general a good 'off the cuff' assessment of quality of the concrete cover.

It is important, however, to appreciate the factors influencing readings. One of our surveys highlighted the effect that a high level of moisture in the concrete could have on Pundit results. A number of readings were taken over the inside surface of the web and top flange of a precast concrete beam. In the area where the concrete was saturated, values in excess of 3.5 km/sec were recorded, yet visually the concrete appeared to be of low quality due to the ease with which water was permeating into it. In the non-saturated area a much lower velocity was recorded (2.2 km/sec) which might have been more indicative of the general quality of the concrete. Calibration for moisture-content effects is sometimes necessary.

Overall Assessment of Present Condition

The factors given above which could seriously affect interpretation of NDT results illustrate the extreme care which an engineer must take when analysing results to assess the state of the structure. However, if such pitfalls are recognized, an assessment of the structure's condition and even the cause of any deterioration can usually be made. With the data so obtained it is possible to make useful predictions on the structure's future performance.

PREDICTION OF PERFORMANCE

From the results of the global and detailed survey of the structure an estimate is made as to the state of the concrete and the amount and significance of any deterioration processes identified.

In order to determine whether areas of the structure need to be repaired, at what timescale repairs might be required, and whether additional inspection is advisable, it is necessary to arrive at a prediction of the rate of penetration of the environment and therefore, the probable timescale when repair and/or additional inspection becomes necessary.

Typically, a structure will exhibit areas requiring repair, areas where a more detailed inspection is necessary, and areas which, in the future, may require attention or may, with perhaps additional protection, perform satisfactorily for the desired design life of the structures.

An important part of the survey process is therefore aimed at advising the client on:

(1) Areas requiring repair now, and areas likely to require repair within the design life.
(2) Areas requiring additional detailed inspection now and areas which are likely to remain sound.
(3) What, in the form of maintenance and additional protection, can be done to extend the life of components beyond the design life without major repair.

Over the past 10 years we have, therefore, had to develop simplified methods of determining rate of penetration of, for example, water, CO_2 and chlorides, and a simple theory of damage [9,10].

Deterioration can be viewed as a two stage process where t_0 is the time taken for the environment to penetrate through to the steel and t_1 is the time taken for the resulting damage to become significant, i.e. require repair for structural, aesthetic or safety reasons. Therefore:

$$D \text{ (Design)} \leqslant t_0 + t_1$$

This theory, and the computation of diffusion coefficients have been discussed elsewhere [3, 10].

When Will the Environment Reach the Reinforcement (t_0)?

Penetration of Chlorides

Assuming a constant surface chloride level and diffusion coefficient, rates of chloride penetration can be calculated and the rate of progress of chlorides with time predicted. Figure 8(a) shows curves calculated for two surface chloride levels (2% and 5% chloride by weight of cement) and two ages (1 and 10 years). Assuming that reinforcing steel becomes active at a chloride level of 0.4% by weight of cement, an examination of Fig. 8(a) shows the following.

Surface chloride level (% by wt of cement)	Chloride diffusion coefficient ($\times 10^{-8}$ cm^2/sec)	Time to activate steel (t_0)	Fig. 8(a)
5	5	1 year at 30 mm cover	Point A
		10 years at 100 mm cover	Point B
2	5	10 years at 75 mm cover	Point C

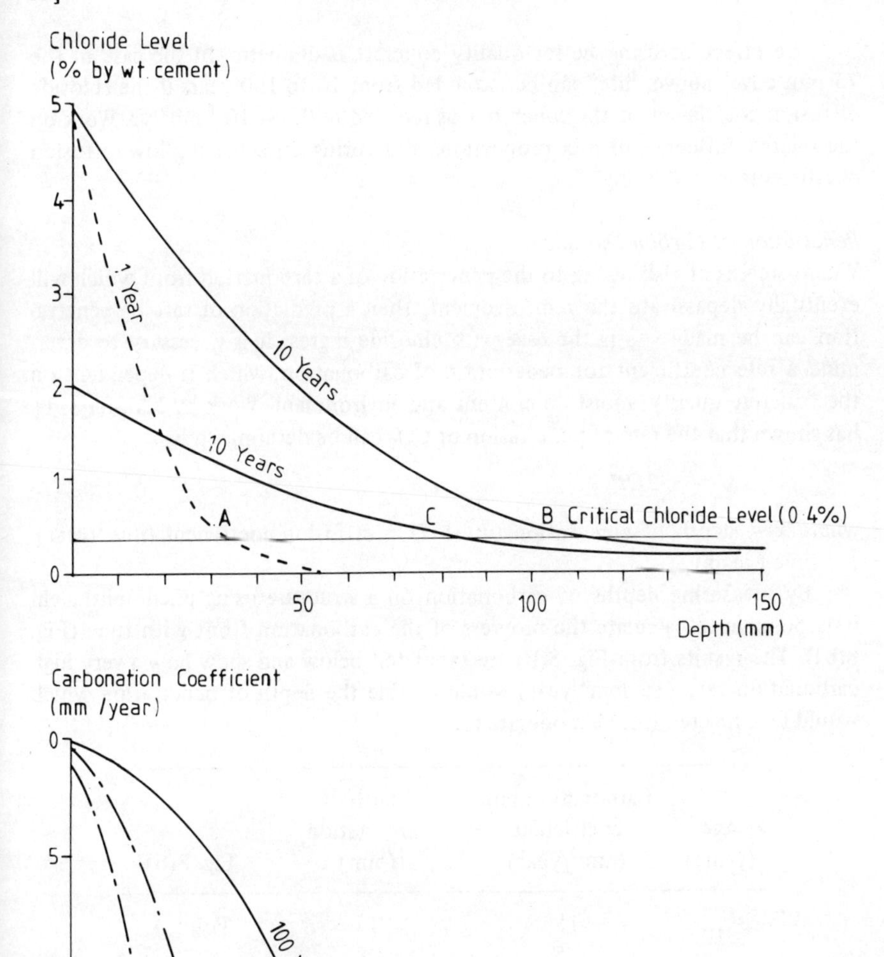

Fig. 8 – Top (Fig. 8(a)). The penetration of chloride into concrete showing the effect of cover and degree of exposure on t_0. Below (Fig. 8(b)). Penetration of carbonation showing the effect of cover thickness and carbonation rate coefficient on t_0.

The effect of using better quality concrete is dramatic. In the case of the 75 mm cover above, 'life' can be extended from 10 to 100 years if the chloride diffusion coefficient of the concrete was reduced to 0.5×10^{-8} cm^2/yr. Work on the relative influence of mix proportions and curing on achieving low diffusion coefficients is continuing.

Penetration of Carbon Dioxide

Where steel is at risk owing to the penetration of a carbonation front which will eventually depassivate the reinforcement, then a prediction of rate of penetration can be made. As is the case with chloride ingress it is necessary to determine a rate coefficient for penetration of carbonation, which is dependent on the concrete quality, moisture content and environment. Work by Klopfer [11] has shown that the rate of penetration of CO_2 can be determined by:

$$x = \sqrt{2Dt}$$

where x = depth of penetration (mm); D = diffusion coefficient (mm^2/year); t = time (years).

By measuring depths of carbonation on a structure using phenolphthalein it is possible to calculate the progress of the carbonation front with time (Fig. 8(b)). The results from Fig. 8(b) are tabulated below and show how a very high carbonation rate (15 mm^2/year) would double the depth of penetration which would be expected from a moderate rate.

Age (years)	Carbonation rate coefficient (mm^2/year)	Depth of carbonation (mm)	Fig. 8(b)
10	15	17	Point A
	5	8	
30	15	30	Point B
	5	15	
100	15	55	Point C
	5	32	

When Will Significant Damage Occur (t_1)?

While it is possible to calculate the time it takes for an aggressive environment to penetrate through the cover to the steel (t_0), we can only estimate the length of time it will then take for significant damage to occur (t_1), assuming of course that the damage is not visible. This estimate is based on experience and will have to take into account a number of factors including:

(a) *Definition of Serious Deterioration*
Cracking in high-rise cladding panels would be regarded as serious, as the next stage, spalling, could endager life. On the other hand, a structure such as the Mulberry Harbour Units at Portland [12], which only now act as a windbreak, might be allowed to reach a state of near collapse before its condition would be considered serious.

(b) *Exposure Conditions – i.e. Environment*
Serious deterioration has been observed in the Middle East after as little as 1 year. Here the conditions are severe, with a combination of high temperatures and an aggressive, humid, marine environment. Similar deterioration may only become apparent after 15 to 30 years in UK coastal structures [3].

The environment affects t_1 in three ways:

(i) *Temperature.* Steel corrosion is a chemical reaction and therefore the corrosion rate is influenced by temperature. Arrhenius's Law of a doubling of reaction rate per 10°C rise is a reasonable guide.

(ii) *Moisture content.* Concrete moisture content is an important factor in controlling the rate of reinforcement corrosion. The conductivity of the concrete will determine the magnitude of any corrosion current set up by areas of varying potential. Concrete's conductivity is approximately inversely proportional to the moisture content.

On surveys, the resistivity of the concrete is measured and an empirical relationship (Table 3) used to determine the possible rate of corrosion provided sufficient oxygen is available.

(iii) *Oxygen.* Lack of oxygen availability limits the steel corrosion rate in permanently immersed or semi-permanently saturated concrete (i.e. underwater and tidal zones), However, in splash or atmospheric zones oxygen is available in sufficient quantities not to limit the corrosion rate.

Other Factors Affecting Both t_0 and t_1

(a) *Cover and its Variability*
Figures 8(a) and 8(b) illustrate the importance of cover for a given concrete quality. For chloride penetration, by reducing the cover from 100 mm to 30 mm we would expect that corrosion could start after one year, not ten.

Cover, like strength, varies on a structure around a mean value. Concrete strength is specified as a characteristic strength, (i.e. 5% defectives at two times the standard deviation). However, concrete cover is normally only derived from durability tables [13] as a single value with a placing tolerance of, say, 5 mm. Even this, less rigorous, tolerance may not be achieved in practice.

Figure 9 illustrates measured cover depths for cladding panels on a high rise structure. The design cover of 25 mm allowed a placing tolerance of 6 mm. In fact, the average cover measured was 14 mm with a standard deviation of 6 mm. Only 16% of the cover was above 19 mm (design cover minus tolerance). This is a very extreme example where covers were measured where spalling had occurred and on core samples. However, problems can occur with precast cladding, especially when the aggregate is exposed and the panel is cast face down without spacer blocks. Such problems rarely occur with cast *in situ* work.

(b) *Shape*
Corner sections are particularly at risk due to biaxial penetration of chlorides, carbonation etc. [14]. Spalling of corners is often the first indication of reinforcement corrosion, followed by more widespread deterioration.

(c) *Reinforcement*
Bar sizes, their distribution and amount can all affect the time at which distress becomes serious, and also affects the type of damage (i.e. pop-out in low cover areas, cracking, spalling and eventually delamination in wall sections).

It is the assessment and interrelationship of all the above factors (definition of serious deterioration, environment, cover, shape and reinforcement) which enables an estimate of t_1 to be made before deterioration becomes visible.

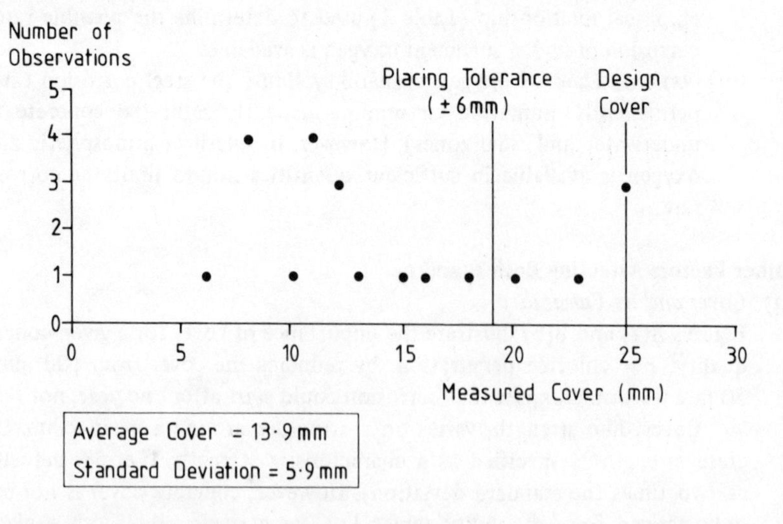

Fig. 9 – Standards allow a placing tolerance around the design cover, but in the case of one structure surveyed, the average cover was well below the minimum allowed for.

PROTECTION AND REPAIR

Having obtained estimates for the life of the structure and its components, it is necessary to apply a grading system to define how much repair is required, what type of repair will be necessary, and which areas should be inspected or re-inspected in the future.

Selection of Repair Procedures

Repair methods are selected on the basis of the type of deterioration found (cracking, localised spalling, delamination, softening etc.) and the reasons for deterioration (CO_2 or chloride ingress or other) together with knowledge of the required life of the structure.

Repairs to corroded reinforced concrete structures generally follow the same set of procedures although the actual methods employed may vary considerably depending on circumstances.

The outline procedure normally adopted is as follows:

(1) Removal of deteriorated concrete and concrete preparation.
(2) Sealing of cracks and/or treatment of the remaining sound concrete.
(3) Replacement and/or cleaning up of reinforcement and treatment of the steel.
(4) Replacement of the cover.
(5) Application of protective treatments to repaired and possibly unrepaired areas of the structure.

Table 4 gives the procedure in more detail.

Selection of Repair Materials

One of the most complex areas in concrete repair is the selection of suitable repair materials, i.e. for reinforcement coating, patch repair, protective coating etc. With a vast number of materials on the market, generally accompanied by little data on important properties such as bond strength, resistance to ultra-violet degradation, chloride ion, water and gas permeability etc. and generally little information on previous performance on similar concrete structures, choosing the correct materials is virtually impossible purely from the information supplied by the manufacturer.

Over the years, therefore, we have had to instigate and conduct test programmes to evaluate ranges of materials and inspect both structures under repair and the completed repair to verify performance. This work has involved investigating the fundamental properties of materials and carrying out basic property tests.

From this experience, we developed the kind of repair procedure outlined in Table 4. Selection of the appropriate repair material needs to be based on a number of factors which are illustrated by the two points below:

Table 4 – Outline Repair Procedure.

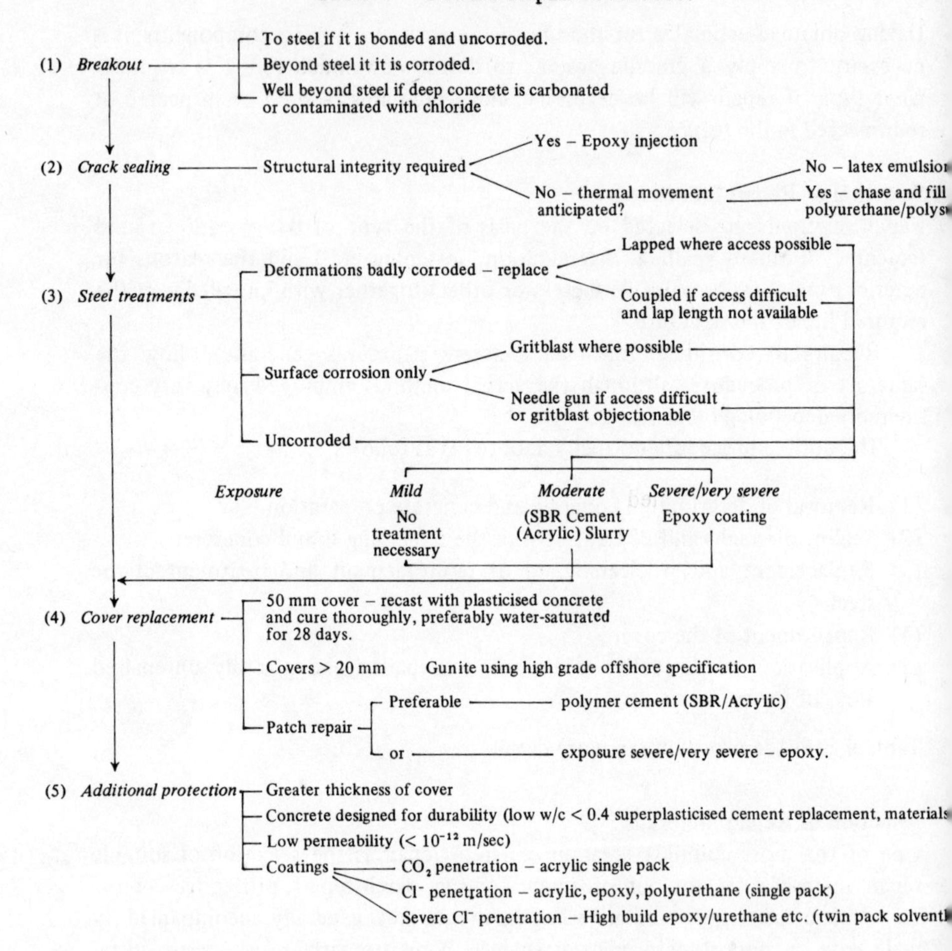

(1) Type of Repair/Mechanical Properties

Where repairs are required on load-bearing members, or where considerable stress or strain is expected, strength, bond strength and rate of bond strength development of the repair material are likely to be critical. Table 5 gives shear bond strengths for control, polymer modified and epoxy mortars. In terms of bond strength, for example, epoxy mortars perform the best. For protective and/or cosmetic repairs, mechanical performance may not be as critical as the ability of the repair material to withstand environmental penetration.

(2) *Exposure of Repair/Physical Properties*
In conditions where the environment is aggressive, the primary consideration is the ability of the repair to prevent or reduce the penetration rate. Table 5 shows that permeability coefficient of repair materials can vary by four orders of magnitude, depending on material type and supplier. Equally, large variations can occur in other physical properties such as chloride and oxygen diffusion rates and ultraviolet degradation.

The above two considerations for repair material selection are not mutually exclusive and are only intended to illustrate the range of factors which have to be taken into account to achieve a long-term effective repair. Material selection on the basis of one factor, such as strength, is inadequate.

As understanding of the many factors affecting repair selection can only be achieved by a systematic analysis of *in situ* repair performance backed up by detailed laboratory property testing.

Table 5 – Patch Repair materials, shear bond strength and permeability

Repair mix	Surface preparation + state	Shear bond strength N/mm^2	Permeability coefficient m/sec $\times 10^{-9}$
Control mortar a/c 3:1 w/c 0.5	Sawn, wet	20.46	9.6
Sytrene Butadiene 1 Polymer content 0.6% by wt cement a/c 3.1:1 w/c 0.47	Sawn, wet Bonding aid used	16.9	0.01
Styrene Butadiene 2 Polymer content 0.12% a/c 2.5:1 w/c 0.35	Sawn, wet Bonding aid used	23.3	0.0002
Acrylic 1 Polymer content 0.18% a/c 3:1 w/c 0.36	Sawn, wet Bonding aid used	21.5	0.04
Epoxy mortar As supplied	Sawn, wet Primed	Failure in concrete, bond intact	Totally impermeable to liquid water

CONCLUSIONS

In the last five years we have inspected some 25 reinforced and prestressed concrete structures in a wide variety of environments of quite different severity. Our experience has led us to the following conclusions:

(1) The combination of integrated inspection techniques, prediction of performance and grading of performance together constitute the most appropriate method of carrying out the condition survey of any concrete structure based on existing technology.

(2) Visual inspection alone, or in combination with one or two poorly understood techniques is not sufficient for concrete inspection work. A full understanding of the performance of the structure is required if repair work is to be effective.

(3) Pitfalls are inherent in the use of N.D.T. techniques, particularly when those are used in isolation. It is necessary to know how and where to apply the technique, its real on site sensitivity and accuracy and the situations where such techniques can be misleading.

(4) Using a simple damage theory, together with information on the rates of penetration of deleterious materials it is possible to assess for a structure whether repair/protection is necessary, and what extent and type of repair should be employed.

(5) At present a large proportion of our survey work has been done on structures where deterioration has become apparent. In these cases the survey objectives have been to establish the cause of deterioration, to recommend remedial measures and to predict the likely future performance of components which have not yet shown signs of deterioration.

(6) A smaller proportion of our work is for statutory or routine surveys in which the objective is to assess the current state of the structure, before extensive deterioration becomes visually apparent, thereby saving the cost of expensive remedial and replacement methods. It is expected that this type of survey work will increase in the future.

(7) Repair methods must be selected on the basis of the type and cause of any deterioration found. Considerable care is necessary in making this selection, especially as manufacturers of repair materials do not provide sufficient information necessary to ensure effective, long life repairs.

(8) Low maintenance concrete structures can be designed by careful selection of the concrete used and by selecting the appropriate treatment to protect the structure in different environments.

ACKNOWLEDGEMENTS

The authors wish to thank the Directors of Taywood Engineering Ltd. for permission to produce this paper, and the staff of the Research Laboratory whose efforts made the paper possible. Special thanks are due to Mike Adams, for his work on accuracy by sample analysis for different testing techniques and for details of computer graphics which he has developed to support our survey programme, and to Geoff Brine for his exhaustive efforts in editing and arranging the contents of the paper.

REFERENCES

[1] Browne, R. D. and Domone, P. L. J., 'The long term performance of concrete in the marine environment', *Proceedings of conference on Offshore Structures.* The Institution of Civil Engineers, London, 1975.

[2] Taylor Woodrow Research Laboratories 'Marine durability survey of the Tongue Sands Tower', *Concrete in the Oceans Technical Report No. 5.* The Cement and Concrete Association, Wexham Springs, 1980.

[3] Browne, R. D., 'Design prediction of the life for reinforced concrete in marine and other chloride environments', *Durability of Building Materials,* Vol. 1, pp. 113–125. Elsevier Scientific Publishing Company, Amsterdam, 1982.

[4] Browne, R. D. and Baker, A. F., 'The performance of structural concrete in the marine environment', In *Developments in Concrete Technology,* edited by F. D. Lydon. Applied Science Publishers Ltd., Barking, 1980.

[5] Davies, O. L. and Goldsmith, P. L. (editors), *Statistical Methods in Research and Production,* (4th edn). Longman, London, 1977.

[6] Van Daveer, J. R., 'Techniques for evaluating reinforced concrete bridge decks', *Journal of the American Concrete Institute,* December 1975, pp. 697–704.

[7] Denaro, A. R., *Elementary Electrochemistry.* Butterworth, London, 1965.

[8] Browne, R. D., Doyle, V. J. and Papworth, F., 'Inspection of concrete offshore structures', *Journal of Petroleum Technology,* November, 1981.

[9] Browne, R. D., 'The mechanisms of corrosion of steel in concrete in relation to design, inspection and repair of offshore and coastal structures', The American Concrete Institute, *Special Publication SP-65,* 1980.

[10] Browne, R. D., 'Testing and design for durability of concrete structures'. Presented at the RILEM Symposium at the Swedish Cement and Concrete Research Institute, Stockholm, June 1979.

[11] Klopfer, H., 'The carbonation of external concrete and how to combat it', One day conference on the repair of concrete structures, Imperial College, London 1981.

[12] Concrete-in-the-Oceans Technical Report, 'Marine durability survey of the Mulberry Harbour Units at Portland'. Confidential to the contributors to the Concrete-in-the-Oceans programme, 1982.

[13] The British Standards Institution, 'The structural use of concrete. Part 1, Design, materials and workmanship', Document 81/13842 (Draft for Public Comment). BSI, London, 1982.

[14] Browne, R. D., Domone, P. and Geoghegan, M.P., 'Deterioration of concrete structures under marine conditions – their inspection and repair', Conference on Maintenance of Maritime Structures, the Institution of Civil Engineers, London, 1977.

[15] The British Standards Institution, *The structural use of concrete. Part 1. Design materials and workmanship,* Code of Practice 110. BSI, London, 1972.

[16] The British Standards Institution, *Methods of testing concrete. Part 6. Analysis of hardened concrete,* British Standard 1881, BSI, London, 1971.

Chapter 14

The investigation and repair of damaged reinforced concrete structures

R. T. L. ALLEN and **J. A. FORRESTER,** Cement & Concrete Association, Cardiff, UK

1. INTRODUCTION

Structures may suffer damage from a variety of causes, such as corrosion of reinforcement, overloading, mechanical damage, chemical attack, freezing and thawing, poor design or poor construction. It is most important that, before any attempt is made to repair damage, the cause should be established as clearly as possible. Secondly, the objective of repair should be decided and, thirdly, a decision must be made as to whether the structure is to be restored as nearly as possible to its originally intended state or whether some improvement is necessary. Only then should remedial works be put in hand.

Deterioration of concrete structures may, in general, be divided into three stages. The first consists of changes in appearance, such as discolouration with local blemishes and staining. The second stage affects surface texture and is marked by scaling and cracking with, sometimes, a general break-up of the concrete surface. The third stage of deterioration is a disruption with major spalling of concrete away from the reinforcement, which may eventually lead to failure of the structure.

These stages of deterioration may be caused by various influences as described above. This paper considers the causes of corrosion of reinforcement and the remedy of the resulting damage.

2. OBSERVATIONS

The first step in assessment must be a close visual inspection for signs of distress or weakness [1]. These may include obvious poor compaction of the concrete, stains or discolouration, cracking, spalling, pop-outs, erosion and softening of the surface. Any such features that may be observed must then be related to the environment of the structure, loading, restraints to movement, and the position of reinforcement. Even if reinforcement drawings are not available, the probable arrangement can usually be estimated in most normal structures and electromagnetic covermeters can be used in cases of doubt.

Cracks should be inspected carefully for signs of cyclic movement. If such signs are present, they usually indicate that the cause of cracking was not corrosion. Research suggests that cracks that formed after the concrete hardened, less than 0.5 mm wide and transverse to reinforcement, seldom causes serious corrosion but wide cracks may, if untreated, lead to corrosion in the future [2]. If brown stains are associated with cracks then corrosion is probable. Cracks following the line of a reinforcing bar usually are caused by plastic settlement or corrosion and, whatever their origin, they are likely to lead to corrosion in the future. The absence of stains on the surface of the concrete does not necessarily indicate absence of corrosion. The positions and orientation of cracks must be considered in relation to loadings and restraints to shrinkage and thermal movements.

Poorly compacted concrete will carbonate rapidly and will not provide normal long-term protection to reinforcement, so such areas must be regarded as likely sites for corrosion in the future. Areas of dark discolouration corresponding to joints in formwork are a sign of grout leakage which may have resulted in permeable cover to reinforcement.

Spalls that are due to reinforcement corrosion often are easily distinguishable from those caused by mechanical damage because the reinforcement is usually exposed. Any damage that reduces the cover, however, is likely to increase the risk of corrosion in the future.

Brown stains on the surface of concrete are often a sign of corrosion, even if there is no visible cracking, but they may be caused by other factors such as impurities in the aggregates. Iron-pyrites is a contaminant that often causes such stains.

Unfortunately one cannot say with certainty that reinforcement corrosion is absent in all cases where there is no visible evidence such as cracking or staining. It has been found that ingress of chlorides into reinforced concrete structures can cause local corrosion and severe pitting of reinforcement without any external sign of expansion [3]. If there is any reason to suspect the presence of chlorides, either as an admixture in the original concrete or from subsequent ingress, further investigation is necessary.

3. CAUSES OF DETERIORATION

The durability of a concrete structure is affected by choice of mix design, care in placing, and curing regime.

Placed concrete starts its life as a porous permeable solid with continuous capillaries. Subsequent hydration fills the pores with saturated lime solution and deposits mainly of cement gel and calcium hydroxide together with some aluminate and ferrite phase hydrates. The result of pore formation is to affect the permeability of concrete to water and gas and the capability for diffusion of water, oxygen and carbon dioxide. However, mixes with a w/c ratio of about 0.4 and greater will not produce sufficient solid hydration products to fill all the initially water-filled space available within practicable times of site curing [4]. The size of pores will also indicate their ability to dry out, so the amount of water retained and, consequently, the electrical conductivity of the concrete will be controlled by pore size and relative humidity. Reinforcement corrosion rates will be affected by conductivity. When the pores dry out then hydration virtually ceases so adequate curing is essential if the pores are to be filled.

A measure of the quality of existing concrete and hence its future performance is the permeability of the concrete and its resistance to diffusion. Site measurements of the former are difficult to make but the ISAT test [5] will distinguish good and poor concretes and aid the diagnosis and provision of advice on subsequent treatment. Measurement of gaseous diffusion of oxygen is also difficult to perform on site, though a practical measure of depth of carbonation can qualitatively indicate the rate of ingress of oxygen.

Chemical attack will disrupt concrete and make it more permeable. This may be a form whereby the hydrate equilibrium salts are leached out as with acid or soft waters, or of a disruptive nature as when the products of chemical reaction, by virtue of their lower density, impose internal stresses on the concrete and cause internal disruption.

Movements and volume changes in concrete can induce tensile stresses. The corrosion of steel when the supply of oxygen is restricted leads to the formation of black, Fe_3O_4 (magnetite) but, when access of oxygen is unrestricted, the usual corrosion product is FeO(OH) (Goethite) [6] with a volume increase of 190%. The strain capacity of concrete varies with age, as shown in Fig. 1, but it eventually settles to the value of about 100 microstrain [7]. This is considerably less than the strain imposed by corroding steel, and the concrete will crack.

There are various ways in which deterioration of concrete can be caused during its early life:

(1) by movement which exceeds the strain capacity; this can occur at any stage in the life of the concrete;
(2) by excessive moisture loss before compaction; this causes premature stiffening and formation of shrinkage cracks;
(3) by plastic settlement and plastic shrinkage cracking;
(4) early age frost-damage which forms ice in the capillary pores; this inhibits further hydration and causes disruption by expansion of the water as it freezes.

In the hardened state deterioration can arise from:

(1) excessive loading;
(2) thermal effects; and
(3) physicochemical effects.

The mechanism of corrosion of steel is to be dealt with in detail by others in this volume. Concrete normally provides a passivating alkaline environment but this can be upset during the life of the concrete. Change in the alkalinity level with time is affected by carbonation which, in turn, is affected by ingress of chlorides. The amount of effect is conditioned by the composition and properties of the concrete.

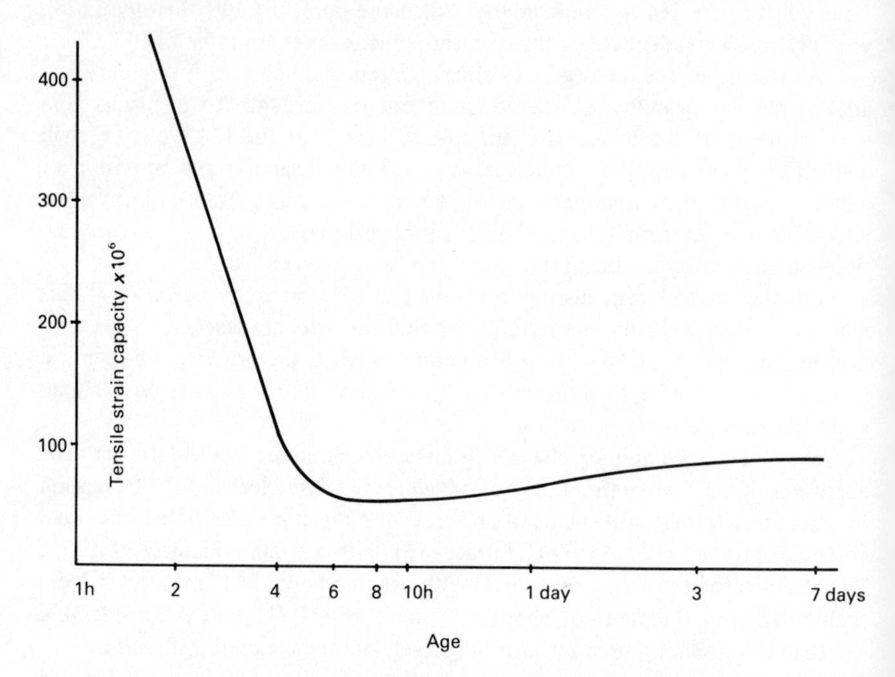

Fig. 1 – Variation of tensile strain capacity of concrete with age.

3.1 Carbonation

The alkalinity of concrete results from the hydroxide produced during the hydration of cement. When concrete is exposed to air, carbon dioxide and water will react with the calcium hydroxide over a period of time to form calcium carbonate. The effect of this carbonation is to reduce the pH value of the concrete sufficiently to destroy the passivating effect on the reinforcement bars. The movement inwards of carbon dioxide is dependent on the pore structure of the concrete and its basicity, i.e. its quality. The rate of carbonation of the concrete is also affected by the relative humidity and is greatest about 55% relative humidity [8].

Studies on rates of carbonation have led to various relationships dependent on various properties of the concrete. All of these relate penetration to some function of the square root of time. Pihlajavaara [9] has produced a useful working relationship which takes into account concrete quality and the environment and this is tabulated in Table 1.

Table 1 – Estimation of carbonation depth in Portland cement concrete. Carbonation depth = $10B\sqrt{t}$. Values of constant B.

Carbonation time (years)	Quality Concrete and storage condition					
	Low strength		Middle strength		High strength	
	Outdoors (moist)	Indoors	Outdoors (moist)	Indoors	Outdoors (moist)	Indoors
	Thickness of carbonation layer					
	$B = 0.6$	$B = 1.0$	$B = 0.2$	$B = 0.5$	$B = 0.1$	$B = 0.2$
1	6	10	2	5	1	2
2	9	14	3	7	1.5	3
5	13	22	4	11	2	4
10	19	32	6	16	3	6
25	30	50	10	25	5	10

3.2 Chlorides

The effect of chlorides in concrete is to overcome the passivating effect of the alkalinity and the mechanism by which this happens is discussed elsewhere in this volume.

The chloride can arise from its addition at the mixer in order to accelerate the strength development of the concrete. It was current practice in the past to allow 1.5% $CaCl_2$ to cement. Now this has effectively been prohibited for reinforced concrete where permissible chloride content is now limited to a maximum level of 0.35% chloride ion by weight of cement [10].

However, chlorides can enter concrete from extraneous sources, such as the use of de-icing salts or through contact with sea water. In such circumstances a concentration gradient is set up inside the concrete, generated by the surface concentration and the diffusion coefficient of the chloride ion. The latter is affected by the pore structure of the concrete, i.e. its quality and to some extent by the aluminate content of the cement. The diffusion coefficient of the chloride ion into cement paste varies, normally between 3 and 7×10^{-12} m^2/s, but it is markedly affected by the accompanying cations. There is an effect on the chloride ion diffusion by the aluminate phase in the cement. Changing from zero to 14% C_3A content can more than halve the diffusion rate, owing to the formation of Friedel salt. This compound, containing chloride and aluminate can, however, be decomposed by carbon dioxide to release chloride ions so the locking in of the latter into the aluminate phase is dependent upon the rate of carbonation.

4. IDENTIFICATION TECHNIQUES

A useful summary of investigation techniques has been compiled by Vassie [11], but some brief notes are given below.

4.1 Permeability

There is no satisfactory site test for permeability of concrete, but the Initial Surface Absorption Test (ISAT) will give an indication of relative quality of concrete. It measures the rate of flow of water into concrete per unit area after stated time intervals from the start of the test and at a constant applied head. The test is described in detail in British Standard 1881: Part 5: 1970 [5].

4.2 Depth of carbonation

This can be assessed by spraying a freshly exposed surface of concrete with an alcoholic solution of phenolphthalein containing a little water [12]. Pink colouration appears on surfaces of materials having pH values of 9.2 or greater. The test must be carried out on surfaces that have not had time to carbonate as a result of contact with the atmosphere. Care must be taken to ensure that surfaces are free from liquid water and that freshly exposed carbonated surfaces are not contaminated with dust from uncarbonated material.

4.3 Chloride content

The chloride content of concrete can be measured on site by using Quantab or Hach indicator strips [13, 14]. Samples of concrete are obtained by dry drilling and the chlorides are extracted by treatment with dilute nitric acid. The special indicator strips can be used to determine the chloride content of the concrete. In order to convert this result to a percentage by weight of cement it is necessary either for the cement content of the concrete to be known, or for further chemical analysis to be carried out. For many purposes it will be a sufficient approximation if a probable cement content is assumed.

4.4 Corrosion of reinforcement

Electrical potential and resistivity measurements can be used in order to detect the existence of conditions under which reinforcement may corrode. These are described in other papers in this volume.

5. REMEDIAL WORK

5.1 Objectives

The objectives of remedial work are to restore adequate strength to the structure, to arrest corrosion of the reinforcement and to ensure, as far as possible, that no further corrosion will take place.

In a large proportion of cases, the amount of steel lost through corrosion will not be great enough to affect the strength of the structure, but it will sometimes be necessary to replace badly corroded reinforcement or to provide additional bars. The immediate environment of the reinforcement must then be returned to a state in which the steel will not corrode. This may be achieved by ensuring that the reinforcement is surrounded by uncarbonated concrete that is not contaminated by chlorides and is of low permeability in order to resist back-diffusion of any chloride ions present in the bulk of the concrete. The reinforcement must be completely coated with cement paste. Alternatively, but less satisfactorily, a coating with low permeability to oxygen and moisture can be applied to the concrete. In the latter case the effect is to slow down the rate of deterioration, and the integrity of the coating is most important.

5.2 Preparation for repair

Whatever the basis of repair, preparation is of fundamental importance. As a first step, all cracked and spalled concrete must be cut away to expose the reinforcement, together with concrete cover in any uncracked areas where there is reason to suspect that the steel may be corroded. Little will be gained by sealing cracks when they have been caused by corrosion of reinforcement, because corrosion will probably continue and cause more cracking and spalling in the future. When corroded steel is exposed, the removal of the concrete should follow along the bar until no further serious corrosion is found. For practical reasons, this may have to be done in stages. Any carbonated concrete in contact with reinforcement should be removed because it will no longer be sufficiently alkaline to protect it from corrosion. It is usually desirable to cut back behind the reinforcment, both to ensure that any carbonated concrete surrounding it is removed and to permit inspection and cleaning of the full perimeter. Wire brushing is seldom adequate for cleaning reinforcement: grit blasting is far more effective, with needle guns or high pressure water jetting as other possibilities. Tests for carbonation are described in section 4.2 above.

It is also necessary to remove any chloride-contaminated concrete in contact with reinforcement if further corrosion is to be avoided. In many cases this is not practicable, but it should be done to the greatest extent possible.

The minimum depth for cutting out concrete should be 10 mm, and edges of prepared areas should be square or slightly undercut. Feather edges must be avoided.

5.3 Execution of repair

The choice is generally between using a repair material based on Portland cement, which will provide an alkaline environment for the steel, and a resin-based material (usually an epoxide) that will exclude moisture and oxygen without, in most cases, providing alkalinity. In either case, careful attention to workmanship is essential. Repairs should be carried out as soon as possible after preparation has been completed.

5.3.1 *Cement-based repairs*

Research into methods of forming construction joints in concrete has shown that the greatest strength of bond is achieved if the older concrete is saturated with moisture but surface dry [15]. Surfaces that are too dry may absorb water from the repair material, while excess water on the surface will increase the water: cement ratio, and consequently reduce strength, of the fresh concrete at the interface.

Bond at the interface may be improved if the surface is coated with a thin slurry of cement grout incorporating a polymer such as styrene butadiene rubber, (s.b.r.) but though some drying of this coat is desirable, it is essential that it should not be allowed to dry completely before the repair material is placed. If circumstances do not permit the patching material to be placed soon enough after the slurry has been applied, there are epoxy resin bonding agents available with an 'open time' of several hours.

Patches of limited area and depth are usually formed by hand-rendering methods. The repair material is normally a cement and sand mortar, and it is usually advantageous for s.b.r or similar polymer to be incorporated in it [16]. The polymer improves adhesion and freeze–thaw durability and reduces the permeability and increases the strain capacity. The effect on modulus of elasticity, and hence on strain capacity, of incorporating polymers in cement mortar is shown in Table 2. This means both that there is less risk of shrinkage cracking and that a given thickness of cover will provide greater protection to reinforcement than the same thickness of unmodified mortar. Repairs are normally built up in layers not exceeding 20 mm thick, each successive layer preferably being applied as soon as the previous one is stiff enough to support it without being disturbed.

Experience is required to judge the best timing. If there is a delay, the previous layer should be keyed as in normal renderings, and a fresh bonding coat applied. On completion, the patch must be moist-cured in order to minimize the risk of shrinkage cracking.

Larger volume repairs involve fixing formwork and placing conventional concrete which often has a maximum aggregate size of 10 mm. Alternatively, grouted aggregate concreting is occasionally used, particularly in underwater work. Polymer admixtures are not normally used when concrete is placed in formwork and it is seldom practicable to use a s.b.r. bonding slurry, because the slurry would usually have to be applied before the formwork was erected and might dry out too much before the concrete was cast.

In these circumstances the epoxy resin bonding agents mentioned above are more suitable. The concrete mix must be designed to minimize bleeding and settlement and, if a gap forms at the top of a repair to a member such as a column or a deep beam, it may be necessary to inject a resin grout in order to seal the crack. Adequate access must, of course, be provided for compacting the concrete against existing surfaces and care must be taken to ensure that the repair is filled completely.

Large area repairs are often repaired by sprayed concrete – usually dry-process gunite. This is an operation requiring skilled operatives, especially when there is much exposed reinforcement [17]. If the steel is to be protected adequately from corrosion, it is essential that it should be completely encased in sound concrete, without any air voids or pockets of sandy material behind the bars. Polymers such as s.b.r. may, if desired, be incorporated in the mix in order to reduce permeability, especially if the thickness of cover has to be kept to a minimum. Gunite repairs are often thin in proportion to their area, and curing is very important.

Table 2 – Concrete materials. Cement – OPC. Aggregate – silica sand Thames Valley gravel.

Experimental					
Mixes	I	II	III	IV	V
Cement kg/m^3	275	263	263	365	364
Water/cement	0.59	0.40	0.61	0.38	0.40
Aggregate/cement	6.9	6.7	6.7	4.2	4.2
Polymer loading (% of concrete)	–	4.71	2.35	3.45	3.17
Air content	–	–	–	3.2%	6.4%
Slump	64 mm	40 mm	60 mm	45 mm	42 mm
Results					
Concrete Qualities					
Mix	I	II	III	IV	V
Relative density	2.32	2.24	2.27	2.22	2.26
Water absorption %	5.9	1.5	6.0	1.5	1.0
Dynamic modulus of elasticity at 28 d (E_d) kN/mm^2	45.6	23.4	27.9	25.0	23.3
Static modulus (E_s) kN/mm^2	40.0	10.3	15.9	12.3	10.1

5.3.2 *Resin-based repairs*

Resin-based repair materials do not usually provide an alkaline environment for the reinforcement, so the passivating layer of oxide, present in reinforced concrete, is not formed. Consequently it is particularly important that the steel should be really clean and that the resin coating should be defect-free.

Resin-based materials are more costly than cement-based ones but material costs usually form a relatively small part of the total repair costs, so the choice of material is usually based on other considerations. Resins are useful in circumstances such as when it is not possible to provide adequate thickness of concrete cover to reinforcement, when rapid curing is necessary, and when increased chemical resistance is required. They have advantages in overhead work, particularly when lightweight fillers are used [18–20].

It is essential that the resin should be formulated to suit the conditions under which the work will be carried out, such as temperature, humidity, heat dissipation and working time. If these factors are not suitable, the resulting repair is unlikely to protect the reinforcement as intended, and it may sometimes be desirable for the resin to be applied by the formulators, who are fully familiar with its characteristics.

6. PERFORMANCE IN SERVICE

Ideally, a repair should have as long a life as the remainder of the structure but in practice this degree of success is not always achieved. In a large proportion of cases this is due to shortcomings in the preparation of the work before the repair material is applied.

If concrete has not been cut back far enough to ensure removal of all carbonated material adjacent to reinforcement, or the full corroded length of reinforcement has not been exposed, fresh corrosion cells may be set up in the future. It is important to ensure that the back of corroded reinforcement is really clean, and that no scale or corrosion products remain on the steel.

Chlorides in reinforced concrete cause particularly severe problems. It may not be possible to remove all chloride-bearing concrete that is in contact with steel and, even if that can be done, it is extremely difficult to remove all traces of chloride from the exposed steel surface. In such circumstances the probability of achieving a permanent repair is very much reduced.

Repairs may fail if they do not provide effective protection to the steel, and the most likely causes of this type of failure are excessive permeability as a result of lack of compaction of the repair material, the use of an unsuitable material or, in the case of cementitious repairs, inadequate curing leading to shrinkage cracking. Insufficient depth of cover often presents problems and, if the surface of the concrete cannot be built out in order to increase cover, some form of surface coating may have to be used. In these cases it is likely that the coating will eventually have to be renewed, depending on the conditions of exposure and the nature of the coating.

In spite of difficulties that may arise, however, it should be possible in a large proportion of cases to carry out repairs that will, at least, have a useful life that will justify their cost.

There is no doubt that the keys to success are correct diagnosis, selection of the most appropriate methods and materials, and careful and conscientious attention to detail at all stages of the work.

7. REFERENCES

[1] *The durability of steel in concrete: Part 2. Diagnosis and assessment of corrosion-cracked concrete.* Building Research Establishment, Watford, 1982. Digest 264.

[2] Beeby, A. W., 'Cracking and corrosion'. *Concrete in the Oceans Technical Report No. 1.* Cement and Concrete Association, Slough, 1978. Publication 15.286.

[3] *Concrete in the Oceans Newsletter,* No. 5, April 1982. Construction Industry Research and Information Association, London.

[4] Powers, T. C., Copeland, L. E. and Mann, H. M., 'Capillary continuity or discontinuity in cement pastes'. *Journal of Portland Cement Association Research & Development Laboratories,* **1** (2), pp. 38–48, May 1959. Skokie, Ill., USA.

[5] *British Standard 1881*: Part 5: 1970. 'Methods of testing hardened concrete for other than strength'. British Standards Institution, London, 1970.

[6] Evans, U. R., 'The corrosion and oxidation of metals: scientific principles and practical applications', Edward Arnold, London, 1960, p. 91.

[7] Draft revision to *British Standard Code of Practice CP110*: 'The structural use of concrete. Part 2. Complementary recommendations for special purposes'. British Standards Institution, London, March 1982. Document 81/15604 DC.

[8] Verbeck, J. G., *The Carbonation of Hydrated Portland Cement.* American Society for Testing and Materials, Philadelphia, USA, 1958. Special Publication 205.

[9] Pihlajavaara, S. E., 'History, dependence, aging and irreversibility of properties of concrete'. Paper No. 60, *Proceedings of conference on structure, Solid Mechanics, and Engineering Design.* Wiley – Interscience, 1971, Part 1, p. 719.

[10] *British Standard Code of Practice CP110*: Part 1: 1972. 'The structural use of concrete'. Clause 6.3.8. British Standards Institution, London. Amended 1977.

[11] Vassie, P. R., 'A survey of site tests for the assessment of corrosion in reinforced concrete'. *Laboratory Report 953.* Transport and Road Research Laboratory, Crowthorne, 1980.

[12] Roberts, M. H. 'Carbonation of concrete made with dense natural aggregates'. *Information Paper IP 6/81.* Building Research Establishment, Watford, 1981.

[13] 'Simplified method for the detection and determination of chloride in hardened concrete'. *Information Sheet IS 12/77.* Building Research Establishment, Watford, 1977.

[14] 'Determination of chloride and cement content in hardened Portland cement concrete'. *Information Sheet IS 13/77.* Building Research Establishment, Watford, 1977.

[15] Brook, K. M., 'Construction joints in concrete'. *Technical Report TR 414.* Cement and Concrete Association, Slough, 1969.

[16] Eash, R. D., and Shafer, H. H. 'Reactions of polymeric latexes with Portland cement'. *Transportation Research Record 542 Polymer Concrete.* Transportation Research Board of the National Research Council, Washington DC, USA, 1975, pp. 1–8.

[17] 'Assessment of fire-damaged concrete structures and repair by gunite'. *Technical Report No. 15.* The Concrete Society, London, 1978.

[18] Tabor, L. J. 'The effective use of epoxy and polyester resins in civil engineering structures'. Construction Industry Research and Information Association, London, 1978.

[19] 'The use of epoxy compounds with concrete'. *Journal of American Concrete Institute,* **70** (9), pp. 614–645, September 1973.

[20] Clifton, G. C., 'Structural use of epoxy resins with reinforced concrete'. *Research Report 79/2.* University of Canterbury, Christchurch, New Zealand.

Chapter 15

Mortar repair systems – Corrosion protection for damaged reinforced concrete

L. H. McCURRICH, C. KEELEY, L. W. CHERITON, K. J. TURNER,
Fosroc Technology Ltd, Leighton Buzzard, Bedfordshire, UK

---

1. INTRODUCTION

The repair and maintenance of reinforced concrete is becoming increasingly important as the stock of older structures increases. Today's economic climate also puts more emphasis on repair rather than new construction. In the past the main methods of repair for spalled or damaged concrete have been:

(1) sprayed mortar or concrete;
(2) hand-applied sand/cement mortar;
(3) poured concrete or mortar with the aid of shuttering.

Whilst sprayed mortar and concrete behind shuttering still have an important role to play for large-scale structural repairs, there has over the last fifteen years been considerable growth in the use of special mortar products. These can be reliably and easily applied by hand-trowelling. In increasing order of cost they have generally been based on the following:

(1) sand and cement mixed on site with the addition of polymer emulsions based on PVA, SBR or acrylic;
(2) pre-packaged cementitious mortars;
(3) resin mortars including acrylics, polyurethanes, polyesters and epoxies.

In this paper we shall look at the properties required of repair mortars, the methods of test for evaluating these and the way in which a recently introduced pre-packaged cementitious mortar meets these requirements.

2. PROPERTIES REQUIRED OF A REPAIR MORTAR FOR REINFORCED CONCRETE

2.1 Function of a repair

Any repair to a structure should generally fulfil certain basic requirements.

(1) Arrest the deterioration of the structure. In particular it is essential that the repair prevents any further corrosion of the reinforcement steel. This

can be achieved by preventing access of oxygen, water and aggressive ions to the steel, [1] or by providing an environment which chemically passivates the reinforcement. It is also possible to electrically protect the steel by use of sacrificial anodes or imposed voltage, in conjunction with the repair work, but such methods fall outside the scope of this paper, and will not be considered further.

(2) Restore structural integrity. There is some controversy over how much load a patch repair actually carries. However, it is generally accepted that for all but purely cosmetic repairs, the material used should have strength properties similar to the substrate.

(3) Provide an aesthetically acceptable finish. It is very difficult to match a repair exactly to the original structure; for this reason a surface coating applied over the complete building is the best way of achieving a uniform appearance. A good surface coating can also provide additional protection both to the repair and to areas not yet showing signs of distress, and is therefore usually well worth considering as part of an overall job.

2.2 Mortar properties

When considering various alternative repair systems, the general requirements outlined above must be translated into material properties which characterise a repair mortar. The following list summarises the most important of these characteristics.

(1) Bond to substrate.
(2) Movement relative to substrate. This depends on shrinkage, thermal movement and behaviour on wetting and drying.
(3) Permeability to water, gases and aggressive ions. As well as intrinsic permeability, the possibility of shrinkage cracking is important, as this can provide a direct path to embedded reinforcement.
(4) Chemical passivation of bar.
(5) Strength.
(6) Ease of application.
(7) Durability of freeze–thaw cycling, chemical attack, and weathering.

3. EVALUATION OF REPAIR MORTARS

3.1 Adhesion/bond

It is essential that the repair mortar achieves strong adhesion to the substrate and that subsequent stresses are not sufficiently great as to cause debond.

A number of techniques are in use to evaluate adhesion and these include the following:

(i) *Direct tensile test*
Standards: ASTM E149–76, RILEM 13MR [2]
These tests generally involve casting fresh mortar against a hardened mortar substrate and pulling in a standard tensile testing machine.

(ii) *Direct pull off test*
Standards: no standards available but various techniques have been published for floor toppings [3] and by manufacturers of proprietary equipment [4].

(iii) *Flexural testing*
Standards: ASTM E518–74, RILEM 13MR.
The mortar under test is used to bond flexural specimens and these are three-point or, preferably, four-point loaded.

(iv) *Direct shear testing*
Standards: RILEM 13MR.
In this type of test the mortar is subjected to a torque, generally applied with a torque spanner, to a suitable dolly bonded to the surface.

(v) *Slant shear*
Standards: BS 6319 (Draft for comment at present).

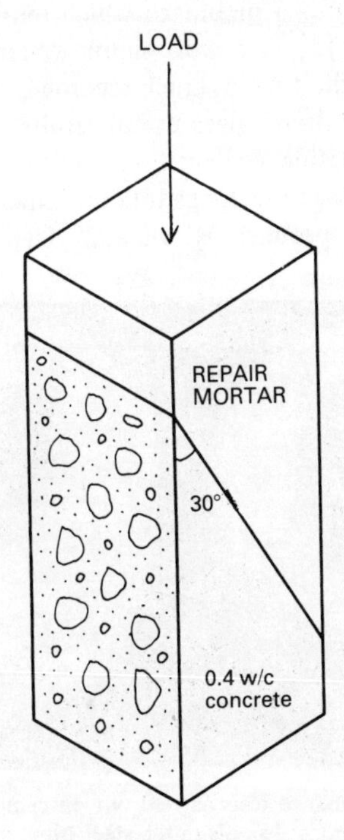

Fig. 1 – Repair prism prepared according to BS 6319: Part IV (proposed). Bond between repair mortar and substrate is evaluated by crushing the prism in a compression testing machine, and recording the failure load.

In this test a prism is produced, with the repair/substrate bond line at 30° to the vertical (see Fig. 1). The prism is simply crushed in a compression testing machine, and the load at which the specimen fails recorded. If bond is good, the sample fails as a monolithic prism, rather than along the bond.

Extensive work has been carried out to evaluate these methods, particularly the slant shear test [5]. It is our view that the slant shear test provides the most reliable method of evaluating the adhesion of mortars.

3.2 Shrinkage, thermal movement and cracking

The usual method of measuring long-term shrinkage and thermal movement is to use standard prisms as defined in BS 1881. These tests give a useful indication of free movement, but cannot easily be correlated to shrinkage cracking owing to the need to relate it to development of tensile strength and stress relaxation which is hard to determine.

An alternative method of evaluating cracking tendency is to carry out restrained shrinkage tests. Various ideas have been tried for doing this, for example casting mortars into a mould to which tensile load can be applied to keep it at constant length [3] or a more simple system where the mortar is cast around a rigid insert and the time to crack recorded. We find that this shrinkage ring test is easy to carry out and gives useful results. We have adopted it as the routine method for evaluating cementitious repair mortars (see Fig. 2). The dimensions of the specimens used are as follows: inner restraining ring $4\frac{1}{2}''$ O.D., outer ring $7''$ I.D. to give a specimen $2\frac{1}{4}''$ wide, $2''$ deep.

Fig. 2 – Restrained shrinkage tests carried out on cement-based repair mortars. The mortar is cast around a $4\frac{1}{2}''$ diameter steel pipe; at 1 day the mould is removed and the mortar is exposed to the given environment. This photograph shows RENDEROC after exposure to 20°C 65% relative humidity for 1 year. No cracking has occurred.

3.3 Permeability

Corrosion of reinforcement can be avoided by excluding oxygen and moisture from the steel. For this reason it is important that a repair should have low permeability. In the case of cementitious mortars, low permeability will also slow carbonation. A number of techniques are available for measuring permeability and these include the following.

(1) Mounting of specimen of mortar in a special test rig and applying a pressure difference of either liquid or gas across the specimen and measuring flow rate.
(2) Measurement of gas diffusion through the specimen. Some very interesting methods have now been established for doing this which make use of modern analytical techniques. The time taken for a gas to diffuse through concrete is measured. This includes work being carried out by D. Lawrence [6, 7].
(3) Surface absorption methods such as the ISAT test described in BS 1881.

For repair mortars the most practical and useful method has generally been taken to be the ISAT technique, although it would seem useful to also measure the rate of oxygen and CO_2 permeability as these have such an important bearing on corrosion protection for reinforcing bars.

3.4 Chemical passivation of embedded steel

Several methods have been used to assess the corrosion risk for a reinfrocement bar in various environments [8. 9]. Polarisation techniques, for instance, have been employed, but one of the simplest methods, both to perform and interpret, is the accelerated corrosion test development by K. Treadaway and A. Russell [10]. In practical terms this follows the development and stability of the passivation film on a reinforcement bar embedded in a solid matrix with an imposed corrosion current. It has been used extensively to test admixtures [11] following the identification of the problems associated with the use of calcium chloride accelerators. As the test requires the solid matrix to conduct electricity, it is not possible to investigate directly the effect of resin-based mortars on reinforcement.

Most cement-based repair materials will show characteristic passivation of the steel, but it is still worth checking prepacked cementitious repair mortars using this technique to ensure that there are no admixtures or other materials present which could affect the passivation process.

In the case of cement-based mortars it is also important that the rate at which carbonation can proceed is kept to a minimum. This can be checked using the standard phenolphthalein test applied to specimens stored under standard conditions.

3.5 Mechanical strength

In many instances the repair mortar is expected to contribute to the mechanical strength of the structure. To this end it is normal for standard compressive and flexural strength development tests to be carried out, in accordance with BS 4551.

It may also be important to know the Young's modulus of the repair mortar. This is measured by BS 1881: Part 5, or ASTM C469–65. A low modulus will generally be advantageous in keeping internal stresses low and ensuring good long-term adhesion and lack of cracking. It is often argued, however, that the modulus should be compatible with the concrete being repaired if uniform load transfer is to be achieved across the section. This of course is true in simple engineering terms but the argument often forgets that a new insert may experience some shrinkage relative to the original material and this will negate any influence of Young's modulus. It is more important that the repair material should be prone to minimum differential movement and have a modulus as near the parent material as possible which is compatible with this.

3.6 Ease of application

To ensure satisfactory performance it is essential that it is easy to mix the materials in the correct proportions and that the wet mix is of suitable consistency for the job in hand. To this end mixing sand and cement on site creates difficulties in ensuring that the correct proportions are achieved. This can be overcome by pre-packaging the materials.

In the case of polyester mortars the chemistry of the resin system can be designed to tolerate a wide range of resin/catalyst proportions and this provides these materials with particularly easy and reliable mixing and cure performance. Several proprietary brands are given in references [12] and [13].

In the case of the epoxy mortars the proportions of resin and hardener must be accurately maintained and this is achieved by manufacturers supplying the individual components in pre-weighed containers. Full containers must be mixed each time.

With regard to physical properties of the uncured materials, it is now possible to obtain proprietary mortars with a wide range of consistencies. These include mortars formulated with lightweight spherical fillers enabling the material to be trowelled onto soffits or on to vertical surfaces to thicknesses as great as 40 mm without slumping [12, 14].

3.7 Freeze–thaw testing

Methods of evaluating freeze–thaw resistance are given in BS 5075, and ASTM C67–78. In addition to such tests (which evaluate the mortar by itself) it is also worthwhile assessing the affect of freeze–thaw cycling on bond strength.

3.8 Chemical resistance

This is usually restricted to specific requirements of a particular repair job. Performance may be best evaluated by exposure testing, often using elevated temperatures to accelerate any degradation.

3.9 Overall performance

The performance of a repair results from a combination of many individual factors. Although most properties can be measured separately, it is still important to assess the all-round performance of any system. Long-term exposure testing is one alternative, but often it is necessary to accelerate the process. Harsh environments, such as salt-spray and elevated temperatures, in conjunction with freeze–thaw cycling, may be used for relative evaluation of alternative systems under test. Monitoring the potential of a bar embedded in a repair specimen using half cell techniques [15] is a convenient way of following the progress of such tests.

4. SELECTION OF A REPAIR MORTAR

Selection of a repair material for a particular situation must be made on the basis of the properties established using the various evaluation techniques available. Major importance is often ascribed to material costs, but in reality this is usually only a minor contribution to the overall cost of a repair. The costs of access, labour and down time form a large part of any job.

In general resin-based mortars are preferred where thin sections have to be applied. In this case advantage is taken of the extremely low permeability of resin materials together with their good adhesion and lack of special curing requirements.

For larger repairs it would seem generally preferable to use cement-based materials. They are lower in cost than the resin mortars and they have thermal expansion and movement characteristics more compatible with the concrete substrate. They are easier to mix and use in large volumes and also have lower exotherms during curing than the resin materials. Care may need to be taken, however, to ensure that dessication does not occur at early ages, either through rapid water loss into a porous substrate, or water loss from the surface.

Steady advances have been made in the formulation of cementitious mortars, particularly to improve ease of use, reduction in shrinkage stresses and reduction in permeability. These have been achieved by the use of special cements, fillers and admixtures and as an illustration the performance of a recently introduced pre-packaged material is given in the next section.

5. RESULTS ACHIEVED WITH PRE-PACKAGED CEMENTITIOUS REPAIR MORTARS

An example of a pre-packaged cementitious mortar is the product RENDEROC [16]. This product incorporates a number of features to meet the requirements outlined previously.

(1) Spherical low permeability lightweight filler to ensure ease of application and non-slump characteristics when trowelled onto vertical and overhead surfaces.
(2) Special cements to provide shrinkage compensation.
(3) A coefficient of thermal expansion similar to that of concrete and steel.
(4) Admixtures to ensure a low permeability and low water: cement ratio.
(5) Accurate proportioning of all components under factory conditions and supplied to site ready to use in 25 kg bags.

5.1 Bond

Using the split block adhesion test the results given in Table 1 have been obtained.

Table 1 – Bond strength to grade 40 concrete.

Repair mortar	Failure load (N/mm^2)
RENDEROC (Using TITEBOND 'R' acrylic primer)	24
4:1 Sand: Cement (Using slurry primer)	5

5.2 Coefficient of thermal expansion

The coefficient of thermal expansion was measured using BS 1881 prisms cycled over a temperature range of 30°C. The results are shown in Table 2.

Table 2 – Coefficient of thermal expansion

Material	Coefficient of thermal expansion ($10^{-6}/°C$)
RENDEROC	7.3
4:1 Sand : Cement	8.0
Typical concrete	5–13

5.3 Shrinkage

Restrained shrinkage tests have been carried out using the shrinkage ring technique outlined previously. The results of these tests are shown in Figs. 2 and 3.

Fig. 3 – Shrinkage tests similar to Fig. 2 carried out on a material prone to shrinkage cracking; cracks in this case became evident within 7 days.

5.4 Permeability

The penetration of water from the surface as measured by the ISAT surface absorption test to BS 1881 is given in Table 3, together with the results from 'in-house' water droplet test.

Table 3 – Surface absorption measured in accordance with BS 1881 ISAT test, and by an 'in-house' water droplet test.

Material	Surface absorption to BS 1881 ISAT test at 10 minutes (ml/m^2 s)	Time for standard drop to be absorbed (s)
RENDEROC	0.18	65
4:1 Sand:Cement	1.11	Immediate

5.5 Corrosion passivation

RENDEROC is alkaline with a pH of 13 and it can be expected to keep the reinforcing bars in an environment according to the Pourbaix diagram [17] where corrosion will not occur.

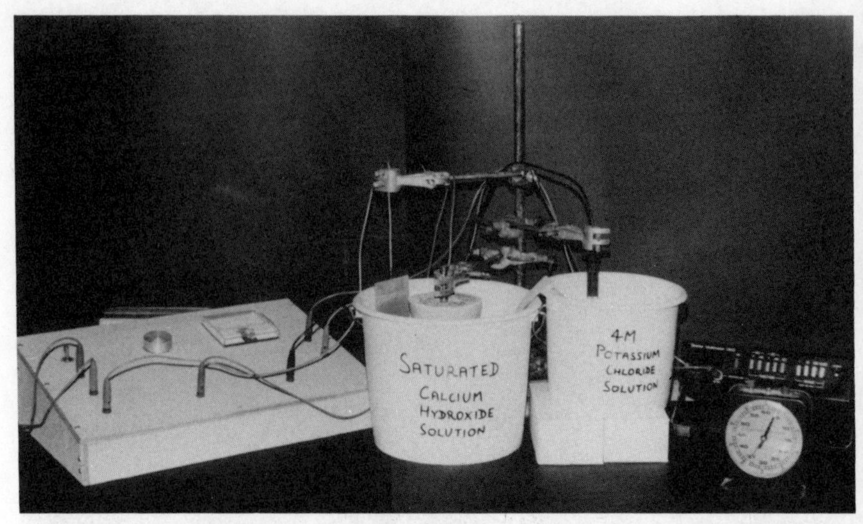

Fig. 4 – Equipment used for measuring accelerated corrosion of steel embedded in repair mortars [10, 11]. The mortar specimen is shown held in the container marked 'Saturated Calcium Hydroxide Solution'. A constant current, corresponding to a current density of 30 $\mu A/cm^2$ at the surface of the embedded bar is passed through the sample, and the potential between bar and solution monitored.

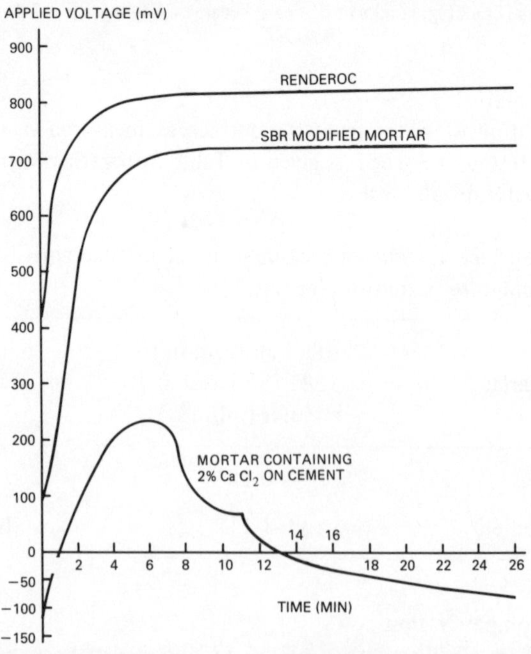

Fig. 5 – Graph showing results of accelerated corrosion tests on a pre-packaged cementitious mortar (RENDEROC) compared with an SBR modified mortar and one containing calcium chloride.

Electrogalvanic tests have been carried out using the constant current system described by Treadaway and Russell. The arrangement is shown in Fig. 4, with typical results plotted in Fig. 5. Figure 5 clearly illustrates that the bar is being maintained in a passive state and corrosion is not occurring.

5.6 Carbonation

The rate of CO_2 penetration into the mortar has been measured by determining the rate of carbonation in a standard condition of 20°C, 65% relative humidity over a period of $1\frac{1}{2}$ years. The results in comparison with sand and cement mortar are given in Figs. 6 and 7. These show the very slow rate of carbonation in RENDEROC compared with typical sand/cement mortar.

Fig. 6 – Results of carbonation test on sample of RENDEROC stored at 20°C 65% relative humidity for $1\frac{1}{2}$ years. The dark area is where the phenolphthalein has turned purple indicating alkali environment maintained. It can be seen that the depth of carbonation is generally in the range 1 to 3 mm.

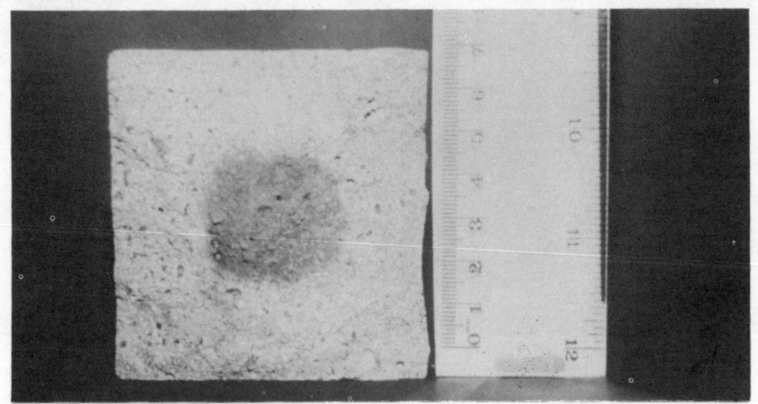

Fig. 7 – Carbonation test carried out on typical 4:1 sand/cement mortar under the same conditions as Fig. 6. In this case carbonation has progressed to a depth of about 20 mm.

5.7 Strength

Typical compressive strength development for RENDEROC is shown in Fig. 8, in which it will be seen that the material continues to gain strength with time and has a strength compatible with that of normal structural concrete.

Age	Compressive strength (N/mm^2)
1 Day	10.4
7 Day	24.5
28 Day	32.0
6 Months	37.5
1 Year	41.5

Fig. 8 – Compressive strength development measured to BS 4551, for RENDEROC at W/P 0.18, wet cured at 20°C.

5.8 Freeze–thaw testing

Evaluation has been carried out in accordance with ASTM C67-78. after 100 cycles, no weight loss was recorded.

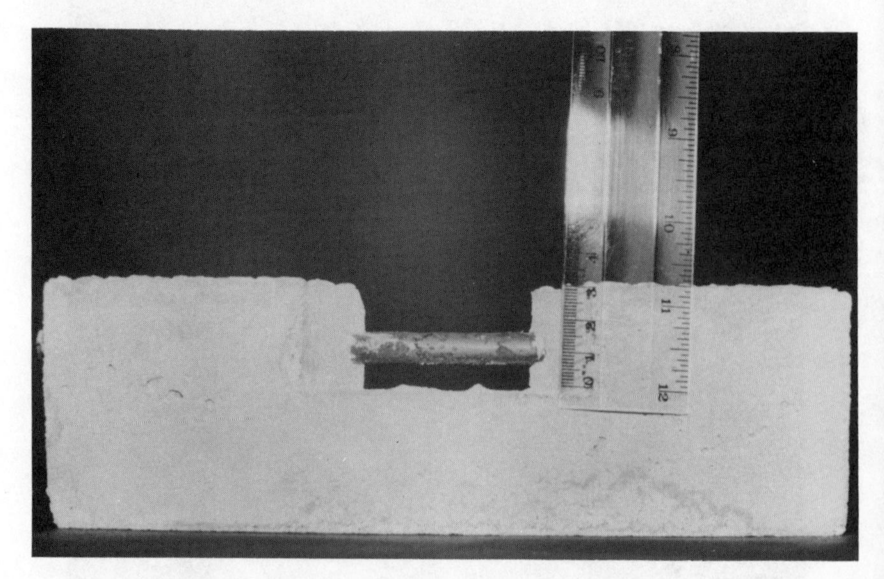

Fig. 9 – Example of specimen used for salt-spray corrosion tests. The single 12 mm bar is cast *in situ* in the concrete substrate, and repair mortar used to fill the pre-formed void. The sample is then exposed to salt-spray and freeze–thaw cycling. The condition of the bar is followed by measuring its potential at 3 points along the specimen using the method specified in ASTM C876–77.

5.9 Exposure cycling

Samples similar to that shown in Fig. 9 are being subjected to the following accelerated weathering regime:

5 days in a salt spray cabinet at 40°C
1 day freeze–thaw using the BS 5075: Part II cycle.
1 day conditioning to 20°C, 65% R.H.

At the end of each 7-day cycle, the condition of the embedded bar is monitored by measuring its potential using the method specified in ASTM C876.

6. CASE HISTORIES

The performance of any material under laboratory conditions must be checked by experienced staff on site and in the next section we review three typical case histories. One shows the performance of an epoxy mortar under high temperature Middle East conditions. The other two illustrate the use of a cementitious repair mortar.

6.1 United Arab Emirates sewage plant renovation (Fig. 10)

This is an example where a high-build epoxy mortar was used to repair the perimeter walls of circular tanks containing the aeration beds of a sewage farm in the United Arab Emirates. Patches of severe attack on corrosion of rebar had occurred owing to intrinsic chlorides in a highly aggressive environment. Epoxy mortars gave the advantage in this case of:

(1) rapid strength development to reduce down time;
(2) low permeability to oxygen;
(3) good chemical resistance;
(4) high-build with no need for shuttering;
(5) no need for special curing under 40°C Middle East conditions.

The objective of the repair was to retain the structural integrity for a limited period but it was envisaged that full arrest of the decay was impracticable.

The repair was completed over one year ago and the epoxy repairs are still performing satisfactorily, although there is some limited evidence of progressive degradation in the unrepaired segments.

Fig. 10 – The aeration beds of a sewage plant in the United Arab Emirates before and after repair with a high-build epoxy mortar. NITOMORTAR H.B.

6.2 Billhay Lane footbridge (Fig. 11)

The footbridge was constructed in the 1930s to provide access over the now disused railway between Wednesbury and Dudley in West Bromwich. The route of the line is being converted into a nature ramble and the footbridge was considered unsafe.

The repair was to both the soffit and support columns which had spalled owing to the corrosion of the reinforcement. The corrosion had occurred as a combined result of carbonation and inadequate cover to steel. The steel was cleaned and coated with NITOZINC (a zinc-rich epoxy primer). The prepared surface was rebuilt using a prepacked cementitious mortar, RENDEROC, to a maximum depth of 125 mm with only limited supporting formwork. The completed repair was cured with a surface-contact curing membrane.

Fig. 11(A) – RENDEROC slurry coat being applied to exposed reinforcement on Billhay Lane footbridge.

This repair was completed in August 1981 and is still operating satisfactorily with no evidence of cracking or debond. The impervious, alkaline nature of the repair has prevented any further degradation of the reinforcement in the areas where the RENDEROC was applied.

Fig. 11(B) – A finished repair in Billhay Lane footbridge column.

6.3 St. Joseph's Primary School, Dundee, Scotland (Fig. 12)

This school, constructed in the 1920s, was suffering from severe degradation of the concrete window and door lintels. The degradation took the form of spalling and cracking at several points, both vertically and horizontally along the lintels.

Inspection of the concrete established substantial degrees of carbonation and high porosity. In several cases reinforcement, which consisted of mild steel $\frac{1}{2}''$ bars, was corroded away completely and had to be fully reinstated.

Fig. 12 (A) – Applying RENDEROC to lintel at St. Joseph's Primary School, Dundee.

The repair was successfully completed with RENDEROC and a 4:1 RENDEROC slurry was used for priming purposes. The soffit was supported where necessary with temporary formwork and the repairs packed into a depth of 100 mm, leaving the final surface 10 mm low and scarifying it to provide a mechanical key. The bulk of the repair was allowed to reach initial set and was then primed with RENDEROC slurry before application of the finishing layer. This 'monolithic' approach avoided any possibilities of delamination whilst allowing maximum accuracy with regard to finished levels. The completed repair was not cured because the subsequent ambient conditions were wet and humid. Recent inspection of these repairs indicates that they are in perfect condition despite exposure to the severe 1981/82 winter.

Fig. 12(B) – Finished repair.

7. CONCLUSIONS

The primary requirement for a repair is to arrest the deterioration of the structure. The most important properties needed by a repair system to prevent further corrosion occurring in damaged reinforced concrete are:

(1) good bond to substrate;
(2) similar thermal, shrinkage and wetting/drying movements to the substrate;
(3) low permeability to water, oxygen, carbon dioxide and aggressive ions (e.g. chlorides);
(4) ability to passivate the reinforcement;
(5) where appropriate, long-term durability to freeze–thaw cycling, chemical attack and weathering.

The paper shows how these properties can in many cases be met by modern pre-packaged cementitious repair mortars such as the one discussed in detail in the paper.

Resin-based systems are generally required where cover to steel is low or the repair is exposed to a particularly aggressive environment.

Three case histories have been given to illustrate typical applications of resin and cementitious repairs.

REFERENCES

[1] R. D. Browne, *Durability of Building Materials,* **1**, (2), 113, Elsevier (1982).

[2] *RILEM Final Recommendations*: 13MR Technical Committee on Mortars and Renderings, Materials and Structures, Sept./Oct. 1982, No. 89.

[3] L. H. McCurrich, W. M. Kay, 'Polyester and epoxy resin concrete'; paper 2 given at the Resins and Concrete Symposium, organised by The Plastics Institute and Institution of Civil Engineers, April 1973.

[4] Elcometer Instruments Limited, Edge Lane, Droylesden, Manchester, M35 6BU.

[5] L. J. Tabor, *Magazine of Concrete Research,* **30** (105), 221 (1978).

[6] D. Lawrence, Cement and Concrete Association, *Departmental Note 4038* (1982).

[7] G. M. Darr and U. Ludwig, 'Determination of permeable porosity', *RILEM Materials and Structures: Research and Testing,* **6**, 185 (1973).

[8] J. D. Gilchrist, *Proceedings of S.C.I. Symposium on Corrosion of Steel Reinforcement in Concrete Construction,* p. 43 (1978).

[9] N. J. M. Wilkins and P. F. Lawrence, *Proceedings of S.C.I. Symposium on Corrosion of Steel Reinforcement in Concrete Construction,* p. 105 (1978).

[10] K. W. J. Treadaway and A. D. Russell, *Building Research Station Current Papers 82* (1968).

[11] L. H. McCurrich, M. P. Hardman and S. A. Lammiman, *Concrete,* **13** (3), 29 (1979).

[12] FERFA Publication *Products and Service Provided by Members* (1981), FERFA, 2A High Street, Hythe, Southampton SO4 6YW.

[13] REEBAFILL Data Sheet, Fosroc Limited, Leighton Buzzard, Bedfordshire, UK.

[14] NITOMORTAR HB Data Sheet, Fosroc Limited, Leighton Buzzard, Bedfordshire, UK.

[15] J. B. Newman, A. M. Bell (Probe Technical Services, Tolpits Lane, Watford, Hertfordshire, WD1 8XA). Private communication.

[16] RENDEROC Data Sheet, Fosroc Limited, Leighton Buzzard, Bedfordshire, UK.

[17] M. Pourbaix, *Lectures in Electrochemical Corrosion,* Chapter 6, Plenum Press, 1973.

Chapter 16

Corrosion monitoring in the industrial-related research area

GUSTAV BRACHER, Sika AG, Zurich, Switzerland

1. INTRODUCTION

Techniques of estimating the corrosion rates of the reinforcement bars in concrete structures range from traditional mechanical to most sophisticated electrochemical measuring systems [1]. The corrosion engineer is confronted with the situation that corrosion monitoring and corrosion measurement is getting more and more complicated and that the cost of corrosion measurement equipment is ever increasing.

On the other hand the reinforced concrete system is not a well defined one, and it always contains a set of not-controllable parameters which cannot be neglected in corrosion monitoring, not even with the most ingenious measuring techniques.

Although a reinforced concrete structure cannot be represented by an easy model – owing to the environment's influences, the not-controllable surface effects and the electrochemical influences of the large net of the electrically connected reinforcement bars – there is a need for studying corrosion phenomena with model systems, e.g. mortar or concrete electrodes. The potentiostat has become one of the most important tools of the corrosion technologist, and it is widely used for measuring corrosion rates, corrosion mechanisms, and surface effects.

A wide range of commercial potentiostats (more or less expensive) are available, but mostly a simple potentiostatic circuit is all one requires, so that the use of a sophisticated and expensive instrument is not necessary.

Additionally, mostly in concrete systems, it is desirable to run several potentiostatic experiments simultaneously in order to get averaged results. The continuous advance in integrated circuit technology has resulted in cost reductions which have led to the design and application of many electronic products. So a simple, inexpensive potentiostatic circuit could be designed by us, although we are not electronics engineers. With the self-made potentiostat, for instance, 24 simultaneous experiments can be performed. The integration of a computer-aided data acquisition leads to a very powerful corrosion monitoring system with various application possibilities in the applied corrosion research area:

– current density curves of steel/concrete/inhibitor-systems to follow the behaviour of different corrosion inhibitors;
– corrosion monitoring of injection mortars for prestressed concrete structures;
– activated chloride penetration to study the effectiveness of various mortar mixtures to seal up the concrete floor or reinforcement against penetration of chloride salts.

2. DESIGN OF THE CORROSION MONITORING SYSTEM

Polarization curves are best obtained with a three electrode probe, the working electrode (normally mortar electrode), the reference electrode and an auxiliary electrode. A defined potential is applied to the working electrode vs the reference electrode, and the resulting corrosion current is measured between the working electrode and the auxiliary electrode, The concept of our corrosion monitoring system was designed with 24 parallel and independent potentiostatic units which are controlled by a minicomputer.

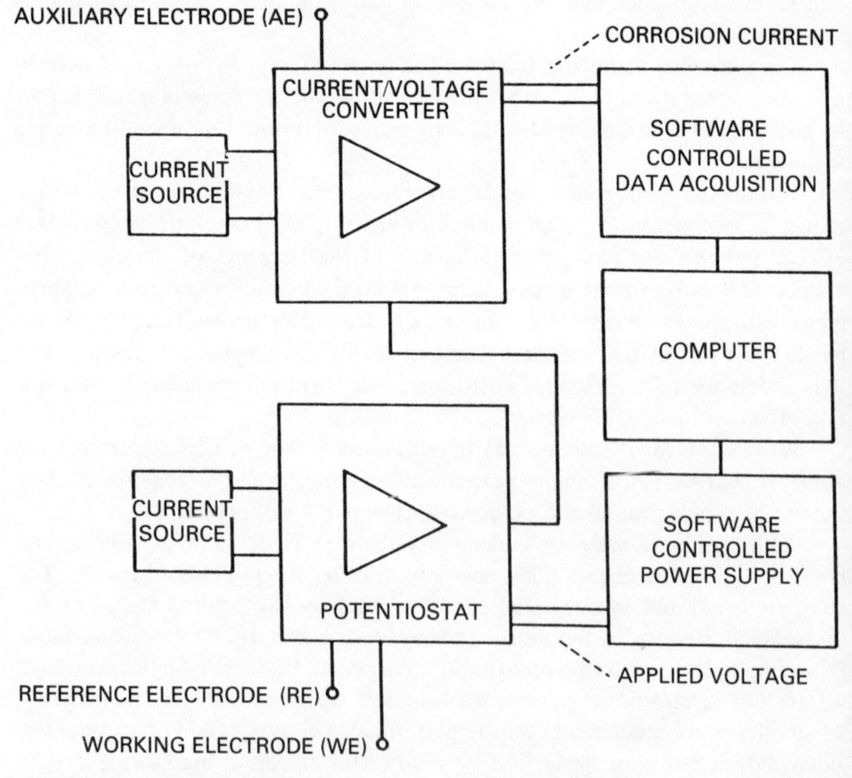

Fig. 1 – Basic scheme of corrosion monitoring system.

The basic scheme is shown in Fig. 1. The computer programs the applied voltage level to the working electrode as a function of time, which is the same for all the electrodes, and controls the data acquisition system to record the individual corrosion currents. For this purpose a current voltage converter for each potentiostatic unit is used. The circuit diagram of a single potentiostatic unit with the corresponding current voltage converter is shown in Fig. 2. The resulting data are stored in the computer system and can be easily averaged and plotted for each set of parameters. By applying this method, polarization curves of mortar electrodes can be obtained and the influence of aggressive and inhibiting substances can be studied.

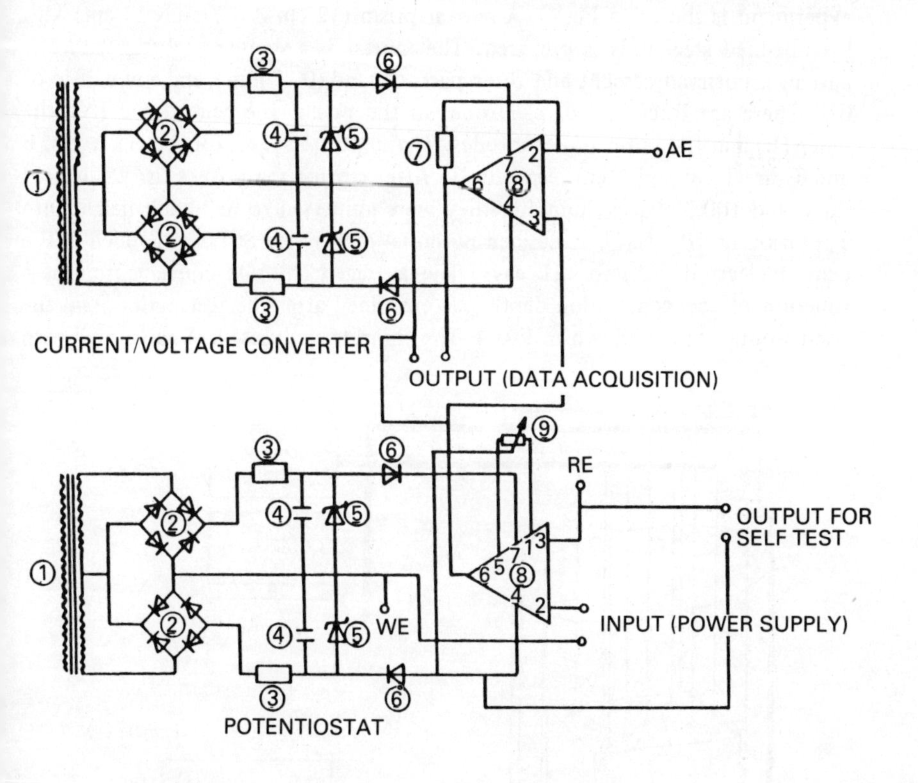

Particle List

1 Eberle BV 3895–5 Primary 220 V Secondary 2x17.5 V
2 Rectifier B 40 C 1500 R
3 R 150 Ω 1/2 W
4 C 25 V 100 μ F
5 Zener Diode ZY 15
6 Diode 1 N 4003
7 R 1K 1%
8 Op Amp LM 741 CN
9 Trimpot 10K (null balance Op Amp)

Fig. 2 – Circuit diagram of a single potentiostatic unit with corresponding current voltage converter.

3. ACCELERATED CHLORIDE PENETRATION

Polarization methods have a serious disadvantage inasmuch as they require IR compensation of the surrounding mortar coating. This is a rather complicated task which cannot be readily performed.

Because chloride is the most harmful aggressor in the corrosion of reinforced concrete structures, and reacts with aluminates of hydrating cements to form calcium-chloro-aluminates, a method of accelerating the chloride penetration was developed to simulate the corrosion process in reinforced concrete structures. This method is a very useful application of the potentiostatic units described in the preceding paragraph. The experimental lay-out for this type of experiment is shown in Fig. 3. A mortar prism (12 cm × 12 cm × 11 cm) with 9 embedded steel rods is prepared. The mortar is a standard mortar with one part of a Portland cement and three parts of sand (0–5 mm), and a w/c ratio of 0.5. There are three sets of electrodes in the prism: the central rod (A), the inner (B) and the outer (C) electrodes. The depths of cover for the electrodes B and C are 1 cm and 2 cm respectively. After storing the probes for 28 days at 20°C and 100% relative humidity they were immersed to half their height into a solution of 10% NaCl. A defined potential of 1.5 V *vs* SCE was applied to the central electrode (A) for 21 days. The averaged chloride concentration as a function of the penetration depth was examined after the test period, and the results obtained are shown in Fig. 4. The chloride analyses were performed with

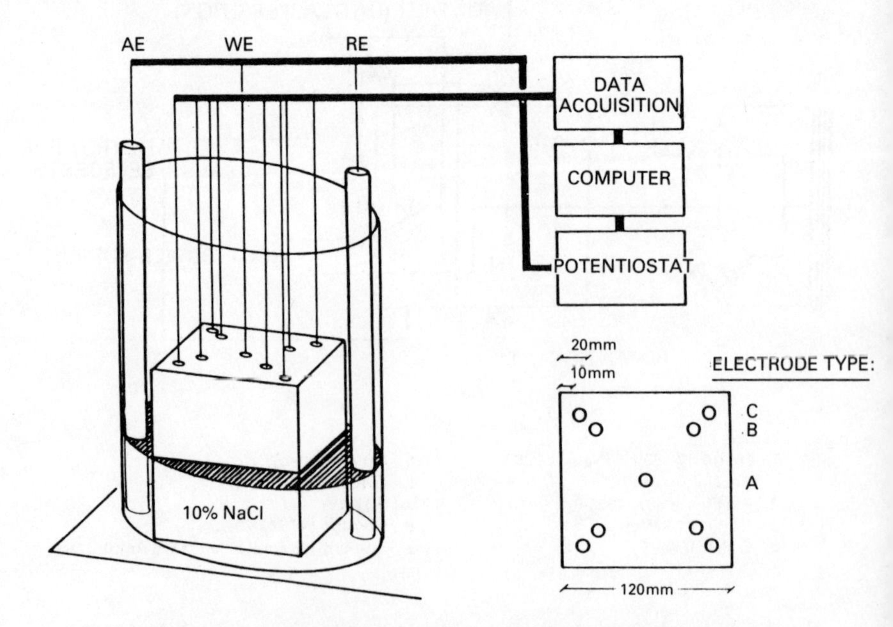

Fig. 3 – Experimental lay-out of the activated chloride penetration.

a ion-sensitive chloride electrode by a calibrated method. A chloride distribution analysis of mortar prisms prepared and stored under the same conditions as the above-mentioned prisms, but without applying a voltage level to the central electrode, clearly demonstrates the acceleration of chloride penetration by applying a voltage level to the central electrode.

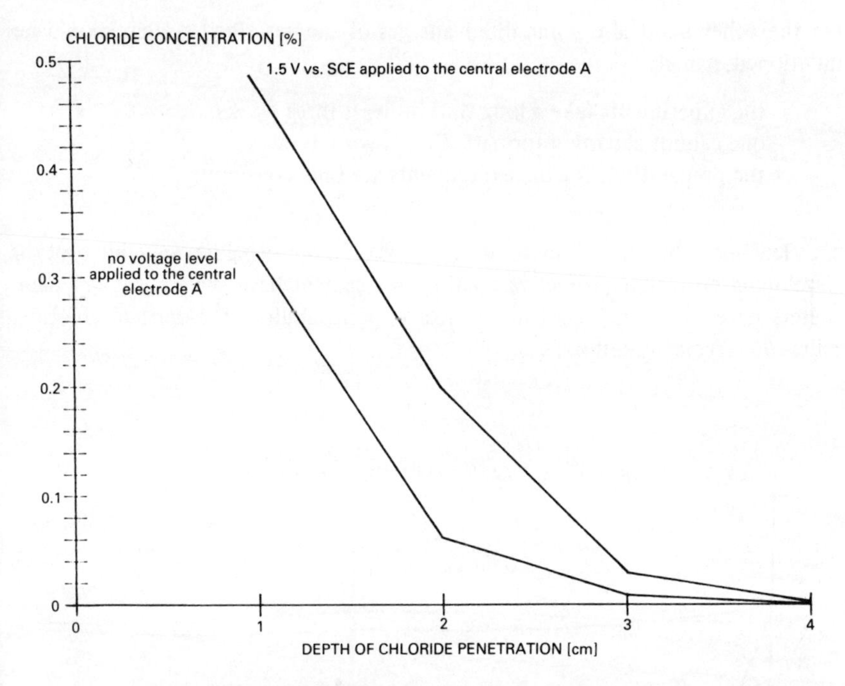

Fig. 4 – Averaged chloride distribution as a function of the penetration depth in the activated chloride penetration experiment.

During the test period the potentials of the electrodes B and C were measured. The potential exposure time diagrams are shown in Figs. 5 and 6 for a non-protected system and a protected system respectively. The corrosion protection consists of a cement-based coating described in this volume by my colleague Mr. Th. Bürge [2].

After the test the iron rods were taken out and inspected visually for corrosion damage. In agreement with further authors [3] corrosion products could be observed in the iron bars when the measured potential was less than −350 mV.

The advantages of this corrosion monitoring system are as follows:

– experimental conditions close to the reality; only acceleration of the chloride penetration process;

– no polarization effects of the examined iron bars;
– different coating systems can be easily compared;
– the results obtained can be verified by visual inspection of the steel rods;
– the chloride distribution is always the same.

On the other hand also some disadvantages of the described system should be mentioned, namely:

– the experiments take a long time, at least three weeks;
– one cannot get any information on corrosion rates;
– the preparations for the experiments are time-consuming.

The above-mentioned method of corrosion monitoring is a valuable tool for developing corrosion protective coatings on cement basis, where a lot of parameters have to be included, like porosity, permeability of water and chloride, adhesion effects, inhibitors, etc.

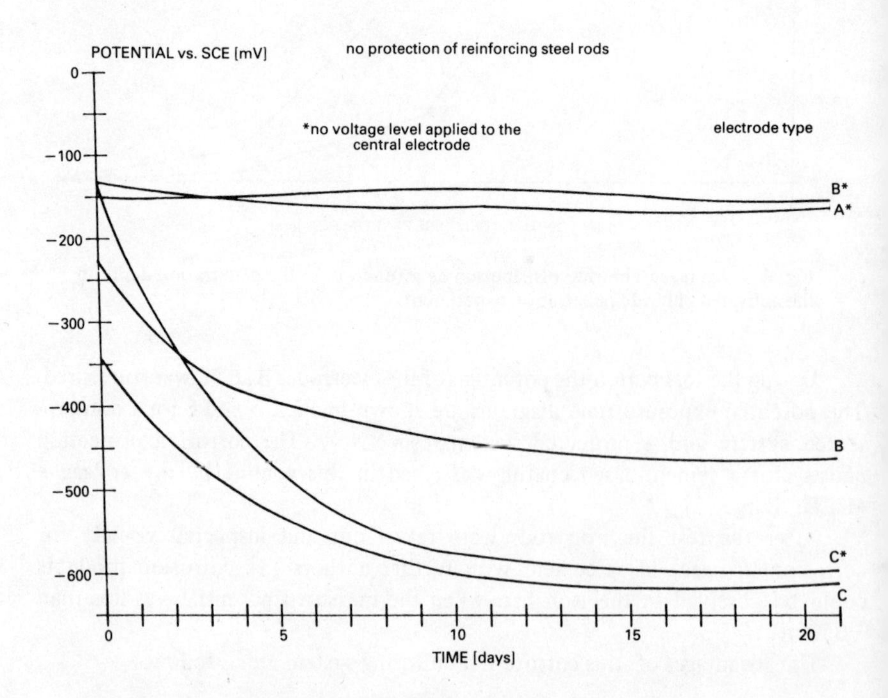

Fig. 5 – The averaged potentials of the different electrode types in the activated chloride penetration experiment.

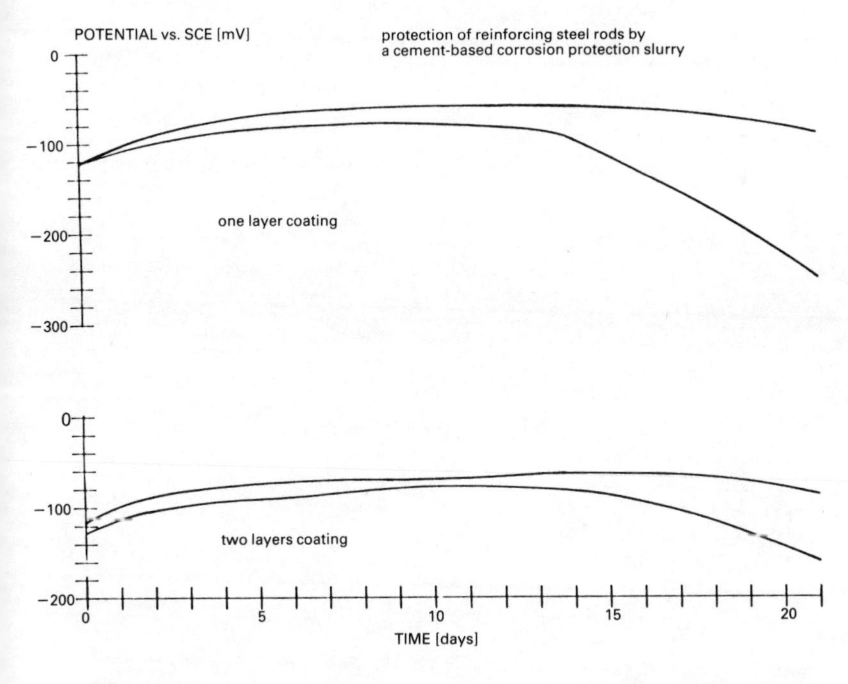

Fig. 6 – The averaged potentials of the different electrode types in the activated chloride penetration experiment.

4. CONCLUSIONS

Corrosion monitoring in the industrial-related research area, as described in this paper, does not lay claim to be highly scientific. The methods described utilize modern technology and are combined with phenomenological effects. Nevertheless the results obtained can be easily used for the development of new corrosion protective systems.

REFERENCES

[1] John, D. G. Novel electrochemical techniques for investigating steel/concrete systems. *Conference 1981: The Failure and Repair of Corroded Reinforced Concrete Structures*.

[2] Bürge, Th., A. Densified silica-cement coating as an effective corrosion protection, in this volume.

[3] Rosenberg, A. M., Gaidis, J. M., Kossivas, T. G. and Previte, R. W. A corrosion inhibitor formulated with calcium nitrite for use in reinforced concrete. *Symposium 1976: Chloride Corrosion of Steel in Concrete.*

Chapter 17

The repair of concrete – A laboratory and exposure site investigation

D. G. JOHN, UMIST, Manchester UK,
A. T. COOTE, K. W. J. TREADAWAY, Building Research Establishment, Watford, UK
J. L. DAWSON, UMIST, Manchester, UK

INTRODUCTION

Reinforced concrete is one of the most widely used structural materials in construction. It displays major benefits of simplicity of manufacture, versatility in application, and, in most circumstances, excellent durability [1]. There have been, however, a number of examples of serious durability problems arising in the medium or even short term as a result of reinforcement corrosion. This can lead to cracking and eventual spalling of concrete cover in extreme cases, if left untreated, to structural weakening [2]. Loss of serviceability as a result of reinforcement corrosion can be a serious problem to the building owner who will, in most circumstances, need to make good any such deterioration.

A growing interest is at present being shown in concrete repairs necessitated by reinforcement corrosion from the points of view of their short-term performance in ease of application etc. and their longer-term performance characteristics of maintenance of bond with substrate, durability and effectiveness in protecting the reinforcement against further corrosion. In the past much of the effort put into the assessment of concrete repairs has been in relation to their physical performance characteristics. Tabor [3] has reported on a system to test bond between substrate and repair and also the effectiveness in bond of crack injection. Hewlett and Morgan [4] have investigated the fatigue performance of cracked concrete injected with epoxy resins and Tabor *et al.* [5] have studied the physical performance of polymer modified cementitious repair systems. Novel repair techniques using glass fibre reinforced cement have also been reported applied, both as patching material [6] and by spraying [7]. Repairs have been applied to a wide range of structures including bridges and buildings [8, 9] but little evaluation of their long-term durability performance seems to have been made. Indeed O'Brien [10] has recently argued for a more coherent study of concrete repair systems especially in relation to the long-term efficacy of repairs in arresting deterioration.

This study has been initiated to develop techniques for the testing of the durability of concrete repair systems with special reference to their ability to provide continued protection to reinforcing steel. An a.c. impedance technique has been used to assess the effect of exposure situations on a range of protective systems. This paper reports results on selected systems within this group to illustrate the effectiveness of the technique. A full report of the performance of all the repair systems examined will be made when further exposure site data is available.

CONSTRUCTION OF PRISMS

For the purpose of the study a standard concrete prism was designed. Factors which were considered during the design stage were:

(i) The prism should be of dimensions that allowed ease of handling in the laboratory and yet be large enough to contain areas for repair that give a realistic assessment of systems, and strong enough to withstand chiselling with a hammer drill.
(ii) The need to provide depth of cover to steel reinforcement as realistic as possible in relation to the need for acceleration of the corrosion process.
(iii) Relative simplicity and manufacture from a mould that could be adopted to make prisms either for repair or to be used as controls, and to incorporate high yield deformed bar or mild steel round bar.

The prism designed to meet the above requirements has dimensions of 250 × 105 × 105 mm with 50 mm of steel projecting beyond the top surface. The version designed to take repair systems is illustrated in Fig. 1. The notched areas simulate areas from which concrete has spalled and are large enough to provide a realistic area for repair. They are formed by placing pieces of hardwood in the middle of the four long edges in the mould. A slot in the hardwood pieces helps locate the steel rod at the correct depth (12.5 mm) within the concrete and locate the surface of the spall well behind the steel to minimise cutting back. Prisms for exposure in the tidal zone have stainless steel sockets cast in their base so they can be secured in position.

Steel reinforcement was cut into 270-mm lengths and cleaned (round bar was de-greased, deformed bar was shot-blasted and de-greased). Round bar for repair systems was masked along the portions to be embedded in concrete by dipping the rods in epoxy resin to provide a barrier between steel and concrete and sprinkling the tacky resin with fine sand to provide a keyed surface. Half the pieces of reinforcement were then pre-rusted; this was achieved by placing them for a few days in a small cabinet that contained an atomising unit which produced a fine mist of water.

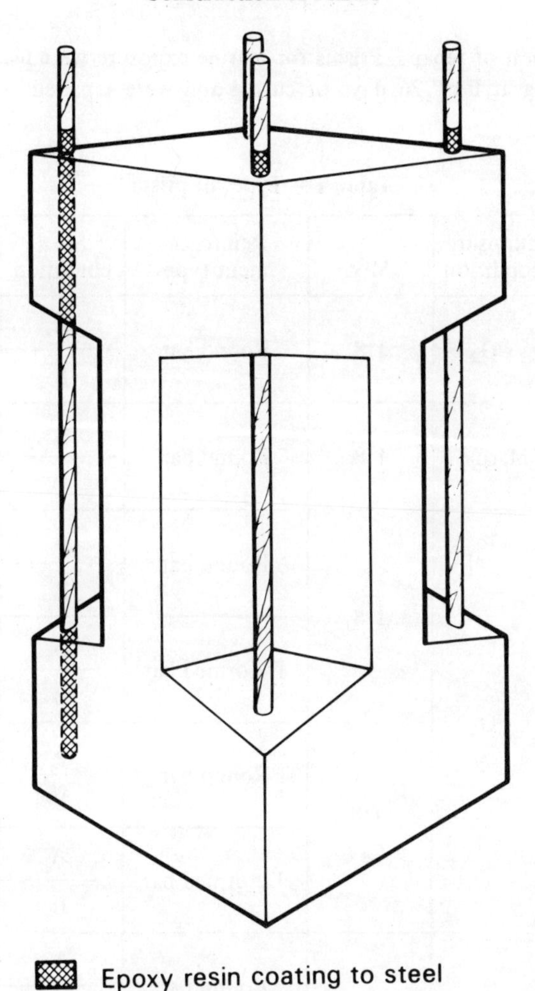

Fig. 1 – Concrete prism showing position of reinforcing steel and spalls for repair systems.

Prisms were cast on their sides in batches of 18; control prisms were cast alongside repair prisms simply by omitting the hardwood spall formers from the moulds. Table 1 summarises the types and numbers of prisms required for all situations and Table 2 gives the constitutents of the three different mixes used. Concrete prisms were demoulded after three days and then closed in polythene sheet for eleven days. Prisms were numbered: those for CO_2 exposure were prefixed C; those for marine exposure were prefixed M. The 4 steel rods in each prism were labelled a, b, c and d according to their relative position to the trowelled surface. Prisms intended for the CO_2 environment were stored indoors until completion of repairs. Prisms for marine exposure were placed in the tidal zone following at least 26 days of curing and were repaired after 40 days of exposure.

Table 1 – Types of prism

	Exposure condition	Mix	Reinforcement type	Steel condition	No. of prisms cast
Repair prisms	CO_2	1:8	Round bar	A	15
				B	15
	Marine	1:8	Round bar	A	30
Control prisms	CO_2	1:8	Round bar	A	3 of each
				B	
			Deformed bar	A	
				B	
		1:8 + $CaCl_2$	Round bar	A	
				B	
			Deformed bar	A	
				B	
	Marine	1:8	Round bar	A	
				B	
			Deformed bar	A	
				B	
		1:10	Round bar	A	
				B	
			Deformed bar	A	
				B	

A = Pre-rusted. B = Bright.

Table 2 – Mix constituents (proportions by weight)

Mix	OPC	Aggregate 5 mm down	Aggregate 5–10 mm	$CaCl_2$	Water Cement
1:8	100	385	415	0	0.775
1:10	100	481	519	0	0.9
1:8 + Ca Cl_2	100	385	415	5	0.775

REPAIR OF PRISMS

The concrete surfaces of spalled areas were cut back using a chisel held in a hammer drill and steel surfaces were cleaned using a mechanically operated wire brush. Prisms for CO_2 exposure contained steel that was nominally either bright or pre-rusted, in fact after wire brushing the two finishes were virtually indistinguishable showing that the pre-rusting method had been inadequate. Marine prisms, on the other hand, were all pre-rusted adequately and showed even pitting over their surfaces.

Details of two repair systems are given in Table 3. Prisms were repaired in an upright position so that the ability of the repair to adhere to the top underhung surface, without slumping, could be assessed. Where repairs were carried out in stages the second application followed one or two days after the first. Prisms for exposure in a CO_2 environment were repaired indoors and the cementitious repairs were enclosed in polythene sheet for 7 days. Marine prisms were removed from the tidal zone for the period of repair (5 days). During this time they were kept outside but cementitious repairs were enclosed in polythene sheet following application of repair systems.

Table 3 – Details of repair systems

Repair no.	Primer (applied to steel and concrete)	Repair system: 1st application	Repair system: 2nd application
1	Slurry of OPC/Water	OPC: Sand 1: Water (1:2.5:0.42) A B	OPC: Sand 1: Water (1:2.5:0.472) A OPC: Sand 2: Water (1:2.5:0.42) B
2	Epoxy resin	Epoxy resin mortar A B	

Sand 1 = sand passing 5 mm aperture; Sand 2 = sand passing 0.6 mm aperture.
A = Prisms for CO_2 exposure; B = Prisms for marine exposure.

EXPOSURE CONDITIONS AND EXAMINATION DURING EXPOSURE

Reference has already been made to exposure situations as one of the test variables. Two exposure conditions were employed. In one, marine exposure, the prisms were supported in the tidal zone; in the other, prisms were exposed in cabinets to an artificially rich CO_2 atmosphere. Both exposure situations are described in detail in Appendix A.

Periodically the prisms were examined, visually for signs of rust staining, cracking and spalling, and also electrochemically using an a.c. impedance technique.

A.C. IMPEDANCE MEASUREMENTS

A.c. impedance data was obtained using either a Solartron 1170 Series Frequency Response Analyser (FRA) or a Hewlett Packard Spectrum Analyser (HPSA). The principle of operation of the two instruments is different; the FRA outputs a single sine wave potential fluctuation (of known amplitude and frequency) to the test cell and the resultant cell current fluctuations are recorded and compared to the perturbating signal, and the impedance for that frequency is obtained. The FRA then outputs a new frequency and the process is repeated; the normal range of frequencies being 100 kHz to 10 mHz. The HPSA, however, outputs a pseudo-random potential perturbation to the test cell and the resulting 'random' current fluctuations recorded. The comparison of potential and current is then made by use of Fast Fourier Transforms (FFT) and a resultant impedance spectrum is obtained; the HPSA requires 2 to 3 runs to obtain the full spectrum and the normal frequency range is 20 kHz to 16 mHz. In both cases (FRA and HPSA) the instrumentation was controlled and the data analysed on a Hewlett Packard HP-85 minicomputer, running IMPED software (copyright CAPCIS/UMIST 1981–1982).

A.c. impedance analysis in general [11, 12] and in particular of steel in concrete [13–17] has been described in detail in previous publications. Normally the data is presented as a Nyquist plot (or Sluyters plot, or complex plane plot) of real and imaginary impedance. Previous work [14] has shown that for steel in concrete the system may be represented by a modified Randle circuit [11, 12]. This circuit, together with a typical complex plane plot of the response, is shown in Fig. 2. The different elements of the circuit being:

R_0 – Concrete resistance ($\Omega\ m^2$)
R_1 – Interfacial resistance ($\Omega\ m^2$)
C_1 – Interfacial capacitance ($F\ m^{-2}$)
R_2 – Charge transfer resistance ($\Omega\ m^2$)
C_2 – Double layer capacitance ($F\ m^{-2}$)
Z_D – Diffusion impedance ($\Omega\ m^{-2}$)

These parameters, their effect on the impedance results and their changes with time are covered below in the description of the results.

Because of limited space no attempt is being made here to cover, in its entirety, all the a.c. impedance data obtained during the programme. However, it is intended to make all the data available at a later date.

The exposure and examination of prisms is continuing: only those prisms which have so far been destructively analysed are considered in the paper.

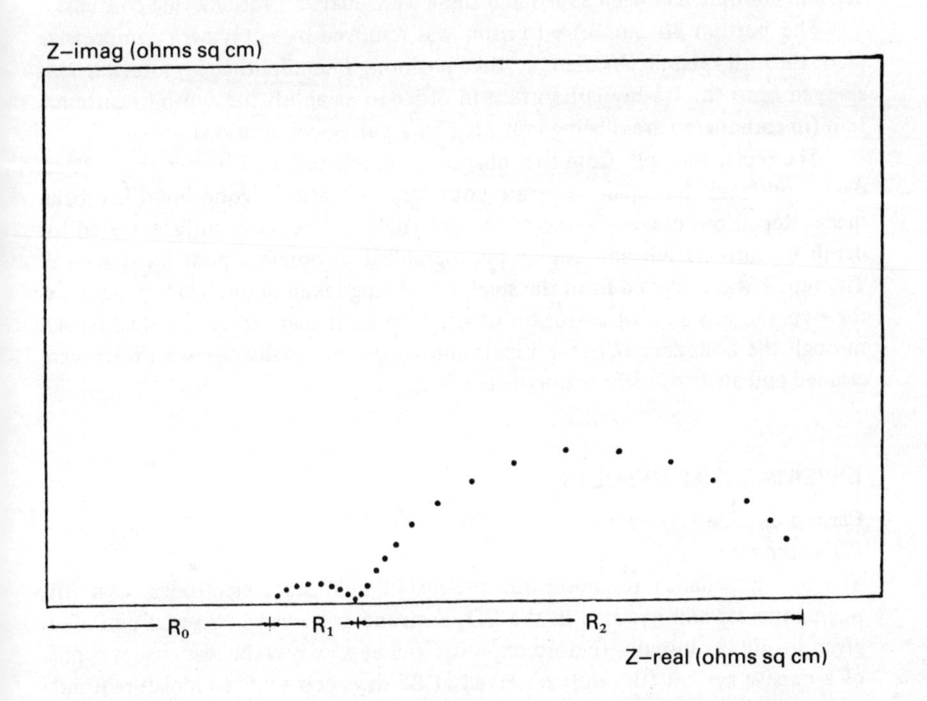

Fig. 2(a) – Typical Nyquist impedance plot for steel in concrete.

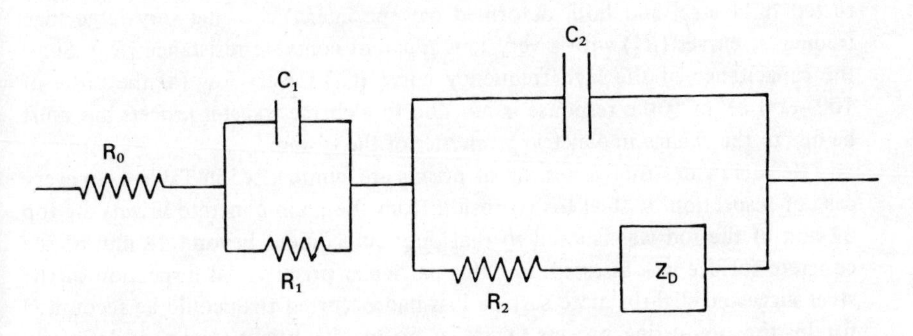

Fig. 2(b) – Equivalent circuit used for computer simulation of impedance data.

DESTRUCTIVE ANALYSIS OF DETERIORATION

At the conclusion of the exposure period marine exposure prisms were cleaned, examined for physical defects and colour photographed from two opposite corners. Powder samples for chloride content analysis were obtained by drilling into both the background concrete and repair systems. Pieces of concrete were broken from the centre portion of the CO_2 exposure plain prisms to which calcium chloride had been added and these were analysed for chloride content.

The bottom 30 mm of all prisms was removed by applying a compressive load through platens located at this position. Phenolphthalein solution was sprayed onto the freshly cut surface in order to establish the depth of carbonation (uncarbonated areas being indicated by a purple colouration).

The repair was split from the substrate concrete using a hammer and bolster; digression from the repair/substrate boundary indicating a good bond had been made. Repair pieces were fractured at right angles to the outer surface, tested for depth of carbonation and colour photographed to obtain a permanent record. The repair was removed from the steel, note being taken of any voidage between the two and any sign of corrosion product, in particular where the steel passed through the boundary between repair and concrete. Finally the steel rods were cleaned and an assessment made of steel loss.

EXPERIMENTAL RESULTS

Carbon dioxide exposure

Plain concrete

The a.c. impedance responses for pre-rusted mild steel electrodes, cast into plain concrete and exposed in the CO_2 environment, over a two-year period are given in Fig. 3. Initially (before exposure and at 85 days) the response is typical of a passive system (the shift observed at 85 days being due to moisture uptake and hence lower resistivity of the concrete). By 178 days and subsequently, however, all the responses changed completely (this also occurred on the pre-rusted mild steel and both deformed bar specimens) showing very large high frequency curves (R_1) with a very low apparent concrete resistance (R_0). Since the capacitance of the high frequency curve (C_1) is very low (of the order of 100–800 nF m^{-2}) the response is not due to a charge transfer process but must be due to the change in dieletric properties of the system.

Results of destructive testing of prisms are summarised in Table 4. In every case of inspection of steel for corrosion from the plain concrete prisms the top 62 mm of the rod was ignored so that only embedment beyond 12 mm of the concrete surface was assessed. For this particular prism visual inspection of the steel suggested slightly more surface loss had occurred than could be accounted for by the pre-rusting process (a similar prism with bright steel rods definitely lost a slight amount of material over the same period). The colour change de-

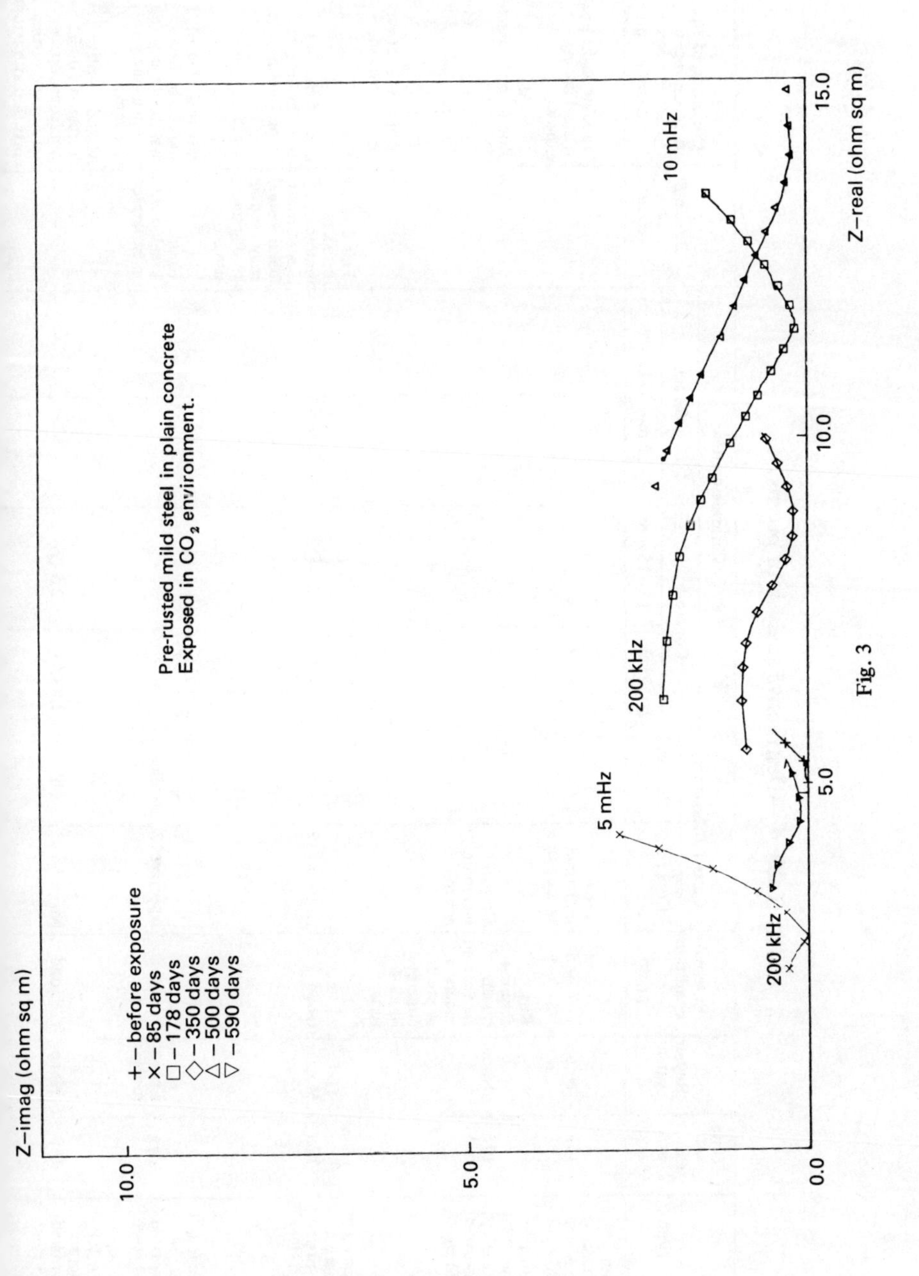
Z–imag (ohm sq m)
10.0
5.0
0.0
5.0
10.0
15.0
Z–real (ohm sq m)
+ – before exposure
× – 85 days
□ – 178 days
◇ – 350 days
△ – 500 days
▽ – 590 days
Pre-rusted mild steel in plain concrete
Exposed in CO2 environment.
5 mHz
200 kHz
200 kHz
10 mHz

Fig. 3

Table 4 – Destructive analysis of prisms.

Label	Type	Time of exposure (days)	Repair system	External condition of prism	Carbonation (mm)		Chloride content				Bonding of repair	Condition of steel in repair region or relevant region in plain prisms
					Concrete	Repair	Concrete		Repair			
							2–12 mm	12–25 mm	2–12 mm	12–25 mm		
C24	Plain 1:8 concrete pre-rusted mild steel	178	None	Good	Too indistinct to measure	–	–	–	–	–	–	Probable slight loss of steel during exposure
C44	Plain 1:8 concrete + 5% $CaCl_2$ pre-rusted mild steel	178	None	Rust staining on surfaces from junction of steel/concrete of all 4 rods	Up to 14 but mainly approx. 9	–	–	2.9 (A)	–	–	–	Rod / Max pit depth / Area*: a 1.5 mm 55; b 1.1 ,, 30; c 0.5 ,, 20; d 1.5 ,, 30. *Approx. % of surface corroded.
C20	Repair 1:8 concrete pre-rusted mild steel	178	OPC Mortar	Good	Approx. 12	1–3	–	–	–	–	Fair d – break at boundary. Some voids behind steel may be due to dry mix used	Loss of steel unlikely but masked by pre-rusting, etc.
C38	Repair 1:8 concrete pre-rusted mild steel	178	Epoxy mortar	Good	Approx. 12	Non alkaline	–	–	–	–	Good. Break occurs in concrete	Any loss of steel masked by pre-rusting etc. c – top of repair one pit (0.1 mm)
M15	Plain 1:8 concrete pre-rusted mild steel	448	None	Good	Nil	Nil	2.8 (A)	2.5 (A)	–	–	–	All rods – pitting corrosion 10–23 mm from top. a d – some pitting corrosion at bottom

M2	Repair 1:8 concrete pre-rusted mild steel	403	OPC mortar	All 4 rods bent causing concrete to spall at top (a c d) and conc + top of repair to spall (b)	Nil	Nil	3.6 (A)	2.7 (A)	2.0 (A)	1.6 (A)	Good. Some Some voids behind steel may be due to dry mix	Loss of steel noticeable where repair had spalled elsewhere corrosion uncertain.
M38	Repair 1:8 concrete pre-rusted mild steel	404	Epoxy mortar	b top corner of prism missing Rust staining through all repairs	Nil	Non alkaline	3.1 (A)	2.4 (a)	0.04 (A)	0.03 (B)	Good. Stronger than concrete	Some corrosion near repair/concrete at bottom (a b c). Elsewhere corrosion uncertain.
M44	Plain 1:10 concrete pre-rusted mild steel	448	None	a top corner of prism	Nil	Nil	3.5 (A)	2.9 (A)	–	–	–	Some pitting corrosion. < 20% of surface corroded but some confusion with pre-rusting likely. Max pit depth approx. 1.5 mm.
M47	Plain 1:10 concrete bright mild steel	448	None	Good	Nil	Nil	3.5 (A)	2.9 (A)	–	–	–	Pitting corrosion less than M44. < 10% of surface corroded. Max. pit depth approx. 0.3 mm.

A = % chloride ion w/w cement. B = % chloride ion w/w repair.

veloped at the boundary between carbonated and uncarbonated concrete following spraying with phenolphthalein solution was too indistinct to allow accurate measurement of the depth of carbonation but it seemed likely that the steel was in partially, if not fully, carbonated concrete. Generally the depth of carbonation of the prisms exposed in the CO_2 cabinet had been greater than expected over this period (178 days) and points to the suitability of this method for artificial carbonation of concrete; at least for concrete of relatively low cement content (1:8 cement:aggregate).

Plain concrete – contaminated with chloride

Figure 4 shows the a.c. impedance response obtained over a two-year period for pre-rusted mild steel electrodes cast into concrete with 5% $CaCl_2$ added. The two obvious differences between this data and that obtained in chloride-free concrete (Fig. 3) are the overall size of the a.c. impedance response (up to 1.5 $\Omega\ m^2$ for chloride-added compared to up to 25 $\Omega\ m^2$ for chloride-free, over an order of magnitude difference) and the presence from the beginning of a definite, well-defined, low-frequency curve, the capacitance of which (C_2) is of the range of 200 to 600 mF m^{-2} which is typical of a double layer capacitance, thus indicating that the low frequency curve is probably due to a charge transfer process and hence corrosion rates (*cr*) may be estimated, using the Stern Geary equation as

$$cr = k.B/R_2$$

where B – Stern Geary constant (B = 26 mV for active system [18]), and k – constant for units.

The corrosion rate ranges from approximately 30 $\mu m\ yr^{-1}$ (1.2 mpy) before exposure of 70 $\mu m\ yr^{-1}$ (2.3 mpy) by 178 days, to 130 $\mu m\ yr^{-1}$ (5 mpy) by 590 days. Unfortunately no information may be gleaned with regard to the nature of the attack and as the destructive results indicate, extensive pitting occurs.

This pitting corrosion was on a scale much greater than any that occurred in the marine prisms (see Table 4, labelled C44). The measured chloride level of 2.9% (chloride ion w/w cement) was a similar value to those obtained beyond 12 mm from the marine prisms. The test for depth of carbonation indicated that at least some of the steel was likely to be in a non-passive environment; there had been no carbonation of any of the marine prisms.

Concrete repaired with OPC and epoxy mortars

Figures 5 and 6 give the a.c. impedance response obtained over a two-year period of pre-rusted mild steel in contact with OPC and epoxy mortar repairs respectively (exposed in CO_2 environment). The data obtained on the OPC mortar repair is of a similar form to that obtained from a plain concrete prism (Fig. 3) but is different in absolute magnitude. The relative difference in impe-

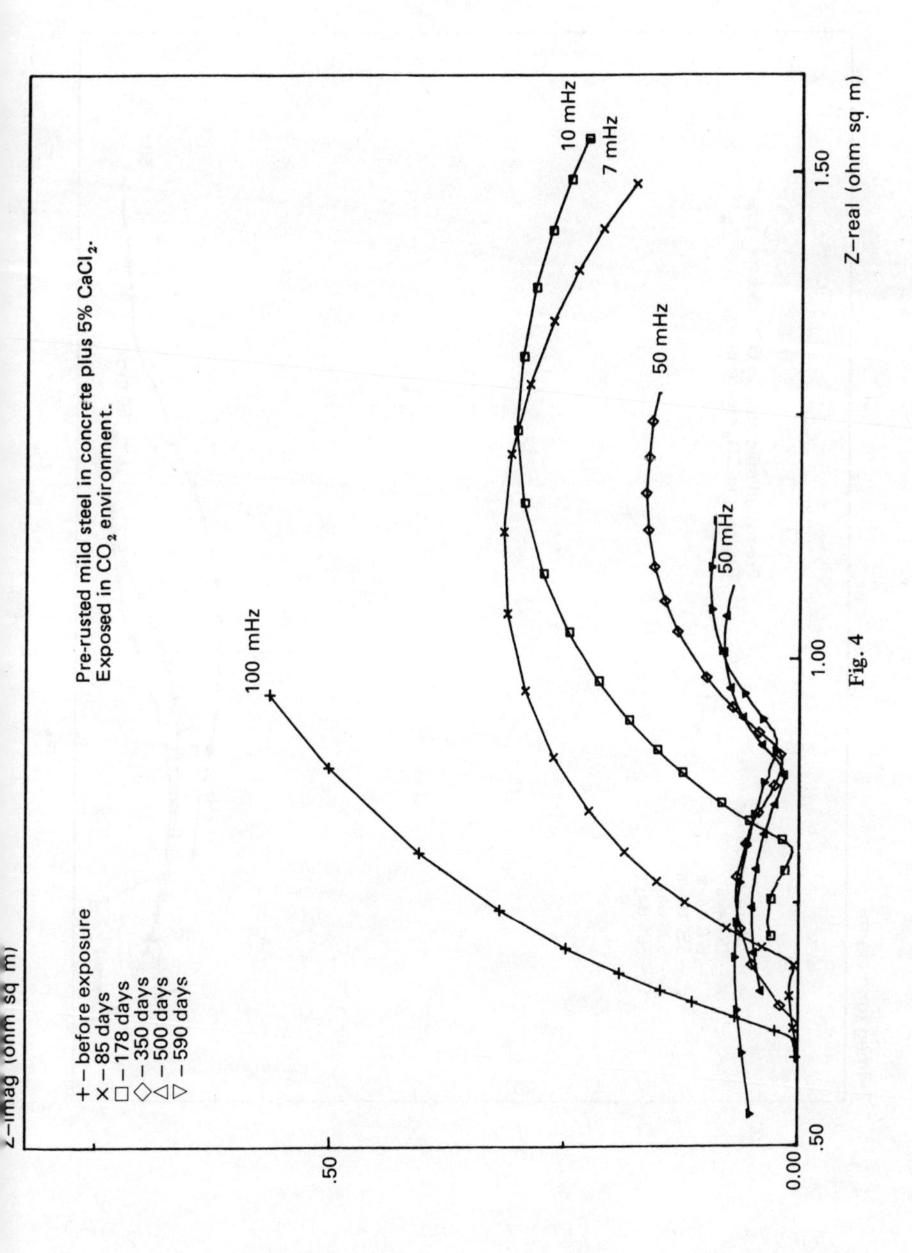
Pre-rusted mild steel in concrete plus 5% CaCl₂.
Exposed in CO₂ environment.
+ – before exposure
× – 85 days
□ – 178 days
◇ – 350 days
△ – 500 days
▽ – 590 days
Z–imag (ohm sq m)
Z–real (ohm sq m)
100 mHz
10 mHz
7 mHz
50 mHz
50 mHz
.50
0.00
.50
1.00
1.50

Fig. 4

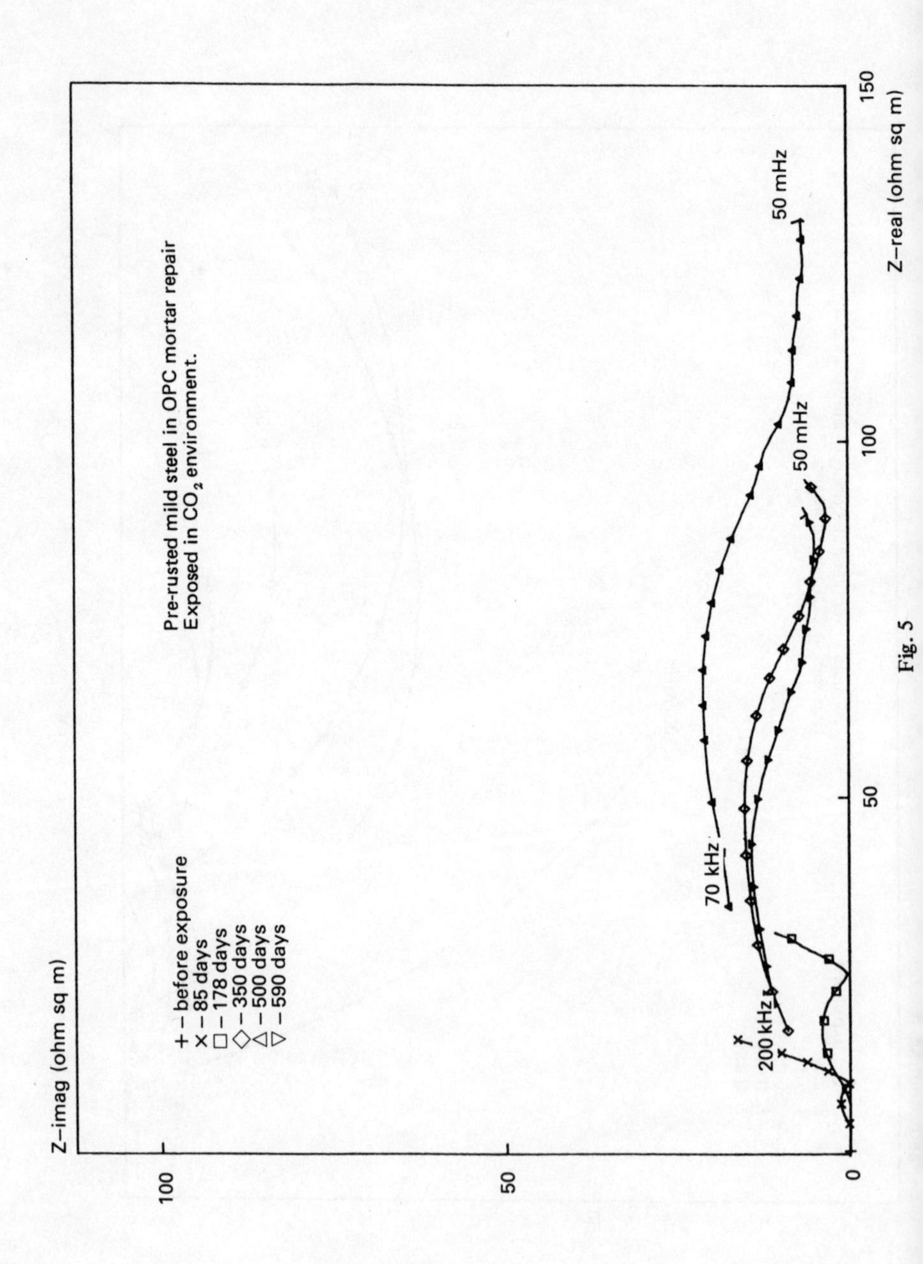
Pre-rusted mild steel in OPC mortar repair
Exposed in CO_2 environment.
+ – before exposure
× – 85 days
□ – 178 days
◇ – 350 days
△ – 500 days
▽ – 590 days
Z–imag (ohm sq m)
100
50
0
50
100
150
Z–real (ohm sq m)
200 kHz
70 kHz
50 mHz
50 mHz

Fig. 5

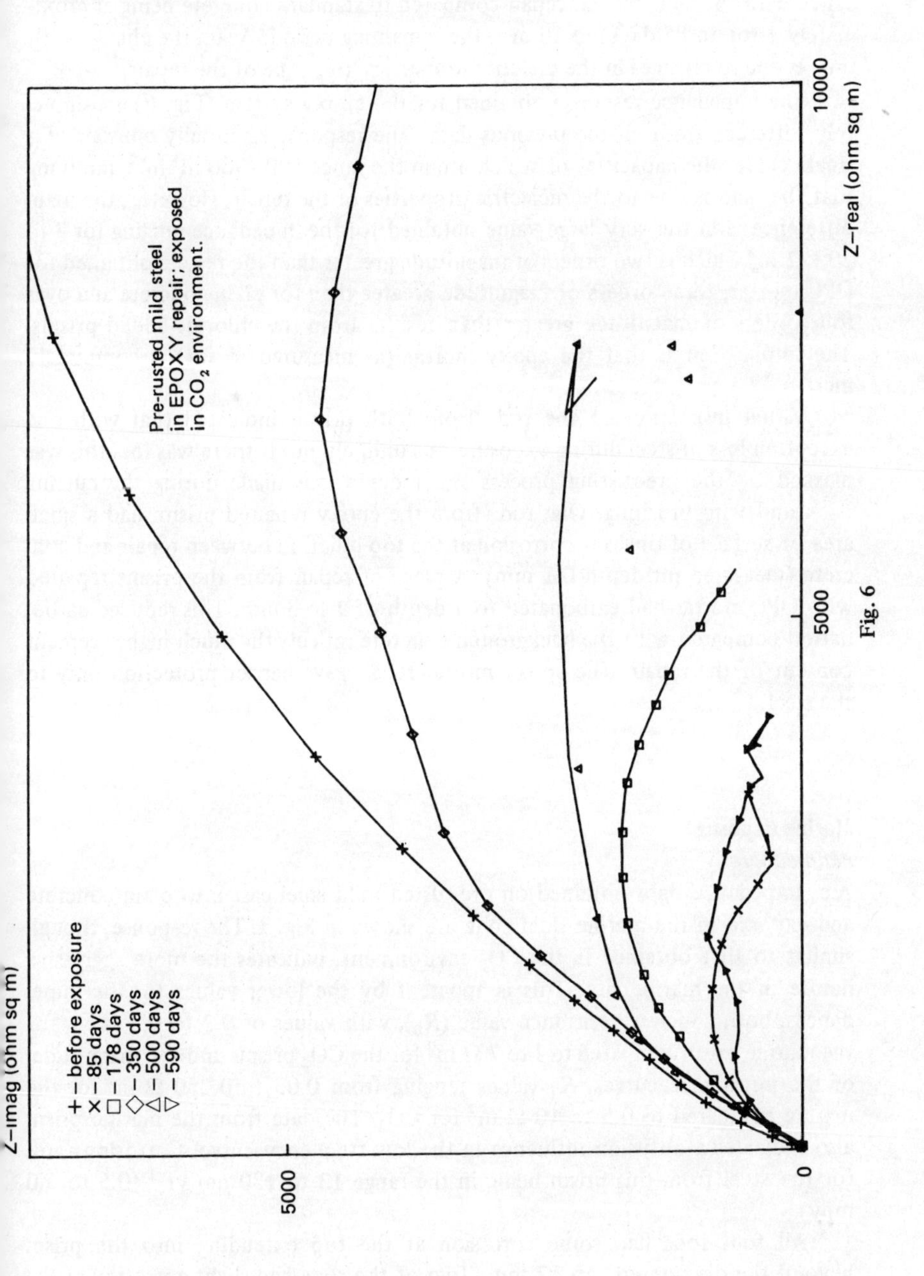
Pre-rusted mild steel
in EPOXY repair; exposed
in CO_2 environment.
+ – before exposure
× – 85 days
□ – 178 days
◇ – 350 days
△ – 500 days
▽ – 590 days
Z–imag (ohm sq m)
Z–real (ohm sq m)
0
5000
10000

Fig. 6

dance data for OPC mortar repair compared to standard concrete being approximately 1 (up to 85 days) to 10 over the remaining period. Again the change with time is due to changes in the dieletric properties, this time of the repair.

The impedance response obtained for the epoxy system (Fig. 6) is completely different from all the previous data. The response essentially consists of a single curve, the capacities of which are in the range 100–900 nF m^{-2}, implying that the data is due to the dielectric properties of the repair. However, the main difference is in the very large value obtained for the impedance ranging for 3 to 20 kΩ m^2 which is two orders of magnitude greater than the results obtained for OPC mortar, three orders of magnitude greater than for plain concrete and over four orders of magnitude greater than results from the chloride-added prisms. The implication is that the epoxy mortar (as measured so far) is completely inert.

Visual inspection of the rods from both prisms indicated that with one exception loss of steel during exposure was unlikely but if there was loss this was masked by the pre-rusting process and indentations made during the cutting back and wire brushing. One rod, from the epoxy repaired prism, had a small area of surface of obvious corrosion at the top junction between repair and concrete (measured pit depth 0.1 mm). A piece of repair from the prisms repaired with OPC mortar had carbonated to a depth of 1 to 3 mm. This reduced carbonation compared with the background concrete reflects the much higher cement content of the repair. The epoxy mortar repair gave barrier protection only to the steel.

Marine exposure

Plain concrete

A.c. impedance data obtained on pre-rusted mild steel cast into plain concrete and exposed in the marine tidal zone are shown in Fig. 7. The response, though similar to that obtained in the CO_2 environment, indicates the more aggressive nature of the marine site. This is apparent by the lower values for the impedances, both concrete resistance value (R_0), with values of 0.2 to 0.5 Ω m^2 for the marine prism compared to 1 to 7 Ω m^2 for the CO_2 prisms and the magnitude, of the impedance curves, R_1 values ranging from 0.03 to 0.250 Ω m^2 for the marine compared to 0.5 to 10 Ω m^2 for CO_2. The data from the marine prism also indicates a diffusion influence in the low frequency curve. Corrosion rates for the steel from this prism being in the range 13 to 130 μm yr^{-1} (0.5 to 5.0 mpy).

All four rods had some corrosion at the top extending into the prism beyond the discounted top 12 mm. Two of the rods had slight corrosion at the bottom. Elsewhere any corrosion was masked by the pre-rusting process. Nowhere was the corrosion sufficient to enable measurable estimates of loss to be

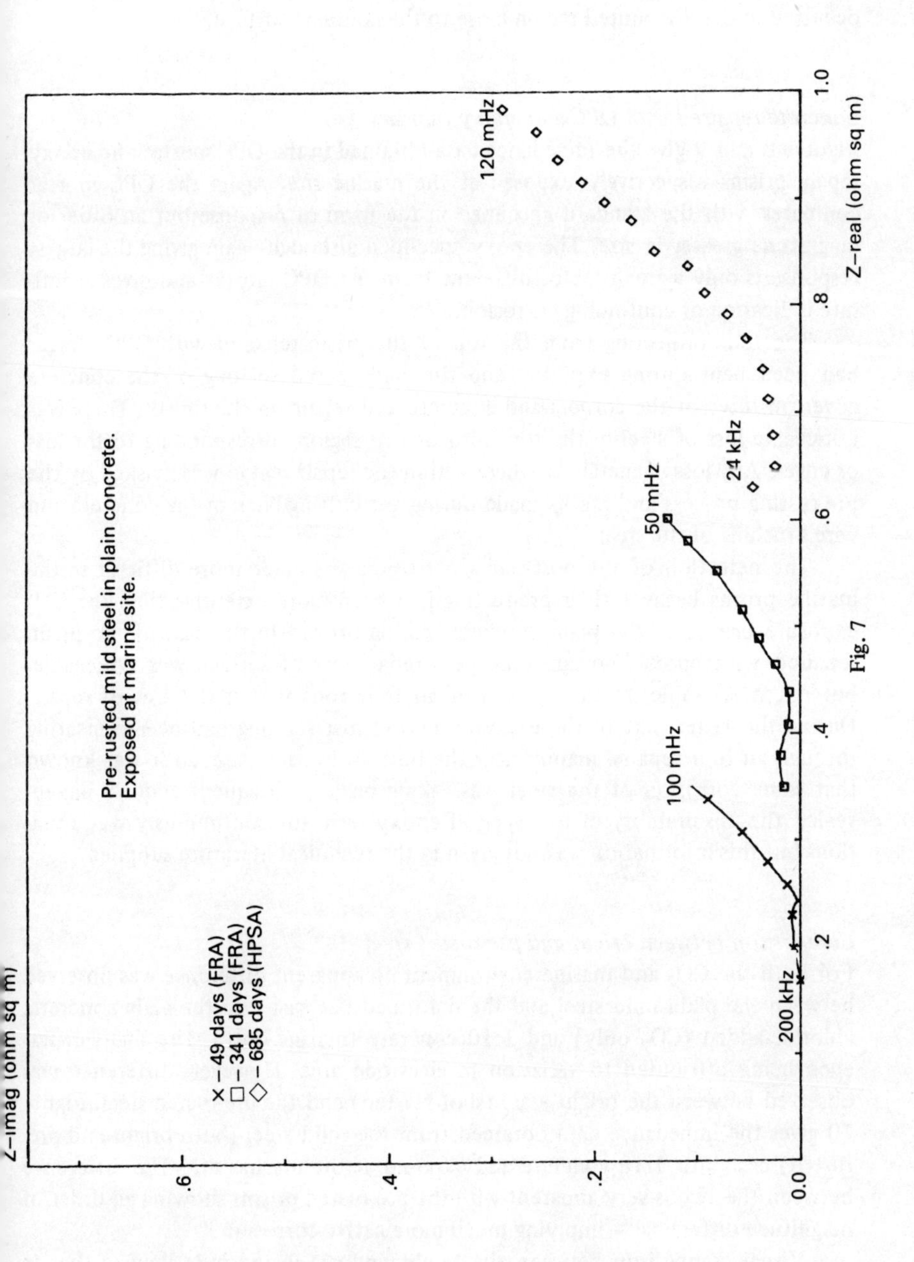
Pre-rusted mild steel in plain concrete.
Exposed at marine site.
× – 49 days (FRA)
□ – 341 days (FRA)
◇ – 685 days (HPSA)
Z–imag (ohm sq m)
Z–real (ohm sq m)
0.0
.2
.4
.6
.8
1.0
200 kHz
100 mHz
50 mHz
24 kHz
120 mHz

Fig. 7

made. Generally the corrosion at the top of the rods was the more severe and it could be that corrosion here was largely initiated by the corrosion that had occurred in the discounted region close to the concrete surface.

Concrete repaired with OPC and epoxy mortars

Figures 8 and 9 give the impedance data obtained in the OPC mortar and epoxy repair prisms respectively exposed at the marine site. Again the OPC mortar compares with the standard specimen in the form of response but an order of magnitude greater in size. The epoxy specimen although again giving the largest response is only a small factor different from the OPC mortar and gives a definite indication of continuing corrosion.

The rods projecting from the top of the prism repaired with OPC mortar had been bent during exposure and this had caused spalling of the concrete cover on three of the corners and concrete and repair on the fourth. There was noticeable loss of steel in the top third of this region corresponding to the loss of cover. Any loss of metal elsewhere within the repair region was masked by the pre-rusting process and marks made during the cutting back of the concrete and wire brushing of the steel.

The detection of any post-repair corrosion was made more difficult in the marine prisms because their pre-rusting had been more extensive than the CO_2 exposure prisms or the plain concrete marine prisms. In the case of the prism repaired with epoxy mortar some post repair loss of surface was noticeable, but not measurable, at the bottom of all four rods within the area of repair. During the latter part of the exposure period rust staining had been appearing through all four repairs, mainly near the bottom in each case, so it was known that some corrosion of the steel was taking place. Subsequent enquiry has revealed the unsuitability of this type of epoxy resin for continuously wet situations but this information was not given in the technical literature supplied.

Comparison between bright and pre-rusted steel

For both the CO_2 and marine environment no apparent difference was observed between the plain mild steel and the deformed bar cast into the plain concrete, chloride-added (CO_2 only) and 1:10 concrete (marine only). The slight difference being attributed to variation in electrode area. However, difference was observed between the bright steel (shot-blasted) and the pre-rusted steel. Figure 10 gives the impedance data obtained from the mild steel (both bright and pre-rusted) cast into 1:10 concrete and exposed at the marine site. The difference between the two is very apparent with the pre-rusted prisms showing an order of magnitude difference – implying much more active corrosion.

Visual comparison between the bright and pre-rusted rods showed that in both cases pitting corrosion had occurred but the latter had corroded to a much

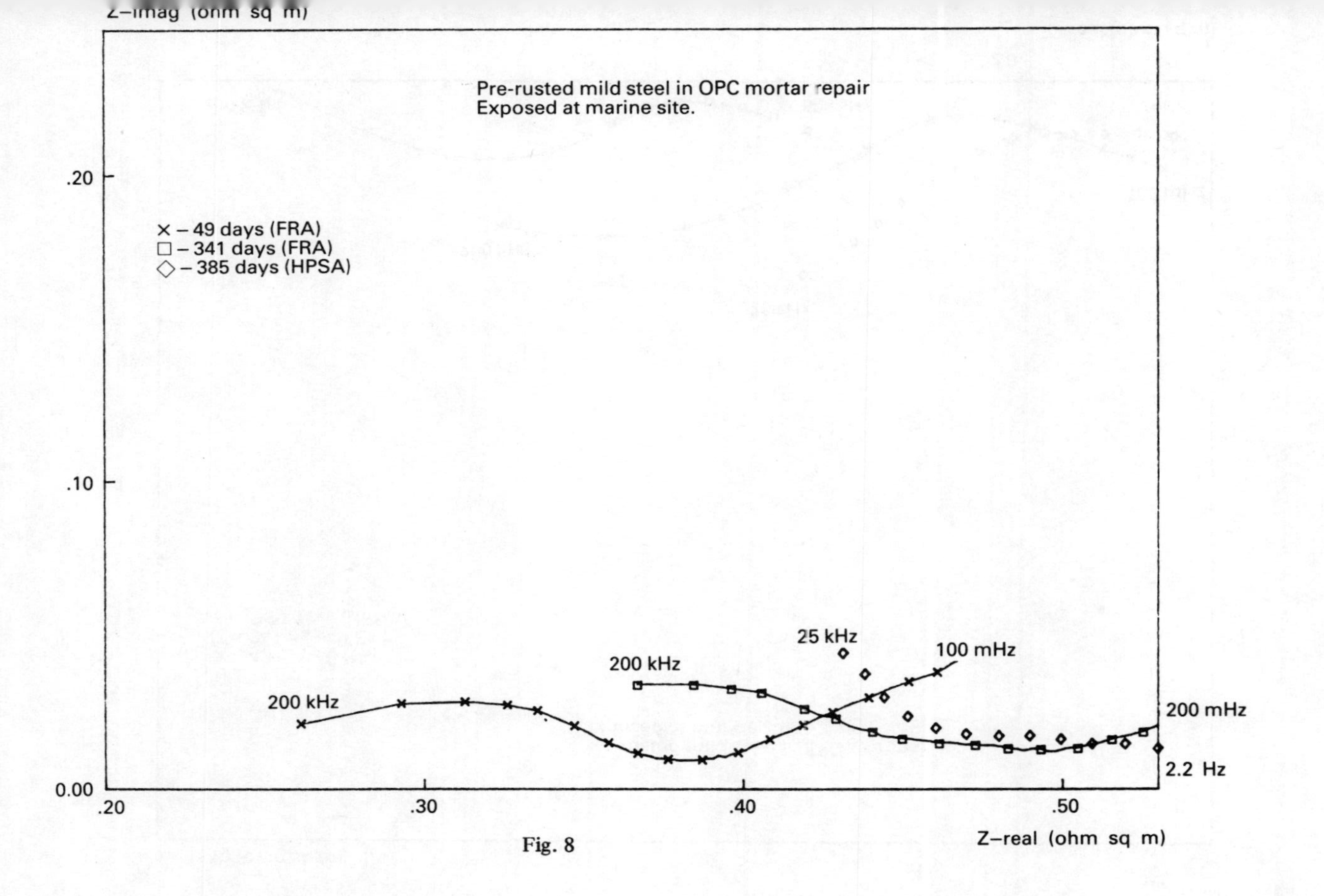
Z–imag (ohm sq m)
Pre-rusted mild steel in OPC mortar repair
Exposed at marine site.
× – 49 days (FRA)
□ – 341 days (FRA)
◇ – 385 days (HPSA)
.20
.10
0.00
200 kHz
200 kHz
25 kHz
100 mHz
200 mHz
2.2 Hz
.20
.30
.40
.50
Z–real (ohm sq m)

Fig. 8

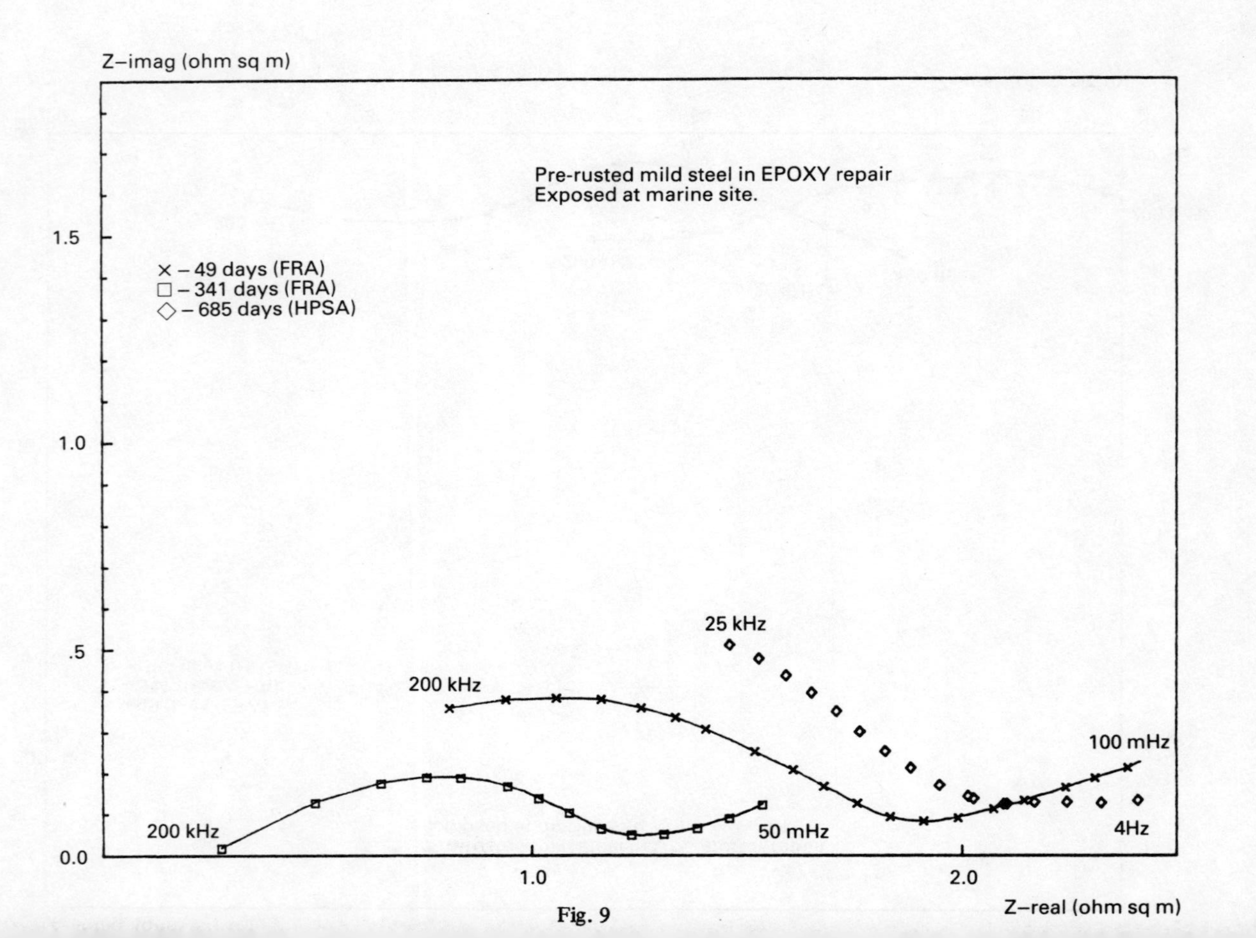

Z–imag (ohm sq m)
Pre-rusted mild steel in EPOXY repair
Exposed at marine site.
× – 49 days (FRA)
□ – 341 days (FRA)
◇ – 685 days (HPSA)
1.5
1.0
.5
0.0
25 kHz
200 kHz
100 mHz
200 kHz
50 mHz
4Hz
1.0
2.0
Z–real (ohm sq m)

Fig. 9

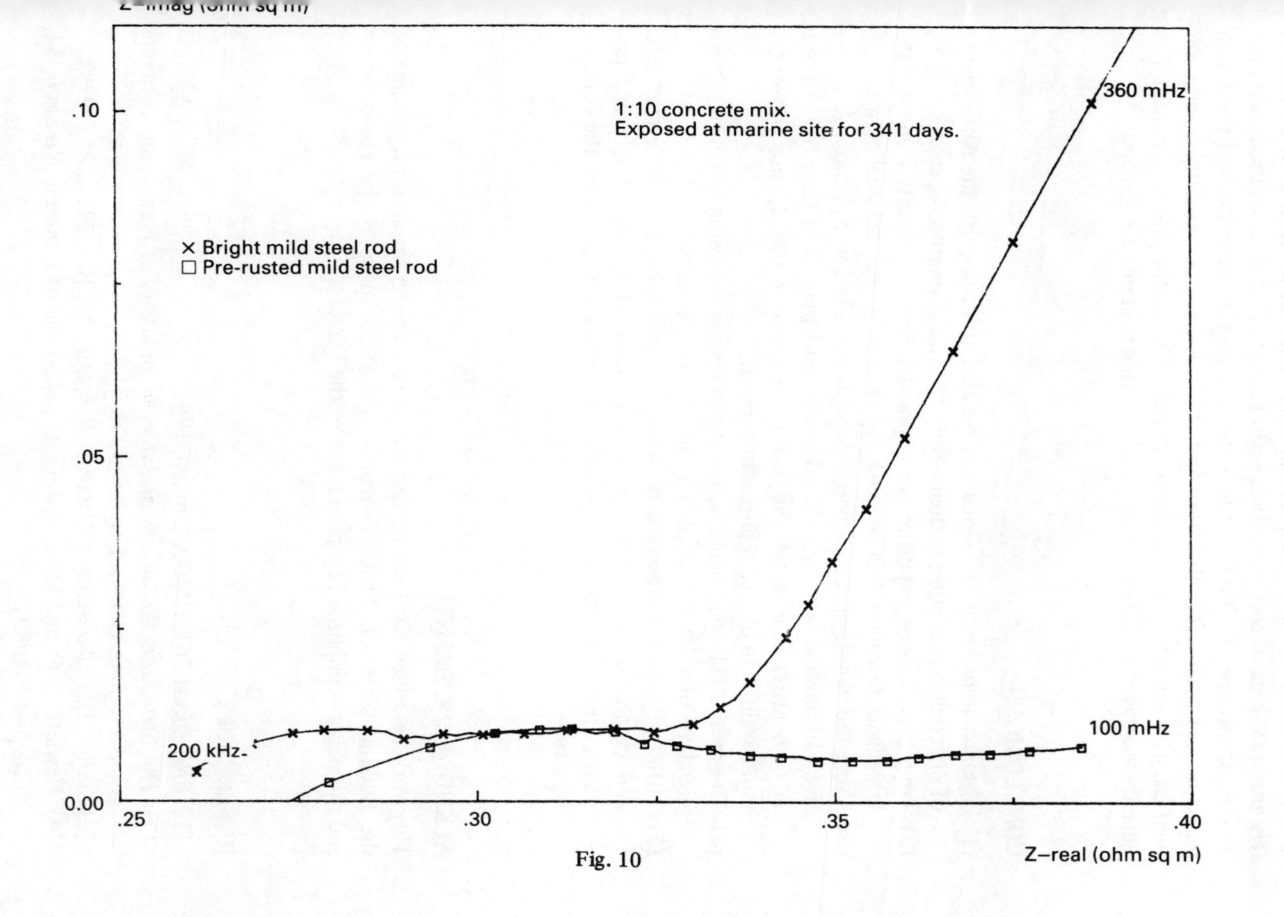
1:10 concrete mix.
Exposed at marine site for 341 days.
× Bright mild steel rod
□ Pre-rusted mild steel rod
360 mHz
100 mHz
200 kHz
.10
.05
0.00
.25
.30
.35
.40
Z–real (ohm sq m)

Fig. 10

greater extent. In the case of the pre-rusted rods it was difficult to assess accurately the total area of corrosion since slight corrosion could be confused with the pre-rusting process. In Table 4 values for the area of corrosion and maximum pit depth are given (labelled M44 and M47). Reference to the table indicates their chloride contents to be higher than the other marine prisms examined, but since they contain less cement their total chloride contents are similar.

CONCLUSIONS

(1) The a.c. impedance technique can be used to distinguish the performance of different repair types in their ability to protect reinforcing steel.
(2) Longer periods of exposure are required before the results from the a.c. impedance technique can be more closely related to actual steel corrosion.
(3) Using the simulated CO_2 environment, as described in Appendix A, extensive carbonation is possible in relatively short periods of time (e.g. 10 mm in 6 months) at least for the quality of concrete tested, thus allowing a rapid evaluation of concrete/repair systems.
(4) The results yet again demonstrate the extensive corrosion obtained when a high level of chloride is used as an additive to concrete.
(5) Little difference is observed between the performance of plain mild steel and deformed bar, but a difference is observed between bright and pre-rusted steel, the extent of the difference being dependent on the concrete quality.

ACKNOWLEDGEMENT

The work described has been carried out as part of the research programme of the Building Research Establishment of the Department of the Environment and this paper is published by permission of the Director.

REFERENCES

[1] *BRE Digest 265*, HMSO, London, 1982.
[2] *The Structural Conditions of Intergrid Buildings of Prestressed Concrete Constructions*, HMSO, London, 1978.
[3] Tabor, L. J., *Magazine of Concrete Research*, **30**, 105, 221–225 (1978).
[4] Hewlett, P. C. and Morgan, J. G. D., *Magazine of Concrete Research*, **34**, 118, 5–17 (1982).
[5] Tabor, M. M., Fowler, D. W. and Paul, D. R., University of Texas, Center for Highway Reasearch, *Report 114–3*, Austin, 1975.

[6] Meyer, A. and Steinagger, H., Repair of a steel reinforced concrete chimney by glassfibre reinforced concrete (GRC). *Proc. Int. Cong. on Glass fibre reinforced cement, Brighton, 1977,* pp. 271–274. Glass Fibre Reinforced Cement Association, Gerrards Cross, 1978.
[7] Schadel, E., Dikeou, J. and Gill, D., *Performance of Concrete in Marine Environment,* AC1 SP-65, Detroit, 1980.
[8] Perkins, P. H., *The Repair and Waterproofing of Concrete Structures,* Applied Science Publishers, 1976.
[9] Cavalier, P. G. and Vassie, P. R., Paper: *Proc. Inst. Civ. Engrs,* Part 1 (1981), **70,** 461–480; Discussion: *Proc. Inst. Civ. Engrs,* Part 1 (1982), **72,** 401–419.
[10] O'Brien, T. P., *Proc. Inst. Civ. Engrs,* Part 1 (1980), **68,** 399–408.
[11] Sluyters, J. H., *Rec. Trave. Chim. Phys-Bas. Belg.,* **79,** 1092 (1960).
[12] Hladky, K., Callow, L. M. and Dawson, J. L. *Br. Corros, J.,* **15,** 20 (1980).
[13] Dawson, J. L., Richardson, J. A., Callow, L. M. and Hladky, K., Paper presented at NACE Corrosion/78 Meeting, Houston, 1978.
[14] John, D. G., Searson, P. C. and Dawson, J. L., *Br. Corros, J.,* **16,** 102 (1981).
[15] John, D. G., Paper presented at 'Failure and Repair of Corroded Reinforced Concrete Structures', Conf., London, 1981.
[16] Searson, P. C., Dawson, J. L. and John, D. G., Paper represented at 'Electrochemical Methods in Corrosion Testing and Research', Manchester, 1981.
[17] Wenger, F., Galland, J. and Lemoine, L., Paper presented at Int. Symp. Behaviour of Offshore Concrete Structures, France, 1980.
[18] Andrade, C. and Gonzalez, J. A., *Werkstoffe und Korrosion,* **29,** 515 (1978).

APPENDIX A

Exposure conditions

Carbon dioxide

Exposure cabinets were constructed to house the concrete prisms in a humid carbon dioxide/oxygen environment. The cabinets consisted of glass tanks 1500 × 700 × 500 mm sealed with a melamine-faced particle board lid and silicone rubber. Two access hatches were incorporated into the lid. The temperature in the cabinets was maintained at 26–29°C using two 100-watt light bulbs controlled by a standard thermostat; a high-speed fan was used to continuously circulate the gases (2.5 m^3/min) in the tank.

The high relative humidity (89–92% RH) was maintained by having saturated sodium carbonate solution in trays at the bottom of the tanks. The solution was stirred by externally located magnetic stirrers. The tanks were first vented with carbon dioxide for two hours and thereafter by a cylinder containing an 80% carbon dioxide/20% oxygen mixture.

Marine

For the marine conditions, a tidal exposure site was used. The site is situated under a jetty in a harbour on the South Coast. The prisms were secured by the sockets cast into their bases on to heavy elm timbers that spanned between, and were securely fastened to the concrete whalings of the jetty. In this position the prisms were completely covered at high tide and at low tide the water level was well below the securing timbers. During stormy weather some of the prisms became damaged by heavy floating debris and so the site was protected by a fence of reinforcement mesh fixed to the columns of the jetty.

Part 4
AVOIDANCE AND DESIGN RECOMMENDATIONS

Chapter 18

Criteria for the cathodic protection of bridge decks

RICHARD F. STRATFULL, Corrosion Engineering Inc, USA

During the removal of a bridge as a result of corrosion of the reinforcing steel caused by de-icing salts, sections of the deck were salvaged to evaluate various cathodic protection criteria. These included the 100 mV and 300 mV change in half cell potential, −0.77 and −0.85 volt CSE half cell potentials, and, evaluation of the half cell potential and constant current values obtained from the beginning of the Tafel slope on the *E* log *I* curve.

The effectiveness of each criteria was monitored by means of electrical resistance probes (bars) that changed in electrical resistance when affected by corrosion. They were embedded in salt-contaminated concrete at the level of the top and bottom mat of reinforcing steel.

The cathodic protection currents were applied to the concrete by cutting ⅜″ (9.3 mm) wide and at a depth of ½″ (12.7 mm) longitudinal anode grooves on spacings of one foot (305 mm) center to center. The grooves were then filled with an electrically conductive strand of carbon filament and electrically conductive polymer. Measurements were obtained to determine the current distribution when the anodes were electrically energized on a one-foot (305 mm) and two-foot (610 mm) center to center spacing.

Procedures are given for obtaining an *E* log *I* curve using an oscilloscope so that the measurements are not adversely affected by turning the current on and off to obtain polarized potentials.

INTRODUCTION

The first reported use of a cathodic protection system on a bridge deck was in 1974 [1]. Although the cathodic protection system has been successfully operating for about eight years, a question of whether or not the cathodic protection criteria used was appropriate or economical has been raised [2].

As a result of this concern, and the raising of the same question by the United States Department of Transportation, Federal Highway Administration

(FHWA), the latter department decided to award a research contract to explore the effectiveness and applicability of the existing cathodic protection criteria used on bridge decks.

In 1981, the FHWA awarded Contract No. DTFH61-81-R-00051 to Automated Management Systems, Inc. The Co-Principal Investigators, Dr. Errol C. Noel and Richard F. Stratfull, undertook the task of exploring the cathodic protection criteria. The applicability of the following cathodic protection criteria were subjected to research.

(a) Break in $E \log I$ curve-constant potential
(b) Break in $E \log I$ curve-constant current
(c) −0.85 volt CSE instant-off potential
(d) −0.77 volt CSE instant-off potential
(e) 100 mV polarization shift (decay)
(f) 300 mV cathodic voltage shift (i.e. difference between before CP static and instant-off polarized potentials).

A cathodic protection system was installed on an approximately 7 foot by 17.7 foot (2.1 by 5.4 m) salvaged section of bridge deck that was removed from the Gleebe Road Overcrossing on the George Washington Memorial Parkway. The bridge was in the State of Virginia and was reported to be constructed in 1957, and, because of deck distress, removed in 1979.

The salvaged section of the bridge deck was moved, stored and subsequently prepared for the cathodic protection criteria testing at the FHWA Fairbanks Laboratory at McLean, Virginia.

The research was completed in 1982, and the following are the results of that program.

PREPARATION OF TEST SLAB

The salvaged section of the bridge deck was moved from its outdoor storage location and placed in its permanent outdoor test site. Then the slab was cleaned by a heavy duty rotary hammer, chisels, and steel brushes. After this the slab was sand blasted.

Samples of the concrete at different depths below the concrete surface were obtained by means of collecting pulverized samples at appropriate depths. Each time a concrete sample was collected, the hole in the concrete was vacuum cleaned and the rotary hammer bit and spoon were cleaned with a wire brush and then washed with alcohol. The chloride ion for each sample was determined by the test procedures listed in *Report No. FHWA No. FHWA-RD-77-85 and in AASHTO Specification No. T620*. The results of the chloride analysis are shown in Table 1.

Table 1 – Chloride content of cores (pounds per cubic yard–concrete)

Core No.	Depth below concrete surface–inches							
	$\frac{1}{16}$–$\frac{1}{2}$	$\frac{1}{2}$–1	1–1$\frac{1}{2}$	1$\frac{1}{2}$–2	2–2$\frac{1}{2}$	2$\frac{1}{2}$–3	3–3$\frac{1}{2}$	3$\frac{1}{2}$–4
1	2.41	2.24	2.10	2.14	2.24	1.64	1.91	1.62
2	9.52	4.81	2.92	2.77	2.24	1.35	0.56	0.33
3	11.97	5.76	2.81	2.46	0.77	0.82	0.70	0.32
Avg. Cl.	7.97	4.27	2.61	2.46	1.75	1.27	1.09	0.76

Both the top and bottom surfaces of the slab were traversed with an 'R Meter' or Pachometer, to determine the amount of concrete cover over the reinforcing steel. In addition, the length and the surface area were obtained and calculated for the embedded reinforcing steel. The results of 102 concrete readings resulted in the determination that the average depth of cover was 1.92 inches (48 mm) with a standard deviation of 0.35 inches (9 mm).

The calculated amount of reinforcing steel in the salvaged bridge deck slab for the top mat was 28.88 square feet (2.68 square meters), and, 50.02 square feet (4.65 square meters) for the bottom mat of steel. The total surface area of reinforcing steel in the slab was 78.9 square feet (7.33 square meters).

To be sure of electrical continuity, all reinforcing steel was welded together. This was accomplished by progressively welding a number 3 reinforcing steel bar to all visible ends of the projecting slab reinforcing steel.

Electrical half cell potential measurements were made after the slab was cleaned, repaired, and made electrically continuous. Initially, two sets of readings were made seven days apart to determine the reproducibility of the technique. These electrical half cell potential measurements were made both on the top and bottom surfaces of the slab, and, were made on one foot (305 mm) in both the longitudinal and transverse directions. At a later date, a saturated copper–copper sulfate half cell (CSE) was affixed to the top surface of the slab for control of the testing of cathodic protection criteria.

Longitudinal grooves were then saw-cut into the center of the slab on one foot (305 mm) centers. The grooves were cut to have a width of $\frac{3}{8}$ inch (9.3 mm) and a depth of $\frac{1}{2}$ inch (1.3 mm). The average length of the grooves was 16.71 feet (5.09 m) with the shortest being 16.58 feet (5.05 m) and the longest 16.77 feet (5.11 m).

Then, an electrically conductive carbon filament strand was placed in the bottom of the slot. To make an external electrical connection, about four inches

(102 mm) of niobium-coated and platinum-plated copper was placed in physical contact with the carbon filament strand.

Afterwards the wire and filament were placed in the slot, and FHWA patented electrically conductive polymer was poured to fill the slot to the top surface of the concrete. The electrically conductive polymer has a specific electrical resistance of about 2.4 ohm when it is in the pourable state. The platinum-plated copper wire had a diameter of 0.031 inches (0.79 mm). Ottawa sand was then cast by hand over the top of the electrically conductive polymer to enhance skid resistance on future installations at actual bridge decks.

As constructed, the anodes in the grooves had a specific electrical resistance of 0.096 ohm cm. The average lineal resistance of these anodes were 2.32 ohms per foot (7.63 ohms per meter). By means of a 400 cycle alternating current ohmmeter, it was found that the gross electrical resistance between the reinforcing steel and the slot anodes was 13.38 ohms per square foot (1.24 ohm/sq. m.) of anode surface area.

To monitor the performance of the cathodic protection criteria, five corrosion test probes were used. By use of commercially available equipment, changes in the electrical resistance of the probes due to corrosion loss is measured and transformed into a thickness of metal loss.

When the probes were originally installed, they were placed at the top and bottom level of the reinforcing steel mats and backfilled with the original specification concrete for the salvaged slab. Chloride was added to the probe backfill concrete at the rate of 15 lbs/cu yd (8.9 kg/cu m). This amount of chloride ion was added so as to provide a highly corrosive environment for the corrosion probes. Two of the probes were installed at the level of the bottom mat of steel and three were installed at the level of the top mat of steel. A 51 ohm shunt resistor was placed in series with all probes so that the amount and direction of current flow measurements could be calculated from measurements made with a 4½ digit voltmeter. The electrical resistance probes were in the shape of a bar. They were 0.307 inches in diameter (7.80 mm) and 5.0 inches (127 mm) long. The galvanic voltage between the reinforcing steel and the probes is shown in Table 2. In all cases the probes were anodic to the reinforcing steel. At a later date, Probe 5 became inoperative.

Table 2 – Initial galvanic voltage between probes and reinforcing steel

	Probe 1	Probe 2	Probe 3	Probe 4	Probe 5
Dec. 1981	0.147	0.214	0.180	0.240	0.362
Jly. 1982	0.117	0.220	0.166	0.195	–

Note. In all cases, probes were anodic to reinforcing steel.

CATHODIC PROTECTION CRITERIA

A review of the cathodic protection literature indicated that there is a lack of detail as to the method of complying with the criteria, and, even some question of the validity of certain criteria. The criteria of −0.85 CSE should be attained with the current on, and compensation shall be made for IR drop as a result of the current flow. One difficulty is: how to compensate for the IR drop as a result of current flow when it is not spelled out in the Recommended Practice. Additionally, no direction is given as to how to compensate for errors due to IR drop in the pipe as a result of the current flow. In this present work, technically, compliance with the criteria was made by using an oscilloscope to measure the potential when the full wave rectifier had a zero moment of current output during its 120 half wave pulses of direct current.

−0.77 volt criterion

With one exception, the cathodic protection criterion evaluated during this investigation was proposed for underground or submerged pipelines. As a result of a research project concerning the cathodic protection of steel in bridge deck concrete, a corrosion control criterion was recommended [5]. On the basis of tests in solutions, it was recommended that the half cell potential of all of the reinforcing steel be at least −0.77 volts to the saturated copper–copper sulfate half cell (CSE) to protect against general and crevice corrosion. For this project, the −0.77 volt CSE was to be a polarized potential measured at the instant the current is turned off, which would insure that the measurements are free of IR drop as there was no current flow.

Table 3 – Cathodic protection criteria test results

	$E \log I$				I_0	100 mV	300 mV	0.77	0.85
Date	I_{corr}	E_{cp}	I_{cp}	B_c	Probe	I_{cp}	I_{cp}	I_{cp}	I_{cp}
De. 81	0.45	0.42	2.10	0.23	1.5	1.2	7.2	29.5	46
Jul. 82	0.60	0.40	1.70	0.34	1.3	0.73	5.1	20	37
Sep. 82	0.45	0.33	1.50	0.21	1.08	1.32	7.5	27	34

Notes. Icorr:mA/sq. ft conc. surf., E_{cp}: volts CSE, I_{cp}: cathodic protection current, mA/sq. ft. conc. surf., I_0 ptobe: cathodic protection current to R. steel when probe galvanic current is zero, mA/sq. ft. conc. surf., B_c: cathodic Tafel slope, volts/decade.

Break in *E* log *I*-constant potential and current criterion

One of the first publications that described the *E* log*I* criterion was published in 1957 [6]. The half cell potential and the current for cathodic protection could be determined by evaluating the results of incrementally increasing the current. A curve and a section of straight line relationship was found when the data were plotted on semi-logarithmic paper. However, there is no standard method which describes the technique for running the test and then interpreting the test results. Nevertheless, tests were run and the derived potential and current requirements were obtained and utilized in this research work. In another section of this report, the *E* log *I* test method and other details will be presented. The *E* log*I* criterion for cathodic protection is based upon theoretical electrochemical considerations.

−0.85 volt criterion

Although a Recommended Practice [4] specifies that a minimum of −0.85 volt CSE criterion accomplished with the current on. In this test program the −0.85 volt criterion CSE was determined as an 'instant off' value. Using the instant off method, voltages due to IR drop from current flow were eliminated. In addition, there is no approved or described method wherein the error from IR drop could, or has been, used for measuring the polarized potential of cathodically protected steel in concrete.

The −0.85 volt cathodic protection criterion was first suggested as an empirical value based upon experiences with underground pipeline [7]. For many years this criterion has been successfully used as a cathodic protection criterion primarily for well coated pipelines [3].

300 mV polarization shift criterion

This cathodic protection criterion is based upon the polarization of a submerged or buried pipeline of 300 mV above its static or original half cell potential. The 300 mV is determined by reading the half cell potential immediately after turning the cathodic protection currents off. The difference in voltage between the static potential and the instant off potential must be at least 300 mV at all locations on the pipeline.

Apparently this criterion was based upon field experience as one of the first reports of its use simply states: '... lowering the potential −0.3 volt' [8].

This cathodic protection criterion was reported to generally be used for bare or uncoated underground pipelines [3].

100 mV polarization shift criterion

In general, this cathodic protection criterion requires that a buried pipeline receives an unspecified amount of cathodic protection. Then the cathodic protection current is turned off and the instant off half cell potential is obtained. Then the pipe is allowed to depolarize to a steady state potential. This time period is not defined in the literature. The instant off potential must be at least 100 mV more negative than the steady state potential of the pipe after the cathodic protection current is turned off.

The testing in this program conflicted to a degree with the Recommended Practice in that the cathodic protection instant off was set at 100 mV greater than the original or static potential.

The origin of the 100 mV criterion is not clear but one of the earliest reports states in the conclusions' 'the change in potential or polarization necessary for protection of pipelines is probably always less than 0.1 volt' [9]. It has been reported that this criterion is primarily used to cathodically protect uncoated underground pipe [3].

RECTIFIER OUTPUT

The rectifier was connected to the slot anodes and the reinforcing steel. Then, an oscilloscope was connected between the output terminals of the rectifier. At increasing amounts of voltage, the oscilloscope was used to measure the back e.m.f. The results are shown on Fig. 1, and represents the situation where all five strip anodes on the slab were electrically connected together in parallel.

As will be noted in Fig. 1, the back e.m.f. could represent a substantial percentage of the output driving voltage of a rectifier.

DISTRIBUTION OF CURRENT

The 5-slot type of anodes were approximately equally spaced across the 7-foot (2.13-m) slab on one-foot (305-mm) center to center spacing across the width of the slab. A constant current rectifier that produced a pure direct current was used for the current distribution tests. The 'off', or polarized potentials of the reinforcing steel, were made by turning the current off by hand, and then reading the result on a 4½-digit voltmeter. The 'off' potential was measured transversely to the anodes on a 1-inch (25.4-mm) spacing across the slab.

The amount of current chosen to energize the anodes was initially based on an assumption for providing cathodic protection to the entire slab irrespective of the number or spacing of the anodes. As shown in Fig. 2, potential change – one anode, the current to the reinforcing steel just about goes to zero approximately one-half foot (153 mm) distant from the anode. The result was that the current density of 30 mA per lineal foot (98 mA/m) is a relatively high current density of 30 mA/sq. ft (324 mA/sq. m) of concrete surface. However, even at this high current density, there is little significant distribution of reinforcing steel polarization beyond a distance of one-half foot (153 mm).

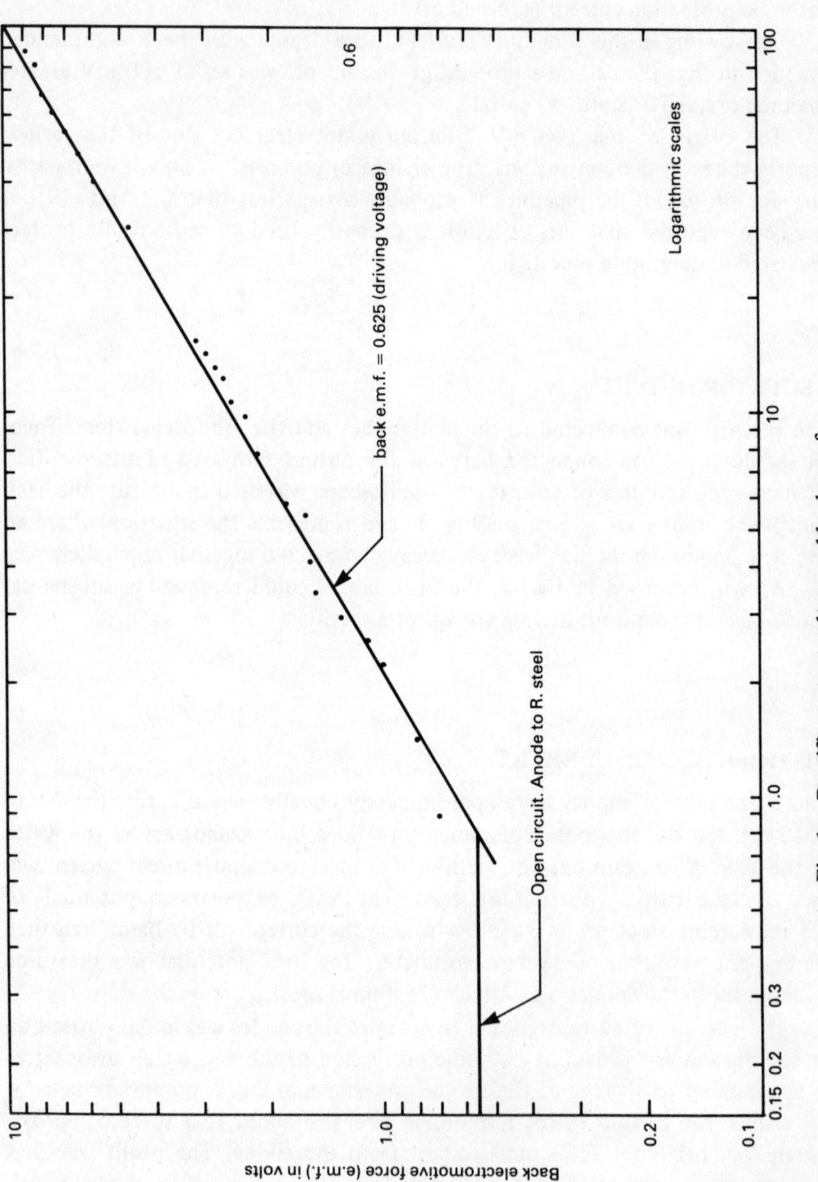

Fig. 1 – Rectifier output voltage and back e.m.f.

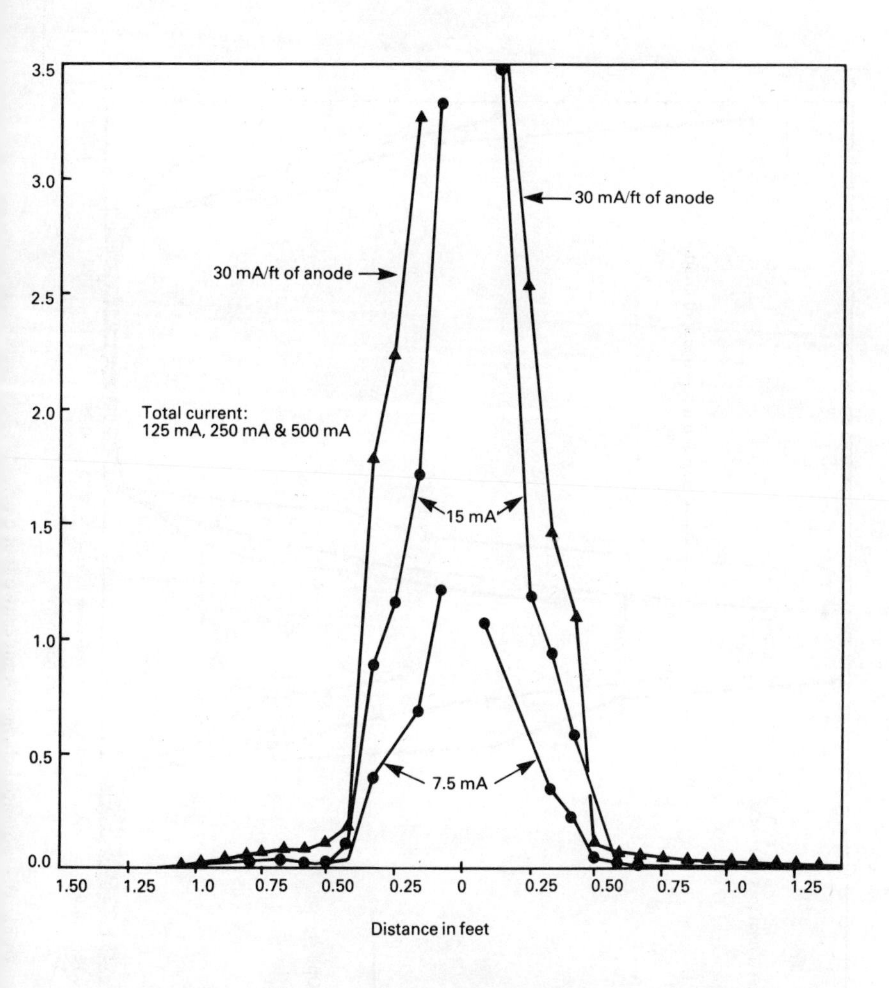

Fig. 2 – Potential charge. One anode.

With the anodes on a two-foot (610-mm) spacing and current densities to the anode of 90 mA/lineal foot (972 mA/m) the distribution of the polarizing cathodic protection current is limited to approximately six inches (153 mm). The limited current throw for three anodes spaced two foot (610 mm) apart is shown in Fig. 3, potential change – anodes two feet c/c.

Fig. 4, potential change – anode one foot c/c, shows that when the anodes were on a one-foot (305-mm) spacing, the current distribution effectively covered the entire slab even at low current densities. It appears that at spacing of the slot anode of more than one foot (305 mm) cathodic protection currents may not fully protect the total top mat of reinforcing steel.

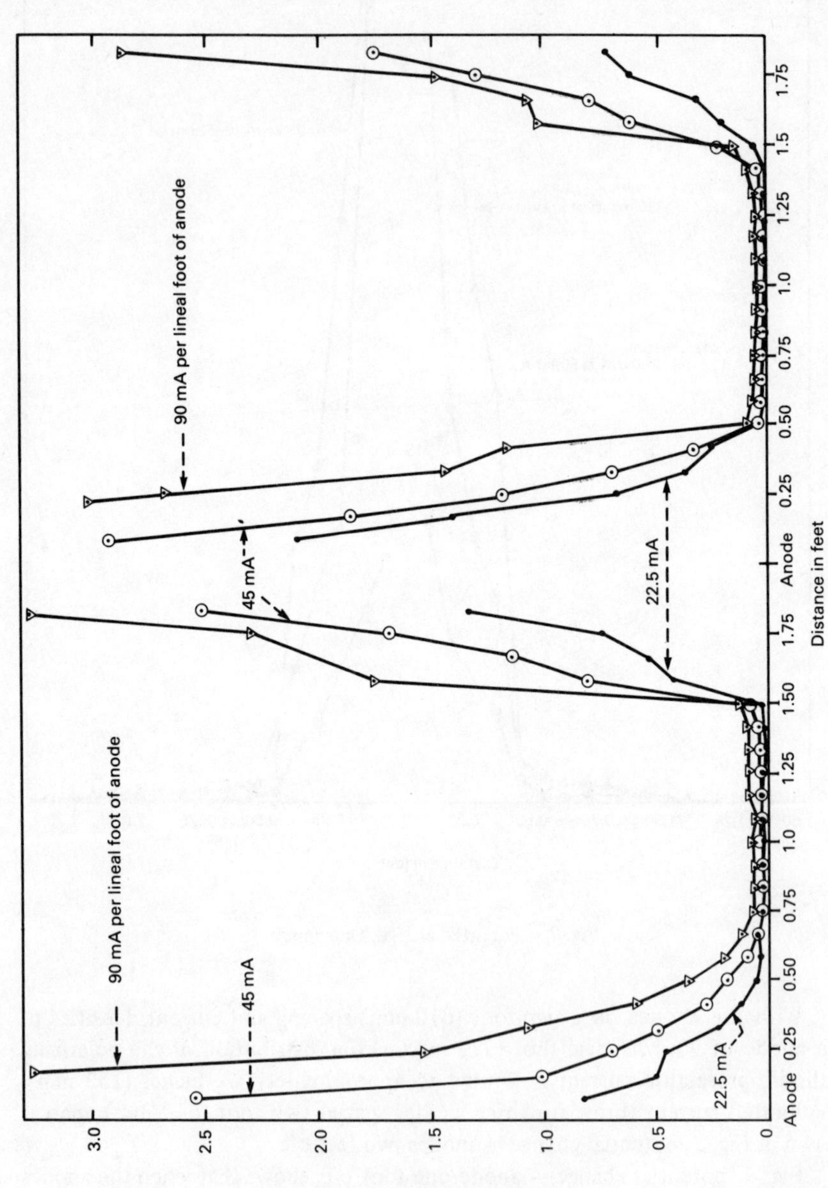

Fig. 3 — potential change. Anodes two feet c/c.

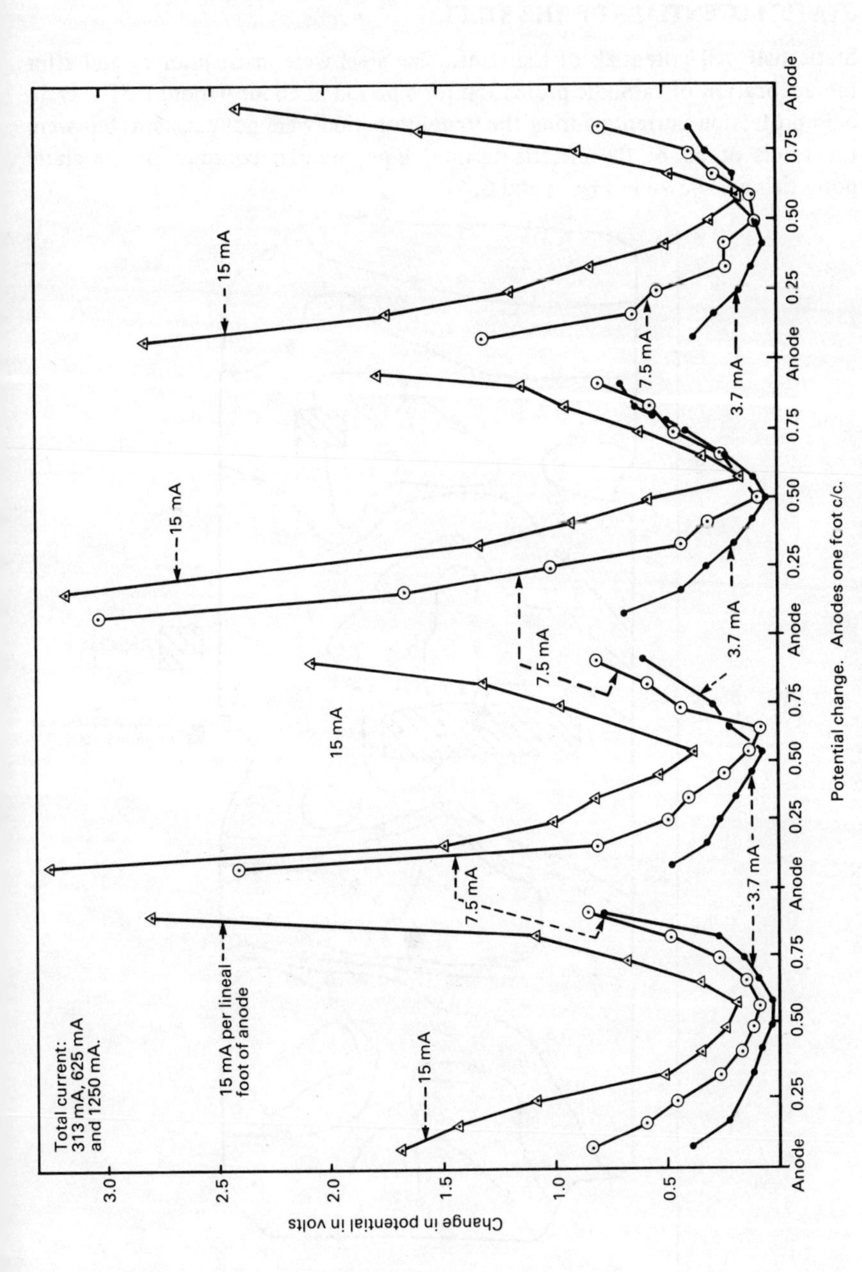

Potential change. Anodes one fcot c/c.

Fig. 4 — Potential change. Anodes one foot c/c.

STATIC POTENTIALS OF THE STEEL

Static half cell potentials of the reinforcing steel were made prior to and after the application of cathodic protection for a period of about 9 months. The cathodic protection currents during the 9-month period were not constant, but were the result of all of the criteria testing. Equipotential contours of the static potentials are shown in Figs. 5 and 6.

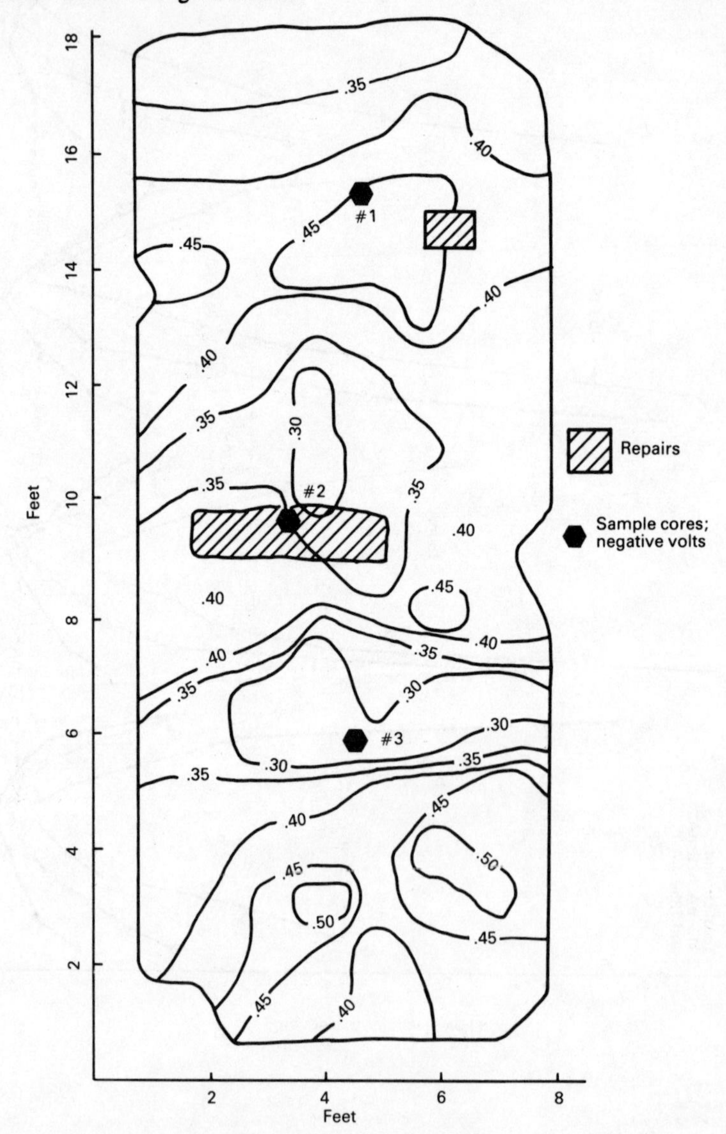

Fig. 5 – Static potentials prior to cathodic protection volts, CSE.

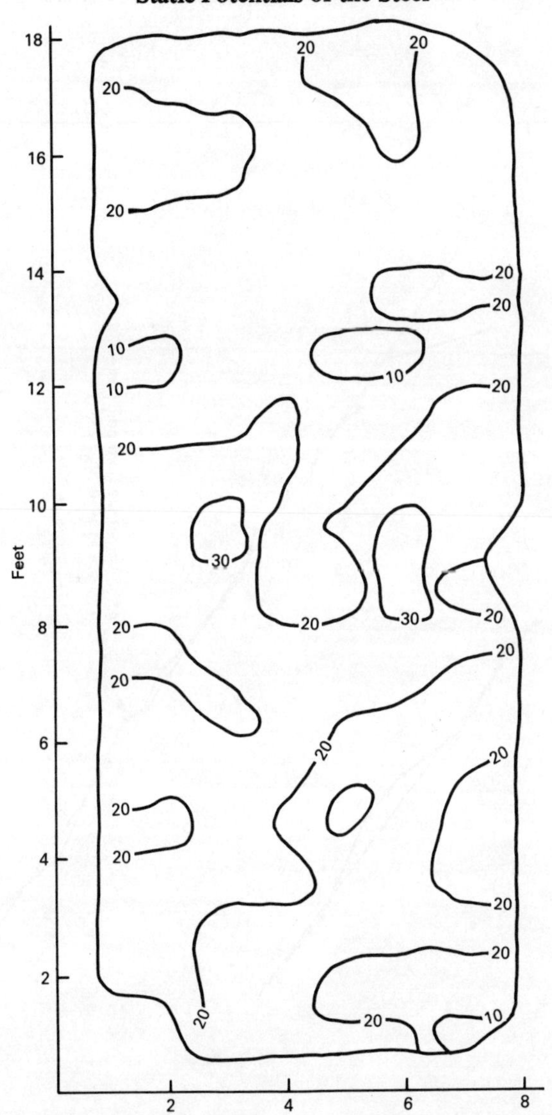

Fig. 6 – Static potential after cathodic protection. Volts, CSE. Slab # G-3. Sept. 14, 1982.

Fig. 7, distribution of static potential – before and after cathodic protection, is a cumulative distribution plot of these potential data. It will be noted on the 'prior' plot that the mean half cell potential is −0.38 volts CSE, with a standard deviation of 0.06 volts. The 'after' plot of the half cell potentials shows that after about 9 months of cathodic protection, that the mean half cell potential of the reinforcing steel was −0.20 volts CSE, and the standard deviation of these data was also 0.06 volts.

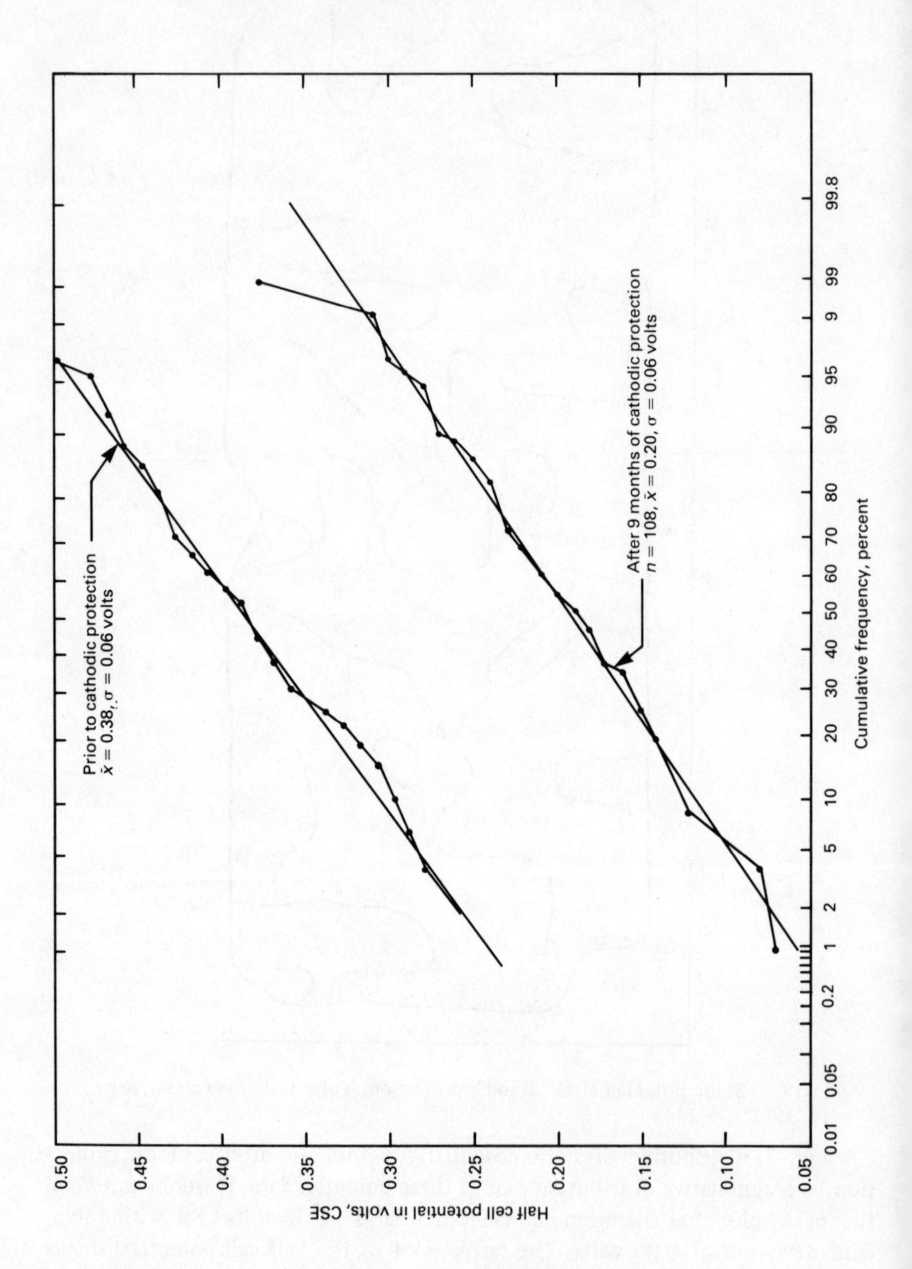
Prior to cathodic protection
x̄ = 0.38, σ = 0.06 volts
After 9 months of cathodic protection
n = 108, x̄ = 0.20, σ = 0.06 volts
Half cell potential in volts, CSE
0.50
0.45
0.40
0.35
0.30
0.25
0.20
0.15
0.10
0.05
Cumulative frequency, percent
0.01
0.05
0.2
1
2
5
10
20
30
40
50
60
70
80
90
95
99
99.8

Although it is known that cathodic protection currents can remove chloride from concrete with time, it is not known if this phenomenon is responsible for the percentage of corrosive potentials (values more negative than −0.35 CSE) to drop from 70% corrosive to less than 1% corrosive potentials.

POTENTIAL OF STEEL IN CONCRETE AND CRITERIA

Fig. 8, distribution of half cell potentials – bridges and laboratory, shows the range of values that may be found [10, 11]. The mean standard deviations are individually plotted for the six bridges on Fig. 8.

In the Recommended Practice [4], the half cell potential criteria for underground or submerged pipe requires that the entire pipeline be energized so that the structure is polarized to a particular potential. For example, when using the −0.85 criteria, all points on the pipe must be polarized to a value at least as negative as −0.85 volts CSE [4]. As it is reported that practical experience had shown that −0.85 volts CSE was the most negative value found in pipeline work, it was chosen as a practical solution [3]. One of the reasons that the −0.85 volt CSE was chosen is that for cathodic protection to be effective, all of the cathodes must be polarized to a negative value at least as negative as the most anodic potential of the anodes.

As will be observed on Fig. 8, the history of half cell potentials both on bridges and in the laboratory indicate that half cell potential criteria of both −0.85 and −0.77 volts CSE are more negative than the most negative values shown for steel in concrete. Therefore, it appears that these latter potential criteria would be appropriate corrosion control criteria for steel in bridge decks providing they are attainable and have no limitations.

One other consideration must be taken into account when utilizing the −0.85 and the −0.77 volts CSE criteria is that the steel in a bridge deck is not necessarily electrochemically the same as steel underground or underwater. Prior to the ingress of chloride to the surface of the steel in concrete, the steel is passivated to a half cell potential that is less negative than −0.35 volts CSE [10]. Therefore, there is the decision to be made when about 10% of the steel in the bridge has active potentials and the other 90% have passive potentials and how this relates to the current requirements for the steel to attain the minimum half cell potential criterion.

The use of the 100-mV and the 300-mV shift criteria also requires that all original or static potentials be polarized the desired number of millivolts as required by the selected criterion. As shown in Fig. 8 for the bridges and laboratory samples, the standard deviation for the data is plotted. One standard deviation on each side of the mean encompasses 68% of the data. Two standard deviations on each side of the mean encompasses 95% of the data. Therefore, even with two standard deviations on each side of the mean, 2½ percent of the

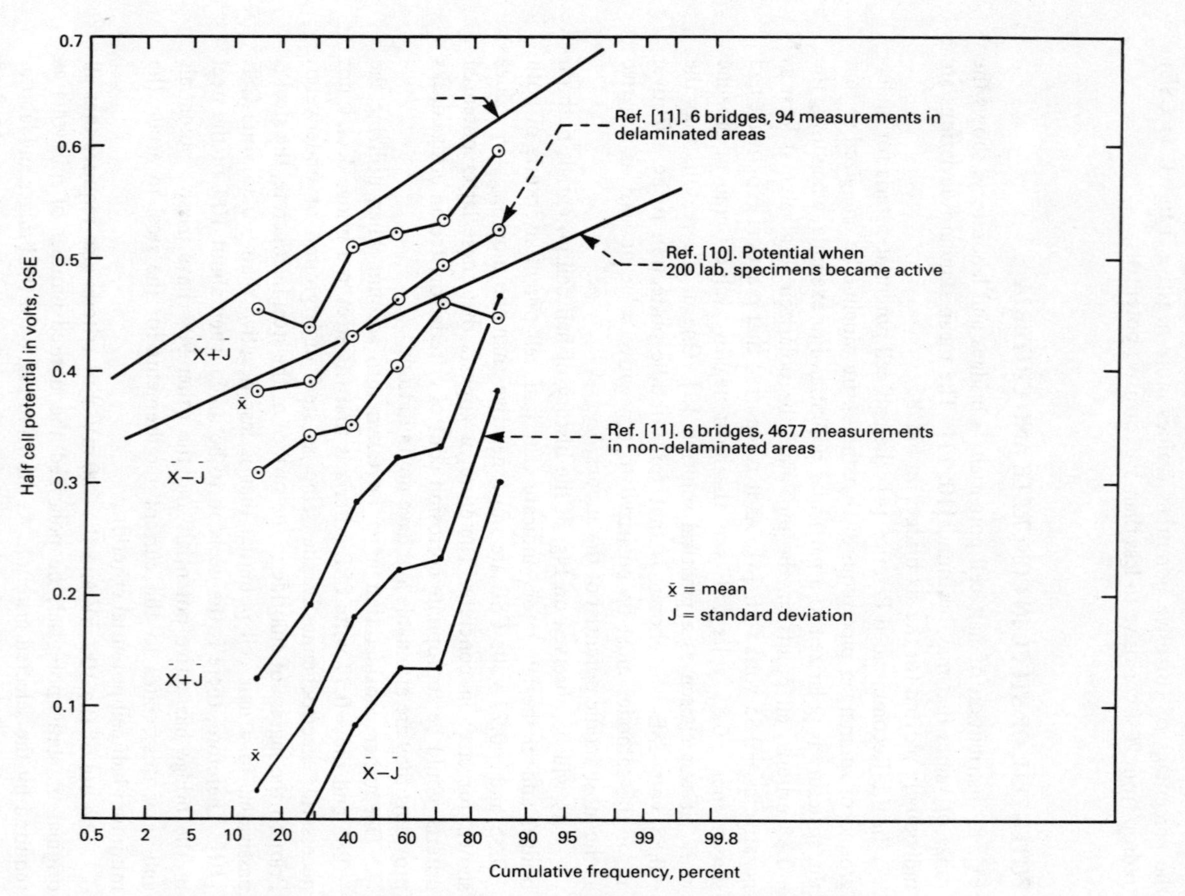

Fig. 8 – Distribution of half cell potentials. Bridges and laboratory samples.

half cell potential readings will be of greater value, and, 2½ percent will be of less value than the two standard deviation limit on each side of the mean value.

In considering the 100-mV and the 300-mV potential shift criteria, it is assumed that these values were used to control the corrosion of uncoated pipelines in localized areas. And, it is further assumed that the values were empirically settled upon as a result of operating economies rather than technical viability.

In Fig. 8, the maximum standard deviation of the half cell potential measurements for the reinforcing steel on individual bridges or for the laboratory samples is approximately 100 mV. This would indicate that 68% of the half cell potentials on a bridge deck and within a 200-mV range, and 95% of the half cell potentials would have a range of 400 mV.

When applying the 'shift' criterion to reinforcing steel in bridge decks, it is apparent that consideration should be given to the relative effectiveness of the application of the current. For example, when comparing the 100-mV shift criterion to the approximately 100-mV standard deviation of bridge deck potentials, then an estimate can be made for the effectiveness of both criteria.

If it is assumed that corrosion is controlled when the cathodes are polarized to the potential of the anodes, and that the maximum potential range found on a bridge deck is four standard deviations while one standard deviation of the bridge deck potentials is 100 mV then the following calculations can be made.

If the 100-mV shift criterion is used on a bridge deck where the standard deviation of the half cell potentials is 100 mV, then the application of current could result in approximately 16 percent of the locations of the measurements, that corrosion will be completely controlled.

With the same foregoing assumptions, when applying the 300 mV shift in potential to a bridge deck, then approximately 84 percent of the locations of the half cell potential measurements will receive sufficient current to theoretically completely stop the corrosion of the reinforcing steel.

Another assumption in the foregoing calculations was that all of the half cell potentials of the bridge deck reinforcing steel was active or more negative than −0.35 volts CSE. However, it will be rare that such will be the case. The influence of having passive reinforcing steel encompassed within the 100-mV and the 300-mV shift criteria is not known.

The *E* log *I* potential and current criteria cannot be compared to the potential data shown in Fig. 8 because these criteria are implemented by test on an individual basis.

E log *I* TESTING

Fig. 9, 'classic *E* log *I* curve,' is developed by applying increments of current and recording the associated half cell potential. Because of the high electrical resistance of concrete, a means for correcting for IR drop must be used. One method is to turn the current off and as quickly as possible obtain the polarized half cell potential of the steel. One problem with this system is that the steel will begin to depolarize when the current is turned off. As a result, there will be a continual shifting and changing of current to correct for the depolarization of the steel during the periods of interrupted current.

The method used during this investigation to obtain half cell potential measurements without disrupting the polarization of the steel to any significant degree was to make the measurement during the period when the full wave unfiltered bridge was in its off cycle.

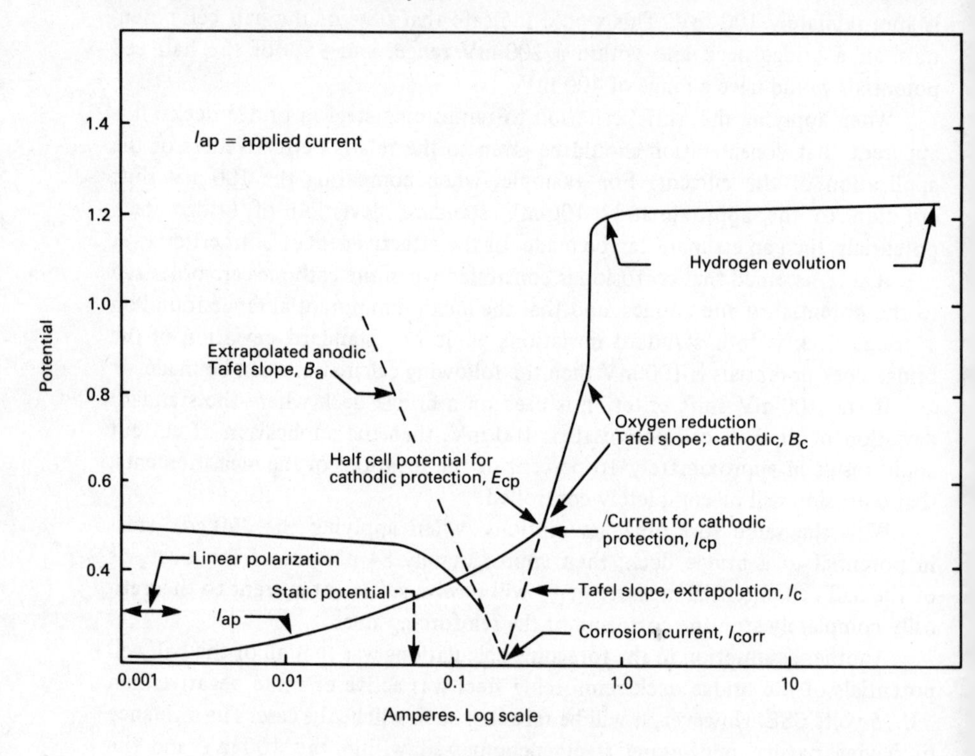

Fig. 9 – Classic $E \log I$ curve.

The schematic for the equipment used to do the $E \log I$ testing is shown in Fig. 10. For the IR free half cell potential measurement, an oscilloscope is used as a zero reading galvanometer. The output signal from a full wave bridge rectifier is a pulsating direct current in the shape of one half of a sine wave. Normally, the voltage would go to zero 120 times a second. When there is a d.c. potential present, such as the half cell potential of reinforcing steel, the pulsating d.c. voltage will not go to zero but will drop and stay at the level of the half cell

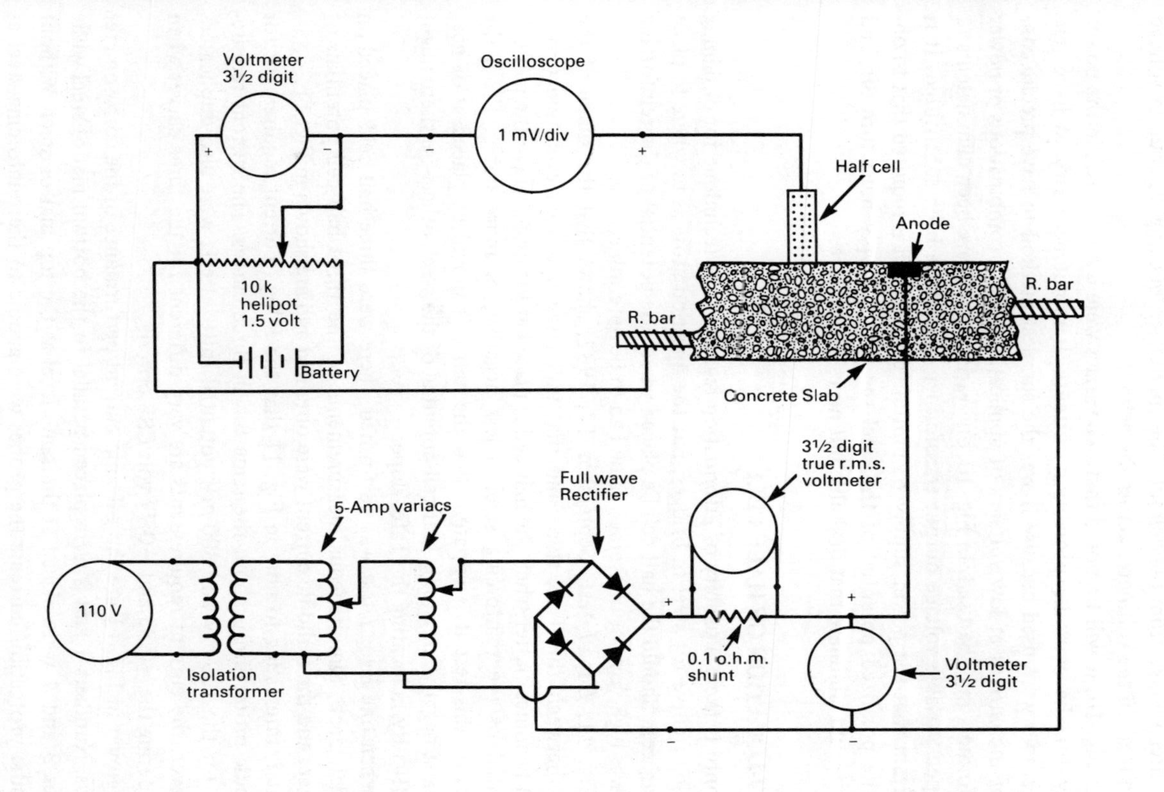

Fig. 10 — Schematic for *E* log *I* testing.

potential of the steel. In this case, the half cell potential of the steel is placed on zero on the most sensitive scale of the oscilloscope by applying a 'bucking' or counter voltage from the helipot. The amount of 'bucking' or counter voltage is then read on the voltmeter across the helipot.

In Fig. 10, it will be noted that two variacs were used in series in the power supply for the cathodic protection currents. It is not necessary to have two variacs, two were used because it was the simplest method to have precise control of the current at low voltages of application. Many combinations of power transformers may be used. In Fig. 10, one variac could have been eliminated and a tapped variable voltage output transformer could have been substituted. It is important that the power supply be unfiltered because it is required that to obtain the polarized potential of the steel the power supply must turn off, and filters would eliminate that desirable but necessary feature.

E log *I*-LOCATION OF HALF CELL

Not only is there a paucity of information regarding the technique for obtaining an $E \log I$ curve for steel in bridges, but the question arose as to where to place the half cell. Should the half cell be placed where the potential of the reinforcing steel was high, low, or of average value had to be determined.

Fig. 11. $E \log I$ test, September 17, 1982, shows that the shape of the curve between the Tafel slope and that indicating hydrogen evolution (−1.2 volts) becomes flatter when the half cell is placed at locations where the half cell potential of the reinforcing steel is less negative, or more noble. From the results of this test it is apparent that the half cell should be placed at or near the most negative or anodic half cell potential of the steel, otherwise there might be difficulty in locating the Tafel slope.

When one $E \log I$ test was performed, there were three half cells placed on the slab deck. Simultaneous measurements of the three half cell potentials of the steel and the cathodic current were obtained and are shown in Fig. 11.

It is interesting to note in Fig. 11 that the $E \log I$ current requirement for cathodic protection is about the same for all three curves. The current requirement for the 100-mV and 300-mV potential shift criteria were also duplicated. However, the current requirements are very different for the three curves when considering the −0.85 and −0.77 volt CSE criteria.

Shown in Fig. 11 are the galvanic current performance of the corrosometer probes. Numbers 1 and 2 were placed parallel to the bottom mat of steel while probes 3 and 4 were placed at the same level as the top mat of steel. Without cathodic protection currents the probes were anodic to the reinforcing steel as evidenced by a measured discharge of direct current. As the cathodic protection currents to the reinforcing steel were gradually increased, the corrosion current discharge of the probes would decrease accordingly. Also shown in Fig. 11 is the amount of average cathodic protection current applied to the reinforcing

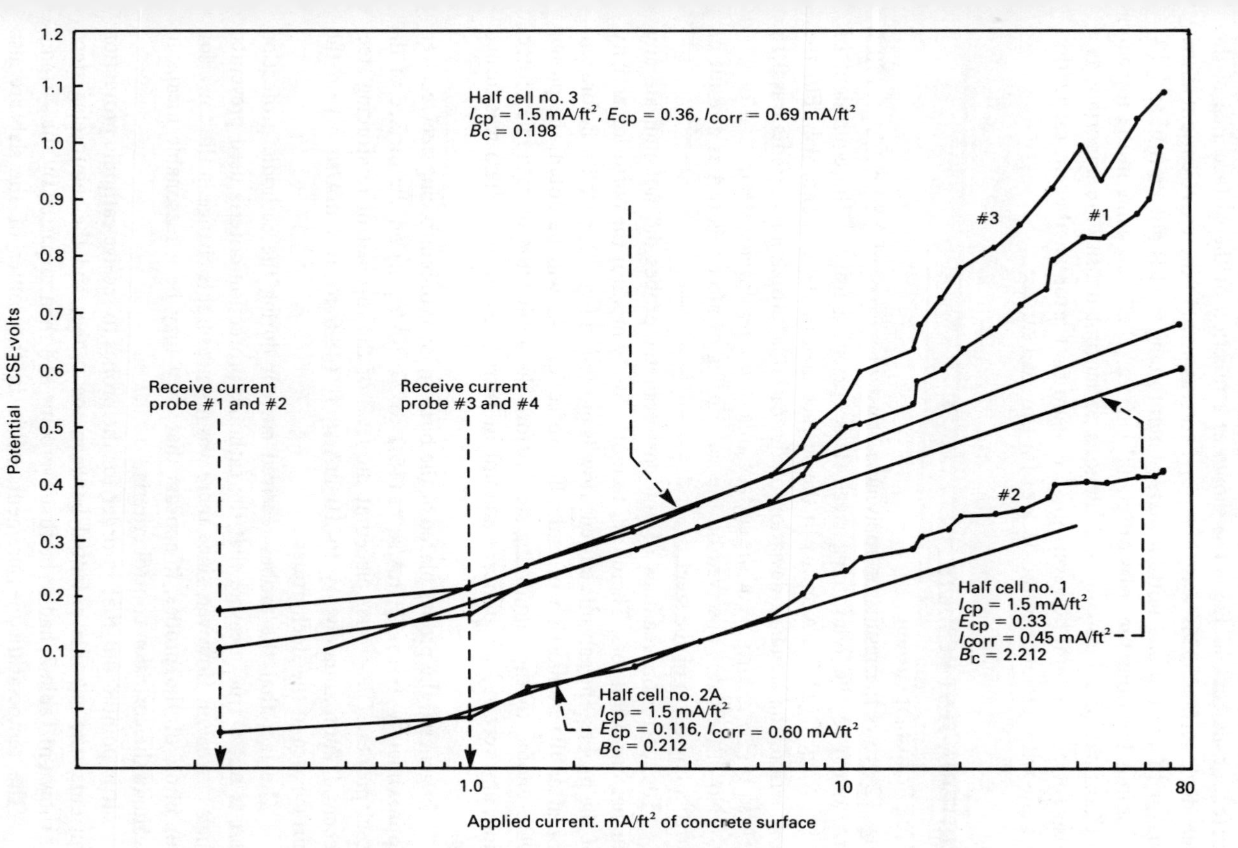

Fig. 11 – *E* log *I* test. September 17, 1982.

when the probe galvanic corrosion currents went to zero. The applied cathodic protection current in Fig. 11 is shown as a function of the surface area of the slab deck surface. This was done for two reasons: (1) the exact current distribution to the top and bottom mats of reinforcing steel is not known, and, (2) at various locations the truss or 'crank' bars vary between being in the top and bottom mat of reinforcing. Calculations of current density are important as a basis for comparison; however, there is significant error whether the calculation is made with reference to the deck or for the steel surface.

CRITERIA TEST RESULTS

E log *I* constant current

Fig. 12 shows the results of applying a constant current of 2.84 mA/sq ft (30.7 mA/sq m) to the reinforcing steel. This current density is the equivalent of 2.1 mA/sq ft (22.7 mA/sq m) to the surface area of the concrete slab. For the test period, the standard deviation of the cathodic protection currents was 0.16 mA/sq ft (1.7 mA/sq m) of the surface area of the reinforcing steel.

No distress was observed on the slab during or after this test as a result of further corrosion of the steel.

The corrosion readings of the corrosometer probes did not indicate any further corrosion loss. Therefore, to more closely monitor the corrosion activity of the probes, the galvanic current flow is plotted in Fig. 12, both for before and for after the application of cathodic protection. As will be noted, the probes were anodic to the reinforcing steel before the application of cathodic protection, and received a significant amount of current when the system was turned on.

In general, the probes placed at the bottom layer of reinforcing steel received approximately 1½ to 2½-mA/sq ft (16.2 to 27 mA/sq m) on the surface of the steel probes. The probes placed at the level of the top mat of reinforcing steel received approximately 4½ to 10 mA/sq ft (48.6 to 108 mA/sq m) on the surface area of the steel probes.

The fact that the probes received current during the cathodic protection test is not a true measure that the cathodic protection criteria used prevented their corrosion. However, since there was no measurable change in the corrosion condition of the probes, it appears that they may be a reasonable means to evaluate the corrosion control criteria.

It is pointed out that in order for the probes to receive cathodic protection currents, the galvanic potential between the probes and the reinforcing steel, as shown in Table 2, had to be first overcome and then surpassed by the system.

The temperature measurements made in the center of the slab are also plotted in Fig. 12. It appears that there is a trend in the data which could imply that, when the temperature is low, the cathodic protection currents received by the probes is higher, and just the reverse when the temperature increases. However, it may be more significant that, when the temperatures were low, 30°F (−1.1°C), the top of the slab was moist with snow or rain as this testing began in the month of February. As a result, rain water ponding on the surface of the concrete probably would enhance the current distribution from the anodes. Therefore the probes, which were centrally placed between the anodes, would receive a greater amount of current when the top of the slab was wet or ponded, such would be the case in the winter.

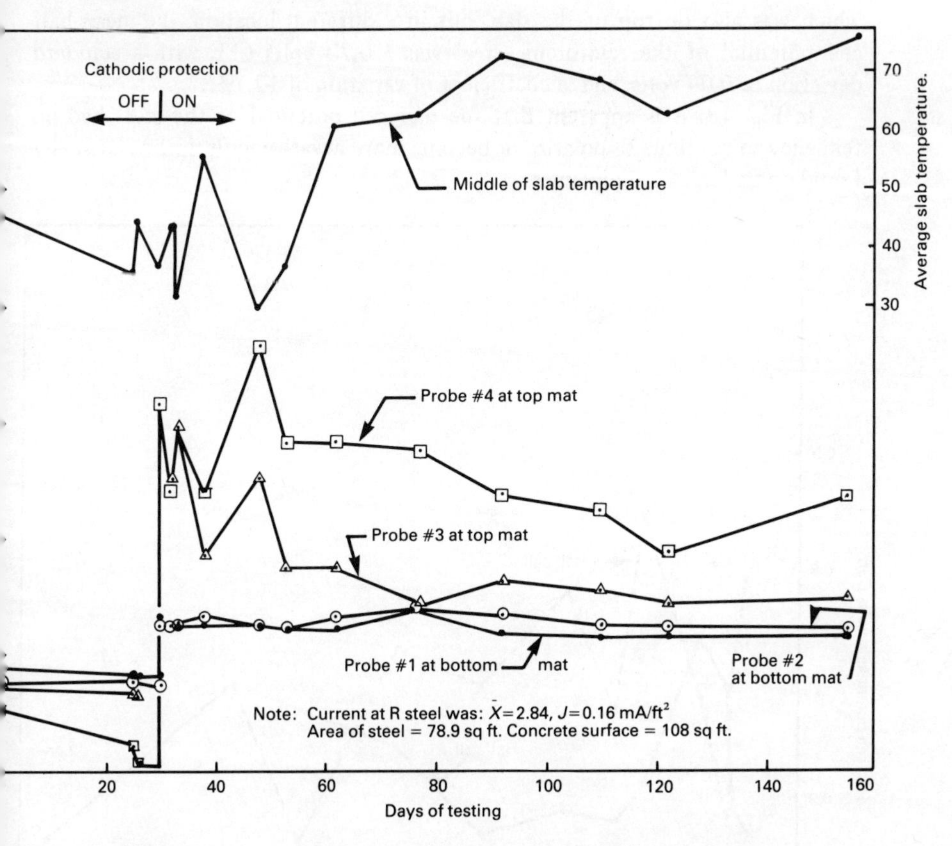

Fig. 12 – $E \log I$ constant current test probe and temperature variations.

Variations on half cell potential

The variation in the half cell potential at two locations during the $E \log I$ constant current test is shown in Fig. 13. In this figure it can be seen that even though the cathodic protection current is held relatively constant, the half cell potentials of the steel varied to a maximum range of 280 mV. At one period of time during the winter, half cell no. 2 indicated that the reinforcing steel had a potential of −0.85 volt CSE. The static, or original, half cell potential of the reinforcing steel using half cell no. 1 was −0.262 volts CSE while based upon the $E \log I$ test, the half cell potential to control corrosion was −0.42 volts CSE as referred to cell no. 1.

During the constant current test shown in Fig. 13, the average half cell potential of the steel using cell no. 1 was −0.63 volts CSE with a standard deviation of 0.06 volts and a coefficient of variation of 9.7%. For half cell no. 2, which was also on top of the slab, but in a different location, the mean half cell potential of the reinforcing steel was −0.70 volts CSE with a standard deviation of 0.09 volts, and, a coefficient of variation of 12.7%.

In Fig. 13, it is apparent that the half cell potential of the steel had no tendency to continue to polarize or become more negative with time as is usually found on underground pipelines.

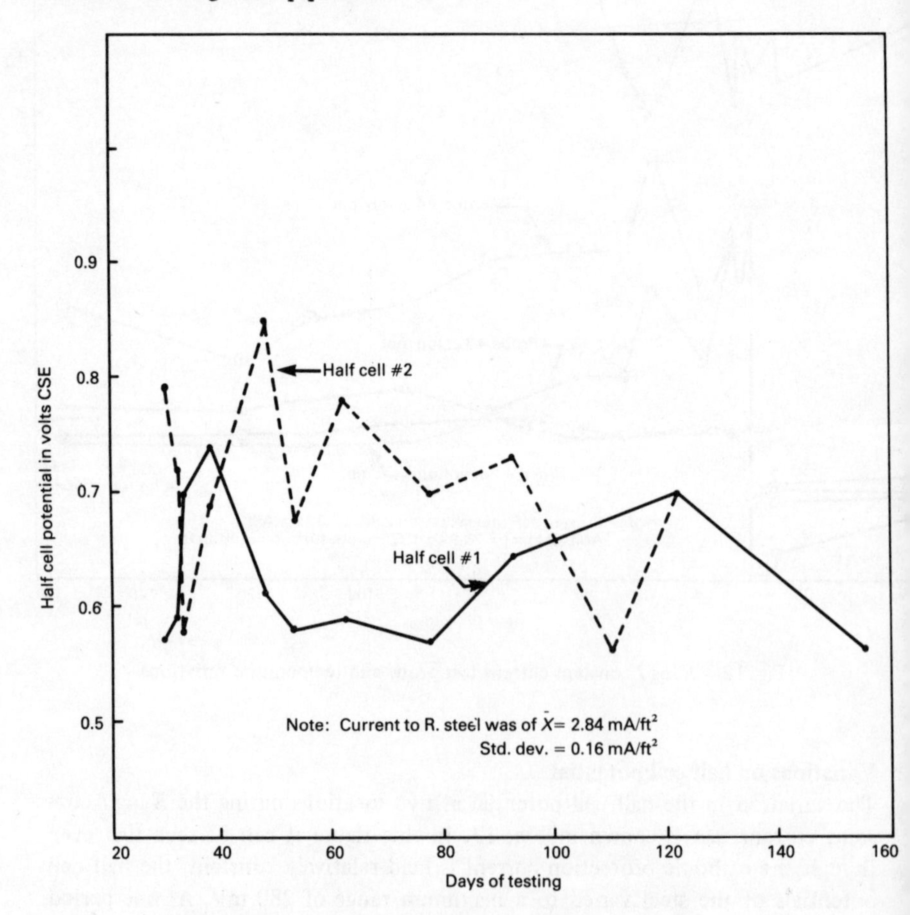

Fig. 13 – $E \log I$ constant current test. Variations in half cell potential.

When a half cell is placed on a bridge deck, it is only influenced by reinforcing steel for a radius six inches (153 mm). This is why the half cell potential of a bridge deck is scanned by making a gridwork of the potentials. Therefore, when a bridge deck is to be placed under cathodic protection with a half cell potential criterion, a half cell placed at one location will only monitor or control an insignificant area of the deck surface.

During the criteria testing of −0.77 volts and −0.85 volts CSE, sixteen locations were designated to measure the half cell potential periodically. These 16 locations were essentially two transverse traverses across the slab where each measurement was centered between the slot anodes. Four additional locations were selected where the half cell was placed for other tests. The results of these measurements are shown in Fig. 14, 'distribution of half cell potentials at 16 locations'. In the case of testing corrosion control criteria, the half cell used to control the output of the potentially controlled rectifier was cell no. 1. Therefore, even though the rectifier was set to control the current output so that the steel beneath cell no. 1 was held at −0.77 volts or −0.85 volts CSE, the steel at other locations were not under this control, and, stabilized in half cell potential according to the amount of current they received. A recently developed potentially controlled rectifier had the ability to measure the half cell potential of the steel with the current off, and regulates the current output to maintain a selected value of potential. Therefore, the potential criteria tested were for off or polarized values.

The data in Fig. 14 does not indicate that the half cell potentials tend to increase with time. Rather, it appears that a very large amount of current would have to be delivered to the reinforcing steel to attain −0.77 and −0.85 volts CSE at all locations. The current requirements for these two latter criteria are approximately 5 to 9 times greater than that used in the $E \log I$ constant current test and the half cell potential values appear similar.

Probe corrosion readings – all tests

In Fig. 15, 'probe corrosion readings – all tests,' the readings for one probe installed at the level of the top or bottom layer of steel is shown. As indicated by the results shown in this figure, before cathodic protection, the probes corroded, and, they did not corrode during any subsequent criteria testing. That the time span between some tests was only about a week may be responsible for the probes having insufficient corrosion to be detectable with the particular probes used.

Behavior of test probes

Although the probes did not display a measurable change in their corrosion loss reading during the criteria testing, they did behave anodically or received cathodic protection currents to various degrees depending upon the cathodic protection test criteria.

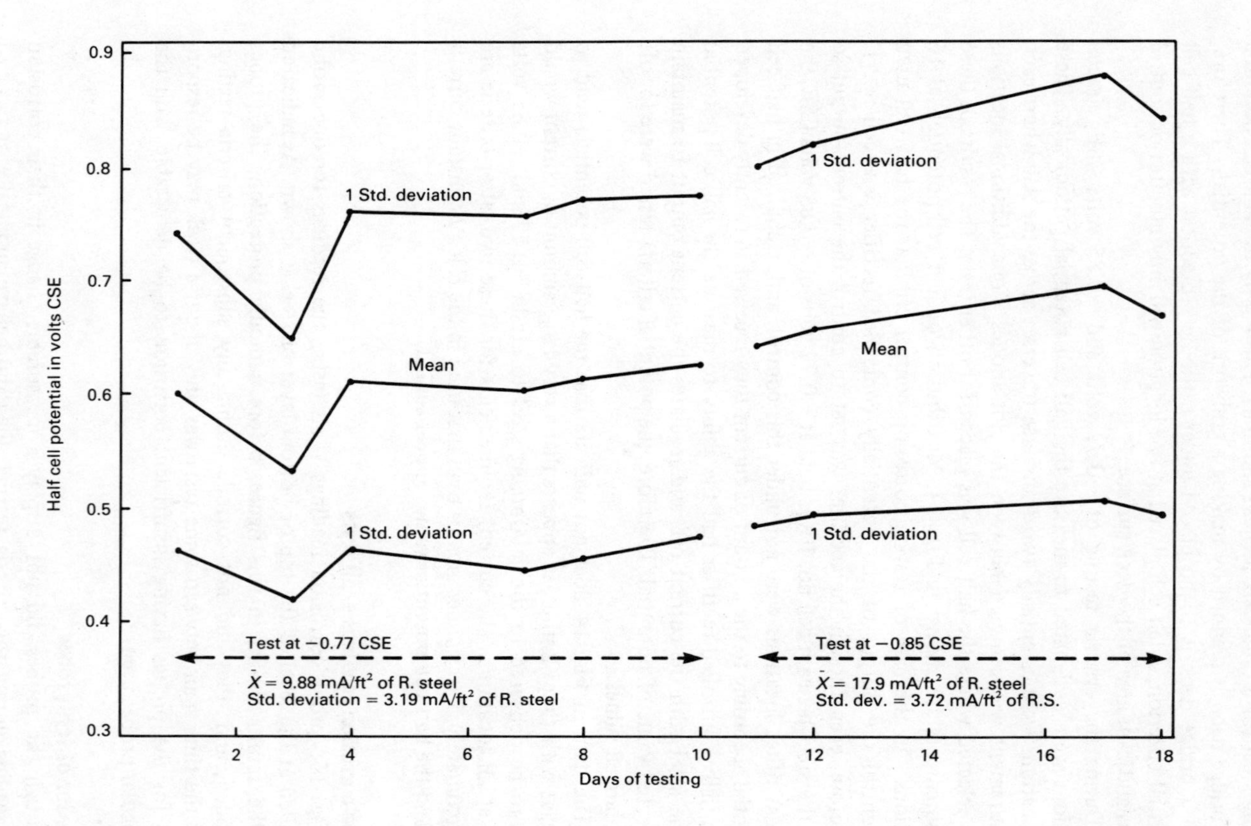

Fig. 14 – Distribution of half cell potentials at 16 locations.

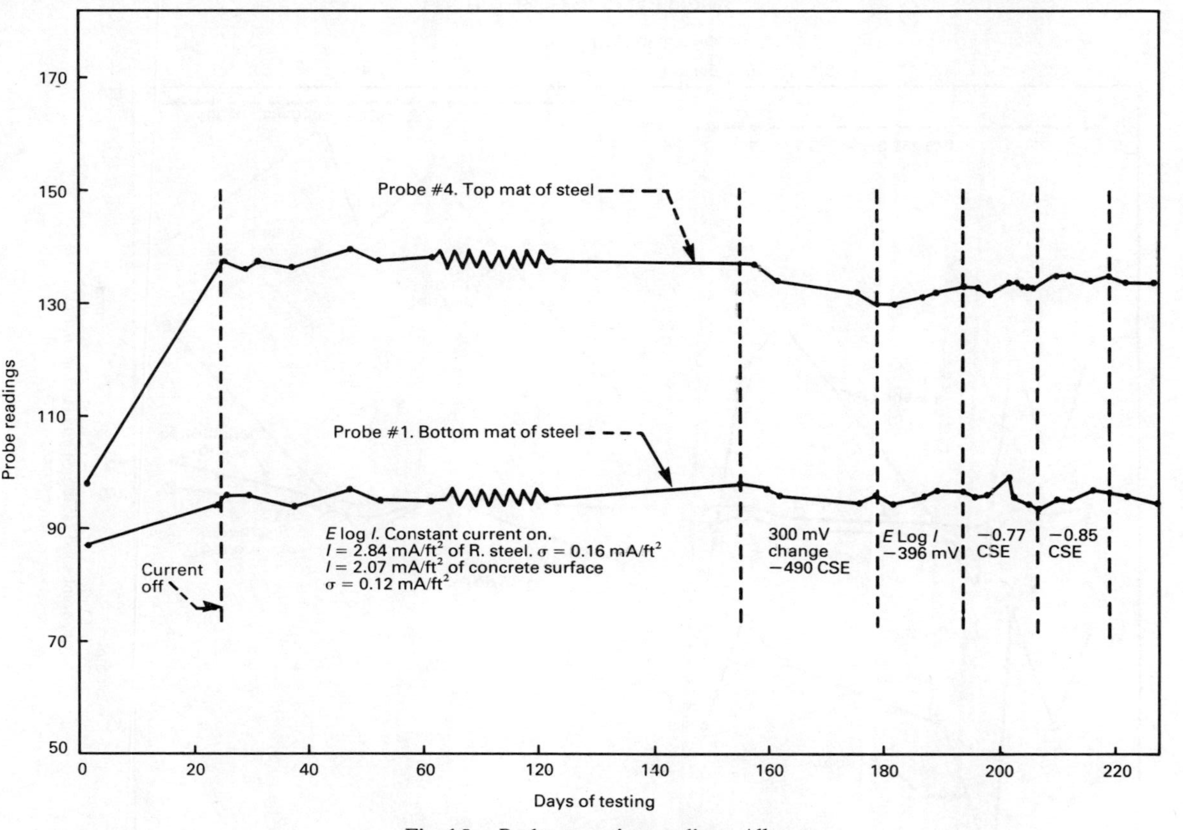

Fig. 15 – Probe corrosion readings. All tests.

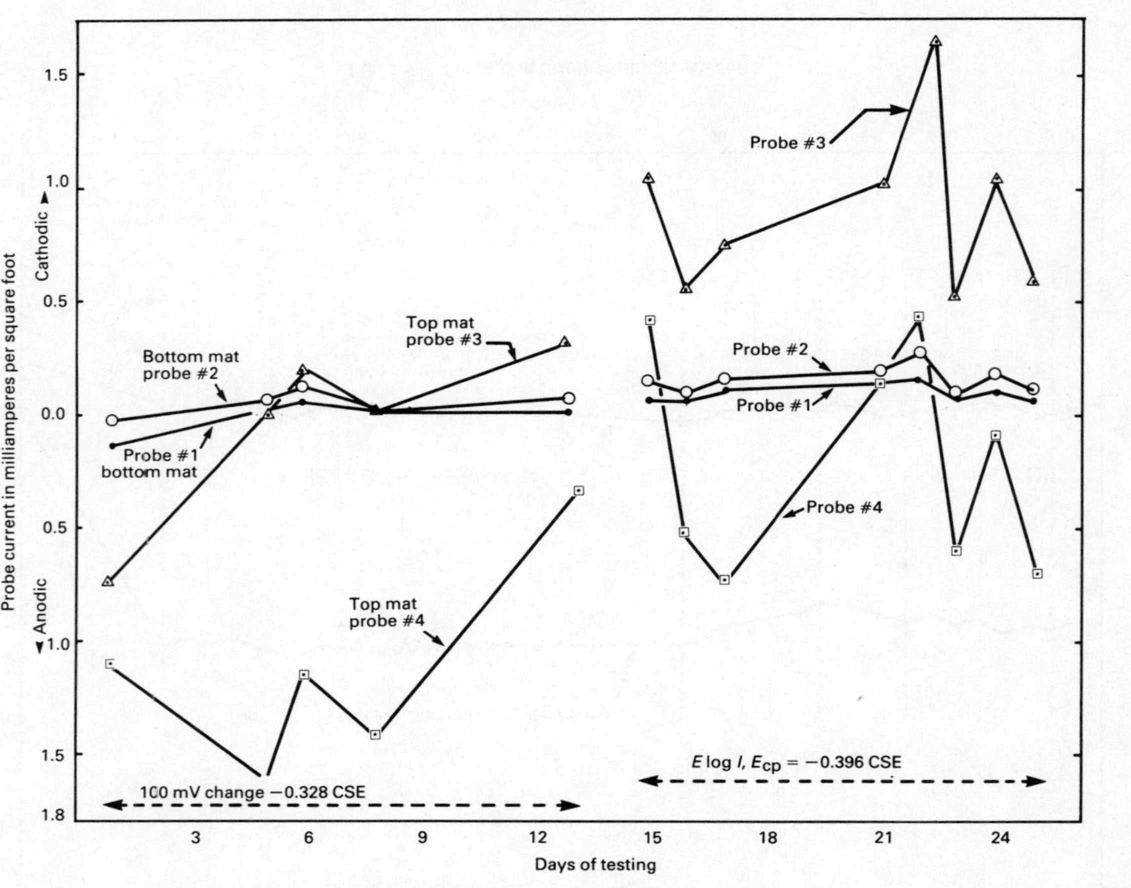

Fig. 16 – Behavior of test probes.

Fig. 16, 'behavior of test probes,' shows that anodic or cathodic behavior of the probes during the criteria testing of 100 mV shift, and the $E \log I$ constant potential test. As will be noted, in Fig. 16 the probes did not always receive cathodic protection currents during the 100 mV shift (a constant potential of −0.328 CSE) or the $E \log I$ constant potential criteria test of −0.396 CSE. Apparently, changes in potential of about 100 mV to 150 mV are insufficient to change corrosion test probes set in highly salt contaminated and corrosive concrete. However, the fact that some probes received a significant amount of current indicates that corrosion loss may be significantly reduced.

Although not shown in Fig. 16, the galvanic current characteristics of the test probes were measured during the 300 mV shift cathodic protection criteria. During the application of the cathodic protection test current, all probes reversed their anodic galvanic current flow and received cathodic protection current. For the probes at the level of the bottom mat of reinforcing steel, the average cathodic protection current received was 0.88 mA/sq ft (9.5 mA/sq m) of probe surface with a standard deviation of 0.36 mA/sq ft (3.9 mA/sq m) of probe surface, the range in current received was from 0.4 mA/sq ft (4.3 mA/sq m) to 1.76 mA/sq ft (19 mA/sq m) of probe surface. For the probes at the top mat of reinforcing steel, the average current received was 2.86 mA/sq ft (31 mA/sq m) with a standard deviation of 1.93 mA/sq ft (20.8 mA/sq m). The range in the cathodic protection current density received by the probes on the top was from 1.18 mA/sq ft (12.7 mA/sq m) to 7.55 mA/sq ft (81.5 mA/sq m) on the surface area of probe.

Variations in current density and temperature

The tests with potential shifts, were based upon the half cell potential of the steel at a selected half cell potential location. The selected half cell potential used to comply with potential shift criteria was normally selected just prior to the test. Since the potentials of the steel in the salvaged bridge deck slab appeared to vary to a degree with the weather and the season, the original or static potential would also change. Therefore, the original static potential may not necessarily be deduced by subtracting the potential shift criteria from the half cell potential used in the test.

As shown in Fig. 17, the current density applied to the reinforcing steel for the cathodic protection criteria testing varied about 40-fold. Within each criterion for cathodic protection it can be observed that there is a considerable change in current density requirement. Contrary to the usual experience with underground pipelines, the current requirement for bridge decks to maintain a constant potential does not generally decrease with time because of the anticipation of polarization.

In Fig. 17, the temperature is plotted for the middle of the slab from the average of three embedded thermocouples. There is no apparent relationship observed between temperature and the current requirements to maintain the

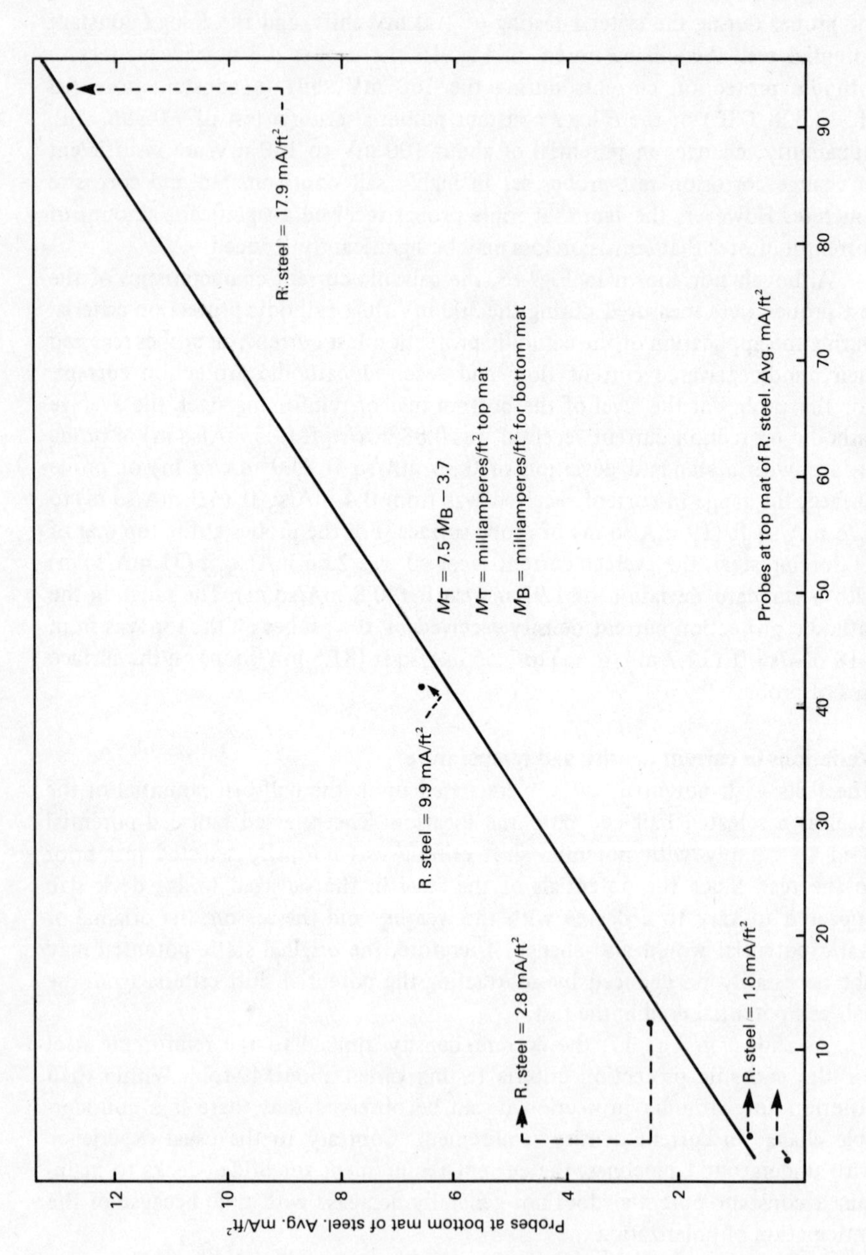

Fig. 17 – Variations in current density and temperature.

cathodic protection criteria. However, it appears that there was insufficient time and changes in temperature during this period of testing that would result in a relationship between the variables.

CURRENT RECEIVED BY PROBES

The corrosion test probes have provided a means whereby an estimate could be made of the relative amount of current received by the top and bottom mats of reinforcing steel. To accomplish this, a regression analysis was made of the amount of current received by the top and bottom probes, and the results are shown in Fig. 18, cathodic protection current received by test probes.

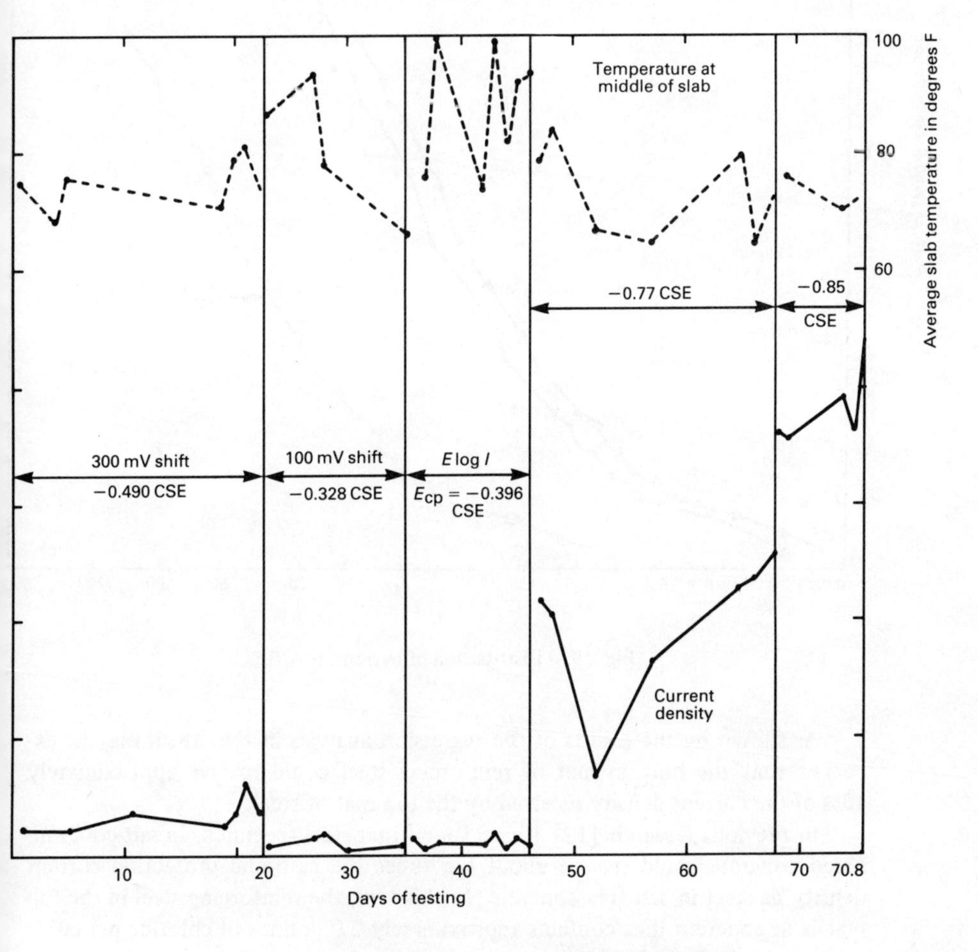

Fig. 18 – Cathodic protection current received test probes.

Also shown in Fig. 18 are some notations of the amount of current received by the reinforcing steel in the slab. Fig. 19, distribution of current, is a complete plot of the current received by the probes and reinforcing steel during one $E \log I$ test.

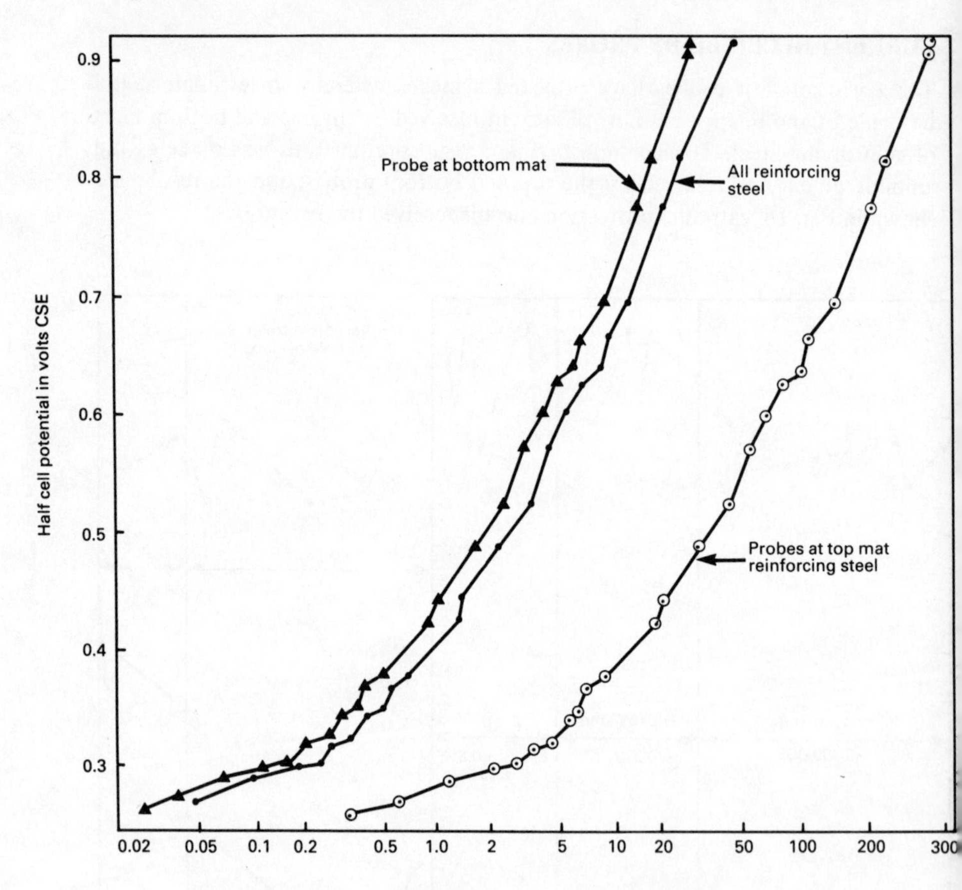

Fig. 19 – Distribution of current. mA/ft².

As shown by the results of the regression analysis in Fig. 18, it may be expected that the bottom mat of reinforcing steel could receive approximately 13% of the current density received by the top mat of steel.

In previous research [12] it was found that steel specimens in salt-contaminated concrete could receive about ten times the cathodic protection current density as steel in salt-free concrete. In this test, the reinforcing steel in the top mat is in concrete that contains approximately 2.0 pounds of chloride per cubic yard (1.19 kg/cu m) while the bottom mat of steel (which comprises about

75% of the surface area of the reinforcing) is in essentially salt-free concrete. The probes were cast into concrete that contains 15 pounds of chloride per cubic yard (8.89 kg/cu m).

The cathodic protection currents preferentially accumulate on the steel in those locations in the concrete with the highest salt content. This may not be a limitation, but a benefit in applying cathodic protection to steel in bridge deck concrete. This is because it has been shown that these currents will slowly remove the salt that has been absorbed by the concrete. Those locations with the greatest salt content (and which are the most likely locations of a high corrosion rate) will receive the highest current density, and, this latter factor would be associated with the most rapid removal rate of salt.

DISCUSSION

With certain exceptions, it is obvious that any moderate amount of cathodic protection applied to a bridge deck will be beneficial. The only question is whether or not it is desirable or economical to completely stop the corrosion of the embedded steel.

For bridge decks, the cathodic protection criteria developed for underground or submerged pipelines would be helpful when using the 100 mV and 300 mV potential shift, but, at the −0.77 volt and the −0.85 minimum polarized potential, it appears that the steel could be severely overprotected with current densities in the range where consideration must be given to the possible slow loss of bond strength of the reinforcing steel.

As noted in this study, polarization of the steel did not tend to increase with time. Instead, the half cell potentials apparently randomly increased and decreased with time. In previous laboratory research [12] it was observed that when steel in salt-contaminated concrete was cathodically polarized under water and then exposed to the air, the polarization would apparently be rapidly lost. However, without additional cathodic charging, when the same specimen was again placed underwater, a significant amount of polarization was recovered.

From this latter experiment and these data it is apparent that the polarization potential of steel in atmospherically exposed concrete may vary according to the oxygen content of the concrete. The pore sizes are highly variable, so much so that some can pass certain gases, while others are of a size that will only pass water vapor, and others capillary water. As a result, the access of oxygen to the steel can highly depend upon the moisture content of the concrete. It is oxygen that enters into the cathodic process of the corrosion cell, and, it is also oxygen that is reduced as part of the cathodic protection reaction.

It is assumed that during the cathodic protection process that when oxygen is consumed at a sufficiently high rate, then concentration polarization occurs, and, the half cell potential of the steel becomes more negative. However, if at the same current density of cathodic protection the concrete dries, then oxygen can become readily available. Then the concentration polarization is lost and the half cell potential of the steel becomes more noble because of the presence of oxygen on the steel. As a result, once the current requirement has been met to cause all the surface of the steel to have oxygen reduction, further quantities of cathodic protection currents may be wasted on the process of reducing oxygen. Therefore, it seems that as the oxygen content of the concrete varies, so will its half cell potential of the reinforcing steel.

Because of the seemingly erratic performance of half cell potential measurements both in the field and in the laboratory, it appears that the use of a half cell potential criteria *per se* may not be the most appropriate means for controlling the corrosion of steel in bridge deck concrete.

Based upon a near maximum standard deviation of 100 mV, a polarization shift of 400 mV should result in corrosion control for 97½% of the bridge deck, as calculated for a four standard deviation data base. However, it is possible that the 400 mV change may result in over protection, not only because of a low oxygen content in the concrete, but because this potential range may not always be found in an actual corrosion cell on a bridge. In all cases, it may be well to limit this 400 mV change so that no steel in the bridge deck will have a half cell potential more negative than −1.10 volts CSE. At half cell potentials somewhat greater than −1.15 volts CSE, electrolysis can occur and the water in the concrete can begin to be broken down into hydrogen. This latter limitation is emphasized by the data shown in Fig. 11 where at the same current density on the steel at one location had a half cell potential of −1.2 and at another location it was about −0.5 volts CSE. The use of a 400 mV shift criterion could result in current densities on the reinforcing steel that might be 10 to 15 times greater than that required by the $E \log I$ constant current criteria.

In Fig. 11, it is shown that at the same average current density to all of the reinforcing steel in the bridge deck slab, the steel with the most negative (active) half cell potential polarizes more readily than the steel that was less negative (passive). Since it is the active (corroding) steel that is to be stopped from further corrosion, it appears that any cathodic protection potential control half cell or testing sites be at those locations where the half cell potential of the reinforcing steel is at or near the location of the most negative (active) potential on the bridge deck.

Also, based upon the performance of the corrosion test probes, it appears that the $E \log I$ constant current criteria is applicable as a cathodic protection criteria for bridge decks. In running these tests, it was apparent that sufficient test points must be obtained in order to develop an accurate curve. Reasonably, the method for obtaining an $E \log I$ curve should be a subject matter within itself. However, sufficient knowledge is available so that a reasonable approximation of the curve can be currently obtained. The appendix contains a method for developing an $E \log I$ curve. The increments of potential change were deliberately made small so that data could be obtained in a reproducible manner.

It was noted during the testing that not only did half cell potentials vary during constant current tests, but current output during constant half cell potential tests also varied. For example, during the constant half cell potential tests shown in Fig. 14, in one case the current from the rectifier had a one standard deviation variation of 32% and in the other test it was 48%. Naturally, if the rectifier were constant current, the same approximate variations in output voltage would be expected. From this it is apparent that cathodic protection rectifiers should have at least double the design power requirements.

Contrary to experience with pipelines where the half cell can be moved to different distances from the pipeline so as to have a more overall 'look' at the half cell potentials, on a bridge deck the 'look' at the half cell potential measurements is limited to a distance of approximately six inches (153 mm). Therefore, this must be considered when planning and maintaining a bridge deck cathodic protection system.

It appears that, for the first few years, an $E \log I$ test should be made on a yearly basis to not only check the performance of the system, but to make any necessary current adjustments. The routine operation of the systems should be checked more often, with some means to detect equipment failure in a short interval of time.

During the testing of the 300 mV corrosion control criterion, it was found that the corrosion test probes at the level of the bottom mat of reinforcing steel received about 30% of the current received by the probes placed at the level of the top mat of reinforcing steel.

During the incremental application of cathodic protection current, the anodic steel in salt-contaminated concrete must first reverse polarity and then begin to receive current. Because of the variations in the initial galvanic voltage between the reinforcing bars as shown on Table 2, the probes will reverse their corrosion current only when that voltage has been nullified. Later, when all the probes are receiving current, then the ratio of current received by the probes at the top and bottom remains relatively constant at 13% for the bottom probes.

SUMMARY AND CONCLUSIONS

1. Based upon the statistical distribution of half cell potentials for steel in concrete, and the criterion that the cathodes must be polarized to a potential at least as negative as the most negative anode, a potential shift of 400 mV should control the corrosion of steel in concrete in 97½% of the bridge deck. It is anticipated that the steel which would not be controlled by cathodic protection would most likely be passive steel that would not be likely to need cathodic protection currents. If the 400 mV shift results in the most negative potential becoming more negative than −1.10 volts CSE, the potential shift shall be limited to that value that caused the −1.10 volt CSE potential at any location. A potential shift cathodic protection

criterion for bridge decks may be implemented by determining the current requirement for the 400 mV shift, and by means of a constant current rectifier, apply that amount of current.

2. That locations at which the half cell potential will be used for current requirement or cathodic protection monitoring tests be at the location where the half cell potential of the reinforcing steel be at or near the most negative half cell potential on the bridge deck.
3. The $E \log I$ constant current criterion appears to be an appropriate means for controlling the corrosion of reinforcing steel in concrete bridge decks.
4. Because of the erratic half cell potential of cathodically protected steel in bridge deck concrete, cathodic protection controlled by a half cell potential criterion does not appear to be a suitable method for corrosion control, especially when such a criterion usually requires that all potentials should be at that value or greater.
5. The corrosion test probes were used to obtain an indication of the relative amount of current received by the top and bottom mats of reinforcing steel. A regression analysis was made of the results of one $E \log I$ test, and, it was found that the probes placed at the level of the bottom mat of reinforcing steel received approximately 13% of the current received by the corrosion test probes at the top mat of steel.

Therefore, it appears that at low current densities, the steel in low salt-content concrete (the cathodes in the corrosion cell) will receive current from the cathodic protection system while the steel in the salty concrete (corroding anodes) will only begin to receive current once their corrosion voltage (potential) to the non-corroding cathodic steel has been neutralized. Thereafter, the steel in the salty concrete can receive a far greater density of current from the cathodic protection system than the previously cathodic steel in the less salt-contaminated concrete.

ACKNOWLEDGEMENT

Special thanks were earned by good work of Dr Errol C. Noel, Co-Principal Investigator, A.M.S. Inc., who administered and directed the moving and preparation of the test slab for the chloride analysis and related field work; to Dr Yash Paul Virmani, Research Chemist, F.H.W.A., who made periodic measurements and implemented the criteria testing on the salvaged bridge deck slab; and K. C. Clear, formerly with the F.H.W.A., but now with Kenneth C. Clear Inc., who initiated the project and provided help periodically. K. C. Clear, Y. P. Virmani, and John J. Bartholomew, formerly with the F.H.W.A., are responsible for the development of the 'slot' anode electrically conductive polymer and the usage of the associated materials.

DISCLAIMER

The contents of this paper do not necessarily reflect the views or policies of Automated Management Systems Inc. (A.M.S. Inc.) of Corrosion Engineering Inc., or of the Federal Highway Administration. This report does not constitute a standard, specification, or regulation. The authors, A.M.S. Inc., Corrosion Engineering, Inc., and the United States Government do not endorse products or manufacturers. Trade or manufacturers' names may appear herein solely because they are considered essential to the object of this report.

REFERENCES

[1] Stratfull, R. F., Cathodic protection of a bridge deck, *Materials Performance,* **13**, (4), 24 (1974).

[2] Stratfull, R. F., Frank Newman Speller Award Lecture, *Materials Performance,* **21**, (6), 38–39 (1982).

[3] Husock, Bernard, Evaluation of cathodic protection criteria, U.S. Department of Commerce, National Technical Information Service, *Report No. AD-A084-783,* April 1979.

[4] Recommended Practice, 'Control of external corrosion on underground or submerged metallic piping systems', Standard RP-01-69, National Association of Corrosion Engineers, Houston, Texas, 1972 revision.

[5] Vrable, J. B., Cathodic protection for reinforced concrete bridge decks, *NCHRP,* 12–13, April (1974).

[6] Stern, M. and Geary, A. L. Electrochemical polarization, *Jour. Electrochemical Society,* **104** (1), 56 (1957).

[7] Kuhn, R. J. Cathodic protection of underground pipelines from soil corrosion, *Proc. Amer. Pet. Inst.* (1933).

[8] Logan, K. H., Comparison of cathodic protection test methods, *Corrosion,* **8** (9), 300–304 (1974).

[9] Ewing, S. P., potential measurements for determining cathodic protection requirements, *Corrosion,* **7** (12), 410–418 (1951).

[10] Stratfull, R. F., Half-cell potentials and the corrosion of steel in concrete, *Highway Research Record No. 433,* 1973, pp. 12–21.

[11] Stratfull, R. F., Jurkovich, W. J. and Spellman, D. L., Corrosions testing of bridge decks, *Transportation Research Record No. 604,* 1976.

[12] Chang, George C., Apostlolos, John H., Myhres, Forrest A., Cathodic protection studies of reinforced concrete, *Calif. Dept. Transp. Report No. FHWA/CA/TL,* March 1981.

APPENDIX – GUIDE FOR OBTAINING $E \log I$ DATA

Introduction

Apparently the $E \log I$ data has this title simply because the polarized potential, is plotted in a linear fashion, and I, the applied current is plotted logarithmically.

These data are the resutl of applying increments of direct current to a structure and then plotting the data and calculating the results as necessary.

The purpose for obtaingin an *E* log *I* curve is to determine the minimum amount of current required to protect the steel in a bridge deck from corrosion.

Equipment reauired

Voltmeters used should have no less than 10 million ohms input impedance, and should be at least a $3\frac{1}{2}$ digit digital voltmeter capable of reading 0.001 volts.

Oscilloscopes used should have a minimum of 1 megohm input impedance and the display shall have a minimum sensitivity of 10 mV per division.

Direct current power supplies shall be unfiltered output, and shall have sufficient power control so that the half cell potential of the steel in the bridge deck can be changed in steps of no greater than −0.002 volts to the saturated copper/copper sulfate half cell.

Direct current power supplies with built in devices for measuring the half cell potential of the steel, shall only measure that potential during the 'off' cycle of the rectifier. These power supplies shall have output power control to the level as previously described for power supplies.

Half cells may be a standard saturated copper/copper sulfate (CSE) electrode or other half cells suitable for use on a bridge deck.

Semi-logarithmic paper with no fewer than three logarithmic cycles may be used. Half cell potentials are to be plotted on the vertical, and, current values are to be plotted on the horizontal axis on the paper. Linear coordinate and note paper should also be available.

A clock or watch which displays elapsed seconds will be required.

Test procedures

1. Attach a wire to the reinforcing steel. This is to designated as ground no. 1.
2. Attach the free end of ground no 1 to the negative terminal of a voltmeter.
3. Place a water-saturated piece of cloth or sponge (approximately three inches square [about 75 mm square]) on the surface of the deck at which the half cell potentials for the *E* log *I* curve will be obtained. This location on the deck will be at or near a location where a previous half cell potential survey detected a maximum or near maximum half cell potential of the reinforcing steel. This location should be within 50 millivolts of the most negative half cell measured on the bridge deck.
4. Place a CSE half cell on the piece of cloth that was placed on the deck during step no. 3.
5. Connect an end of wire to the half cell placed on the deck during step no. 4. This wire is designated as half cell wire no. 1,
6. Connect the free end of half cell wire no. 1 to the positive (+) terminal of the voltmeter.

7. Turn on the voltmeter and read the voltage. The voltmeter will show the reading to be a positive value. However, by scientific convention, these voltages are designated as negative because they are relative to the half cell. Or, traditionally, steel that is buried underground is a 'ground' and, the negative terminal of a voltmeter is normally connected to the ground. In the case of the bridge deck, if the negative terminal of the voltmeter is connected to the steel and the positive terminal is connected to the half cell, the resultant voltage reading between the steel and the half cell will be a negative voltage polarity. So, the half cell potential of the steel is negative to the half cell.
8. If power consumption of the voltmeter is important, turn it off for the present.
9. Connect the free end of some wire that is at least 12 gage to the reinforcing steel. This location should be at least five feet (1.5 m) from ground no. 1 which is attached to the voltmeter. This second ground is designated as ground no. 2.
10. Connect the remaining free end of ground no. 2, which was connected to the reinforcing steel in step no. 9, to the NEGATIVE terminal of the direct current of the direct current power supply.
11. Connect one end of a minimum 12 gage wire to the anode system in the bridge deck. This wire is designated as the anode lead wire.
12. Check the direct current supply and make sure the main power switch is OFF, and all output power controls are set at OFF or minimum output range.
13. Connect the free end of the anode lead wire that was installed during step no. 11, to the POSITIVE terminal of the power supply.

APPLYING CURRENT

From this point on, it is assumed that either an oscilloscope or a commercially available power supply is being used that will measure the polarized half cell potential of the reinforcing steel. Reading the half cell potential when the current is 'on' will result in gross errors due to IR drop, and as such, use of these values will result in erroneous results. Half cell potential measurements obtained by turning the power supply 'on' and 'off' can give results that may be difficult to interpret because the polarized potential of the steel 'drifts' down when the current is 'off', and 'drifts' more negative when the current is 'on'. The use of this latter method should only be considered for use on an existing system in an emergency.

14. Plug the power supply into its power source.
15. Turn on voltmeters, oscilloscope or other powered equipment to be used and let them stabilize according to manufacturer's instructions, if any.

16. Adjust the variable resistor, or helipot, to that point where the digital voltmeter will read the static or original half cell potential of the reinforcing steel. At this point, the trace should be a horizontal line at the zero voltage location on the oscilliscope. Adjust the oscilloscope so that it reads on its most sensitive range, such as 10 millivolts per division. It is important that the helipot be adjusted so that the trace on the oscilloscope is at the zero voltage location. If the digital voltmeter has a reading other than the static potential, and no external cathodic current has been applied, record this new half cell potential. Care should be exercised in preserving the original half cell potential.

There are times when the galvanic current flow from the anode system to the reinforcing steel is sufficient to change the static potential. It is generally a good idea to hook up the oscilloscope circuitry prior to connecting the rectifier. In some cases, a very high resistance, 100K kΩ, variable resistor is put in series with the anode lead wire so the amount of galvanic current can be regulated and measured as a part of the test instead of being a troublesome detail that might affect the static potential of the reinforcing steel.

Note no. 1. Since there are no 'standard' instruments or test method for obtaining E log I data, the polarized half cell potentials can be obtained by a commercially available power supply that has built in circuitry which will produce half cell potential readings obtained on the 'off' moments of the pulsating direct current. An oscilloscope should be used to check the potentials obtained by this instrument to make sure that they are correctly spaced during the cycle, and, are obtained when approximately more than one half of the time period of the 'off' cycle has passed.

When using the oscilloscope, the potential readings should be obtained for that value of polarization that is observed just before it is caused to rise by the power coming 'on'. It has been observed that at high values of current to the reinforcing steel the waveform on the oscilloscope shows that when the power is shut off, the trace can continue down to a lower value than the polarized potential of the steel, and then recovers before the power comes back 'on'. Therefore, if the polarized potential of the steel is read on the far right of the trace, just before the power comes 'on' any distortion in the oscilloscope trace will not affect the value of the readings.

Note no. 2. When running the E log I test, there are in reality two different tests being run. One test is the linear polarization test, and the other is the actual E log I test. Up to a maximum of about 20 mV, change in potential of the reinforcing steel, the plot of potential versus current, will plot as a straight line. The numerical values derived from that linear line can be substituted into an equation and the corrosion current can be calculated.

At polarization values of the steel which are shifted about 20 mV greater than the original or static potential, the data will eventually plot as a straight line on semi-logarithmic paper. This latter straight line is called the Tafel slope, and can be used to determine the corrosion current and the current required for cathodic protection. To obtain these data, current will be applied to the reinforcing steel in incremental steps approximately three minutes apart.

Applying *E* log *I* current

17. Turn all equipment 'on'.
18. Make up a data sheet with column having the headings: Time, Volts and Current.
19. Record the time, and, starting with the undisturbed or static potential, apply sufficient current to change the polarized potential by 2 mV and wait about 2½ minutes.
20. If the polarized potential of the steel has not changed in a ten-second interval, record the potential amount of applied current, and then, at the three-minute time since current was applied, again increase the current until the polarized potential has increased by 2 mV over the last reading, and record the time.
21. Again wait 2½ minutes since the current was increased in step 20. If the potential has remained stable for 10 seconds, read and record the potential and the current. Continue the 2 mV steps until about a 20-mV change from the static or original potential has occurred. From about a 20-mV change in the static potential to approximately 200-mV change, good results have been found when the potential change is in steps of 5 mV. Thereafter, 5-mV potential change steps have also been found to result in good *E* log *I* curves. However, depending upon the polarization characteristics of the bridge deck, that is, if the readings are stable for 10 seconds after the potential increase, it has been found that steps of 10 mV could also give good results after the original 200 mV or so initial shift in potential.

Plotting the *E* log *I* curve

For the values of current used on the bridge deck, divide each by the surface area of the deck. These values should be in the terms of milliamperes per square foot (mA/sq. m). Then, on three or four cycle semi-logarithmic paper, delineate the appropriate range of current measured on the bridge deck, on the *X* axis. On the linear *Y* axis of the paper, plot the half cell potential of the reinforcing steel. Generally, at a location of more than 100-mV change in potential, the potential–current relationship will form a straight line for perhaps an additional 200 mV of potential change. Connect these data points together with a straight edge. Extend this straight line to a point that is less in value than the static or original half cell potential E_{stat}, of the reinforcing steel. In the direction of increasing potential, extend the line so that the length of the line passes through at least one cycle of current. For one cycle of current on this straight line, calculate the voltage change. This calculated voltage change of one cycle of current for this straight line is the cathodic Tafel slope, B_c. The intersections of the cathodic Tafel slope line, B_c, and the static potential line E_{stat}, is the corrosion current, I_{corr}.

With a French curve. draw a line to connect the curved data points. At that point where the curved line departs from the cathodic Tafel slope, is the minimum current required for cathodic protection, I_{cp}. The half cell potential at this same location is the minimum half cell potential for cathodic protection, E_{cp}.

The anodic Tafel slope B_a

The anodic Tafel slope, B_a, may be experimentally obtained by applying increments of anodic current to the reinforcing steel in same manner as for obtaining the cathodic Tafel slope. However, there is always a great deal of reluctance when the reinforcing steel is forced to be anodic (corrode) in any situation. This is especially true when there is a mathematical method which can be used to derive the anodic Tafel slope, B_a, from the data used in obtaining the cathodic Tafel slope, B_c.

The calculation used to derive the anodic Tafel slope is as follows:

$$I_{app} = I_c - I_a \tag{1}$$

wherein I_{app} = The cathodic current applied to the steel. mA/sq. ft (mA/sq. m)

I_c = the cathodic current on the extrapolated cathodic Tafel slope at the half cell potential of I_{app} in mA/sq. ft (mA/sq. m) of deck surface.

I_a = the anode current from which to plot the anodic Tafel slope at the potential of the applied cathodic current, I_{app}. mA/sq. ft (mA/sq. m) of concrete surface.

Graphically solving for the anodic Tafel slope

The plot of the calculated anodic Tafel slope is a mirror image of the true slope. When the true anodic Tafel slope is obtained by testing, for each anodic potential change, the current will increase. This is not the case when the slope is calculated, it is the reverse as the slope is actually negative. If anodic and cathodic slopes were to be plotted, the values of current would be plotted at locations of potential change, and not the actual potential. The point of zero potential change would be the static or original potential. Both the anodic and cathodic slope extrapolations would have their origins at the corrosion current, I_{corr}, intercept on the static potential line.

Divide the half cell potential range between the static, E_{stat}, and the potential for cathodic protection, E_{cp}, into ten equal segments. At these ten locations tabulate the half cell potential and the current applied at appropriate data points. Label these values of applied current I_{app}. At these same locations of half cell potential, record the amount of current that would be found on the cathodic Tafel slope line, B_c. Label these latter values I_c, which is the cathodic current. For each value of half cell potential, subtract the applied current, I_{app}, from the cathode current, I_c. The values which were just calculated are for the anode current, I_a.

Plot each value of the calculated anode current. I_a, at the associated half cell potential on the same semi-logarithmic chart. The anode current data points, I_a, beginning at the corrosion current, I_{corr}, should have a straight line relationship with increasing potential for perhaps 100 mV and then it will form a curve going in the direction of zero current. With a straight edge, beginning at the point where the corrosion current, I_{corr}, intersects the static potential, E_{stat}, draw a straight line through the 'best fit' data points that extends for current; determine the millivolt change for the line. This millivolt change is the slope of the line and is the anodic Tafel slope, B_a.

It will be noted that the calculated points for I_a are in a straight line beginning at the corrosion current, I_{corr}, and then at a more negative potential the data points begin to curve and go towards zero current. At zero current, the anodic current, I_a, data point will have the same half cell potential as the potential for cathodic protection, E_{cp}. Theoretically, when the anodes and cathodes have been polarized to the open circuit potential of the most anodic anode corrosion is stopped. However, the open circuit potential of the most active anode cannot be measured because it has been distorted by polarization as a result of current flow. From a practical standpoint, the difference in potential and cathodic protection currents are small enough to ignore. This is especially true when considering the accuracy of running and evaluating the results of an $E \log I$ test on the bridge deck. If a degree of safety is required perhaps and increase of about 10% over the minimum cathodic protection requirement, I_{cp}, would be all that is necessary.

Linear polarization calculations

Once the values of Tafel slopes, B_a and B_c are determined, the corrosion current, I_{corr}, can be calculated by means of the linear polarization equation. The data for the linear polarization calculation are for those values that have about less than 20-mV change from the static potential, E_{stat}. In the case of the cathodic protection of bridge decks. the corrosion current calculated by the linear polarization equation is used as a means to compare the calculated current to that determined by graphic means. Once a bridge is under cathodic protection, the cathodic protection current must be turned off for a significant period of time to allow for depolarization, before any corrosion current measurements can be obtained.

On linear coordinate graph paper, mark about 30 mV of potential change from the static potential, and mark I_{stat} on the Y axis. On the X axis of the graph paper, mark out current divisions that will range from zero applied current to the current density applied at about 30 mV of half cell potential change from the static potential, E_{stat}.

With a straight edge, draw a line that will average the data points within an area that ranges between zero (the static potential) and about 20-mV change in half cell potential. It would not be unusual for the data points to be linear for only the first 10 mV of potential increase or change from the static potential, E_{stat}.

As in calculating the slope of a line, mark two locations on the straight line and designate the location of least potential with the numeral 1, and the location with the greater potential with the numeral 2. Subtract the half cell potential at location 1 from that at location 2. This value (in millivolts) is now designated E. Subtract the current at location 1 from that at location 2 and this result (in milliampreres per square foot of deck surface) is designated I.

In the calculation of the corrosion current, all terms must be in the same units. Therefore be sure the corrosion current is in the terms of milliamperes per square foot (mA/sq. m) and the Tafel slopes are in millivolts.

The following variables will be required in order to calculate the corrosion current by means of the linear polarization equation:

(1) From the plot on the linear coordinate paper: the change in potential, E, in millivolts, the change in current, I, in milliamperes per square foot (mA/sq. m) of deck surface.

(2) From the E log I plot in the semi-logarithmic paper: the cathodic Tafel slope, B_c, in millivolts, and the anodic Tafel slope, I_a, in millivolts.

The linear polarization equation is as follows:

$$I_{corr} = \frac{I}{E} \times \frac{B_a B_c}{2.3\,(B_a + B_c)} \qquad (2)$$

Wherein: I = the change in the linear current density in milliamperes per square foot (mA/sq. m) of deck surface area for a voltage change E;

E = the change in voltage for the linear current change in current I, in millivolts;

B_a = the anodic Tafel slope in millivolts per decade of applied current;

B_c = the cathodic Tafel slope in millivolts per decade of applied current;

2.3 = a constant.

It has been observed that there is about a 30% difference between the values of corrosion current obtained graphically or by calculation. This difference is not great, and, may be related to the value of the constant or simply to the difference in the methods of obtaining the values.

Chapter 19

Densified silica-cement coating as an effective corrosion protection

THEODOR A. BÜRGE, Sika AG, Zurich, Switzerland

SYNOPSIS

Corrosive attack causes the destruction of concrete structures such as piers, offshore platforms and prestressed marine structures, and of concrete with chloride-containing admixtures, or chloride-containing aggregates.

Corrosion is an electrochemical process, with rust and complex compounds of chloride built up as resultants, the volume of which is higher than that of the reactants. This leads to local crack formation and spalling which allows salt to penetrate farther into the concrete where process of corrosion will be more and more accelerated.

As a result of intense research work undertaken in this field, a method was found which ensures the effective protection of the reinforcement bars against corrosive attack. It consists of a dense, cementitious coat of low porosity which prevents the contact between iron and the corrosive materials. The coating itself is made of a polymer-modified cement-silica fume mortar to which are added bi-carboxylic acids plus organic amines acting as corrosion inhibitors.

Points under discussion are: the effects of the silica fume content on porosity, water, and carbon dioxide permeability, and on the chloride penetration.

INTRODUCTION

Owing to the high pH (>12), concrete normally provides a non-corrosive environment for embedded reinforcing steel. However, the introduction of chloride ions, e.g. from de-icing salt application, into the concrete drastically changes this situation.

The present paper discusses a method of protection against corrosion of reinforcing bars, prestressing cables, steel structures and all these when located within structures.

Structures which are fabricated from concrete, such as piers, harbour walls, offshore platforms, prestressed concrete located in salt water, concrete having additives which comprise chloride or aggregates which comprise chloride, are subject to damage caused by corrosion.

Corrosion of reinforcing bars happens specifically in concrete structures which

- are located in salt water (sea water),
- are manufactured with sand which contains salt,
- are treated with de-icing salts,
- are located in industrial areas (acidic air pollution),
- are manufactured with chloride containing accelerators.

Corrosion is an electrochemical process. The reaction products of rust and complex compounds with chloride have a larger volume than the original material. This leads to the formation of local cracks and chip-offs; accordingly, additional salt can penetrate and thus the speed of corrosion will accelerate as time goes on.

Certain specific chemicals can prevent or delay such a process. These chemicals are added to the concrete in form of an admixture during the production thereof and supposedly prolong the durability of concrete structures. Such additives are already marketed and sold in the USA and in Japan. Generally, their main or sole component is calcium nitrite. Furthermore, a large number of patents or published applications exist, such as the CA-PS 802 281, US-PS 3 427 175, DE-OS 3 005 896, JP-OS 33 940, US-PS 3 210 207, US-PS 3 801 338. Owing to their limited efficiency other inhibitors, such as benzoates, chromates, phosphates, amines, etc., have hitherto not found a large technical application.

The drawback of the currently available corrosion-inhibiting admixtures for concrete is that such admixtures are distributed throughout the concrete which leads to extremely high dosages of such additives. Owing to their high toxicity, highly concentrated nitrite solutions or solid nitrite admixtures are not welcome at construction sites. It is a known fact that the effect of such admixtures regarding the inhibition of corrosion is related to their dosage and simultaneously to the concentration of salt in the concrete structure. In the case of the formation of large cracks, owing to improper concreting, a reinforcing bar which is located at such area will not be protected against corrosion, in spite of an anti-corrosion additive, because the salt at such an area may be present in a high concentration.

Owing to its highly alkaline medium ($pH > 12$) concrete possesses corrosion protective features because reinforcing bars are passivated in such a pH range. In the case of reaction with carbon dioxide the pH value will decrease to a range which is no longer corrosion protective. However, concrete seldom features such a density which does not allow a penetration of chlorides, water or carbon dioxide; so concrete will rarely protect iron against corrosion over an extended time span.

DESCRIPTION OF THE PRIOR ART

In practice, the following three methods are currently followed:

- A coating of the reinforcing bars by synthetic resins, specifically epoxides, leads to complete protection against corrosion where such a layer is sufficiently thick and has not been damaged during the placing or mounting, respectively, of such reinforcing bars; this is, however, almost impossible to achieve. The drawbacks are: high costs, no adhesion of concrete to the protective layer, electrical insulation, and, in damaged areas, the generation of local cells which lead to a locally accelerated corrosion.
- Electroplated, usually galvanized iron, is excellently protected against corrosion. It is, however, known that the zinc coating will be dissolved because of the alkalinity of the cement. Therefore, the protective effect is limited with regard to durability.
- A third known method is the so-called cathodic corrosion protection. It entails an application of a predetermined voltage which inhibits the corrosion of iron. The drawbacks of this method are, however, that different voltages must be locally applied depending on the prevailing concentration of salt and voltage drop within the concrete. Practically, however, such a method can neither be used nor controlled. An unsuitably applied cathodic protection against corrosion may lead to the loss of the adhesion of the iron to the concrete or even to an accelerated corrosion.

DENSIFIED SILICA-CEMENT COATING

Hence, it is the general object of this paper to provide an improved method of preventing contact between corrosion promoting substances and reinforcing bars. With the application of a silica-cement coating on the reinforcing bars a solution to the problem could be found. The coating is basically a two-component product. The liquid component is a dispersion of an alkali-resistant polymer dispersion. The powder component [1] is a blend of cement, silica fume, quartz sand and corrosion inhibitors.

For the corrosion inhibitors an active combination of long chain bi-carboxylic acids and organic amines could be found. This coating provides a low porous [2], waterproof layer which inhibits the penetration of chlorides and carbon dioxide and reacts in an alkaline way in order to passivate the reinforcements. The polymer dispersion provides an excellent bond and a good adhesion of the coating to the reinforcing bars as well as to the concrete.

LABORATORY TESTS

Porosity and permeability

In order to produce permeability out of porosity, there must be an interrelation between the pores of the material which must form a reticular canal system, for it is quite obvious that the presence of pores which are closed on all sides does not lead to air or water permeability. Apart from this, the size and width of the pores within a coherent pore system have a decisive influence on the permeability, because the narrower the pores are the higher must be the pressure to force a liquid – generally water – through the pores.

Porosity tests have been made on cement-silica fume pastes with a constant water–cement ratio of 0.30. 100 parts by weight of cement were blended with 0, 10, 20, 30 and 40 parts by weight of silica fume.

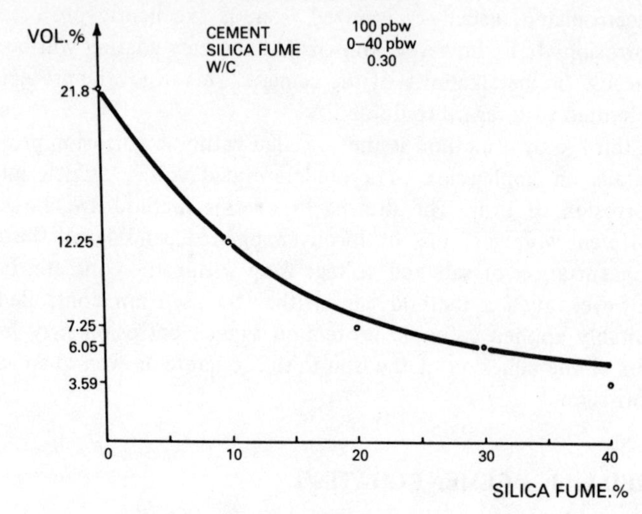

Fig. 1 – Total pore volume.

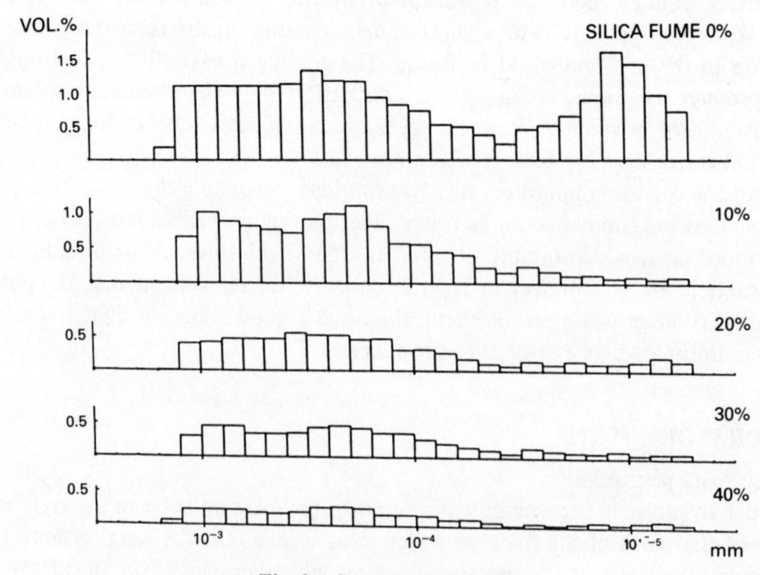

Fig. 2 – Pore size distribution.

After 28 days storage at 20°C and 100% relative humidity the specimens were oven-dried to remove all the capillary water before determination of the porosity. Fig. 1 shows a plot of the total porosity against the content of silica fume and in Fig. 2 the pore size distribution is given for each mix. Total porosity is a function of the silica fume content, the higher the silica fume content the lower the porosity. It can be seen, that the smaller capillary pores between 10^{-4} to 10^{-5} mm are reduced more than the bigger pores between 10^{-3} and 10^{-4} mm [2].

Permeability tests have been made according to DIN 1048 [3] at a water pressure of 0.5 bars for two days, at 1.0 and 2.0 bars for one day. It could be shown, that one coating is not enough always to get a good waterproofing effect. The deviation of the results was too big. Two coatings seem to be necessary to be safe; a third coating does not improve impermeability (Fig. 3).

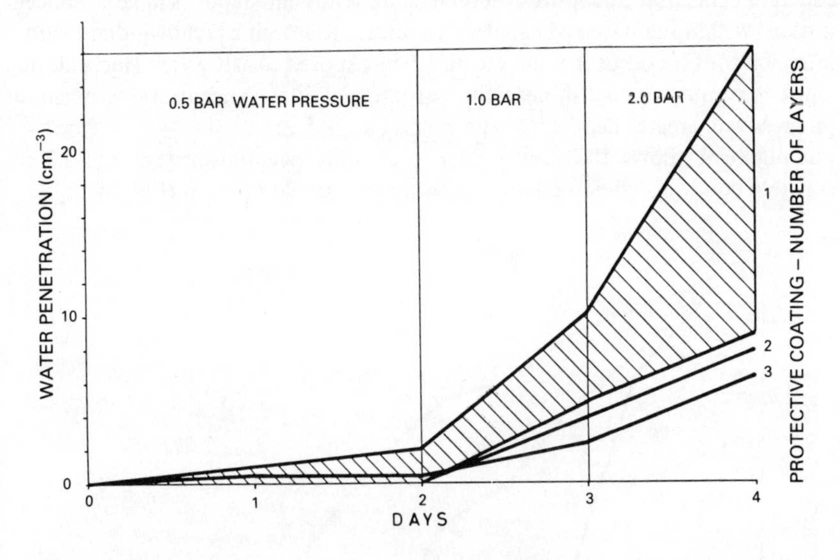

Fig. 3 – Water permeability.

CHLORIDE PENETRATION

Chloride ions are carried into concrete in solution by water. When water penetration is retarded by an effective coating, chloride ion intrusion and deposition is prevented.

Concrete is a naturally porous material. Its porosity is mainly related to the water–cement ratio (W/C). Because of its porosity, concrete absorbs large amounts of chloride ions. The absorption rate is generally a function of capillary structure and exposure conditions.

The actual water demand for hydration of concrete is usually about 0.25 by weight of cement. However, extra water must be added to make the concrete workable. Normal field concreting practice provides concrete with a water–cement ratio ranging from 0.35 to 0.50. The excess water, needed for workability, but not needed for hydration, is called evaporable water. When evaporable water dries it leaves behind an interconnected pore and capillary structure in the hardened concrete. Higher water–cement ratio concretes have more pores and capillaries than lower water–cement ratio concretes, and typically allow greater penetration of chloride ions.

Concrete capillaries vary from approximately 15 to 1000 angstroms in diameter. Chloride ions are less than 2 angstroms in diameter and can easily pass through the smallest capillary.

Chloride penetrates concrete as a water-transported ion. The ion navigates concrete capillaries to depths of several centimetres and deposits in large concentrations within the pore and capillary structure. Research has shown that distinct intrusion profiles occur in concrete members exposed to salt water. High chloride ion concentrations occur near the surface and low concentrations occur at progressively greater depths into the concrete.

Figure 4 shows the results from a chloride penetration test into 12 cm concrete cubes stored partially in 10% sodium chloride solution (Fig. 5).

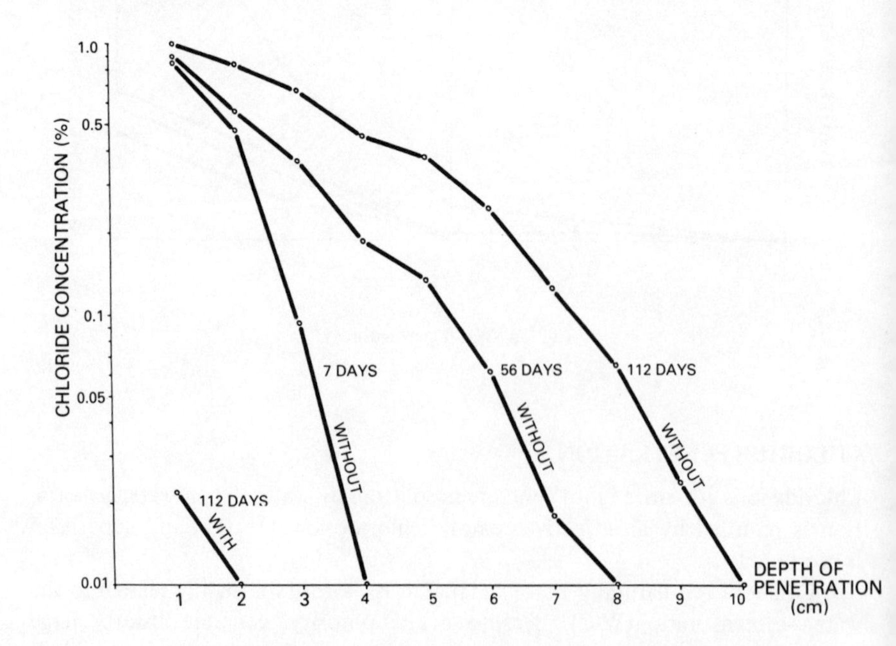

Fig. 4 – Chloride penetration (effect of protective coating).

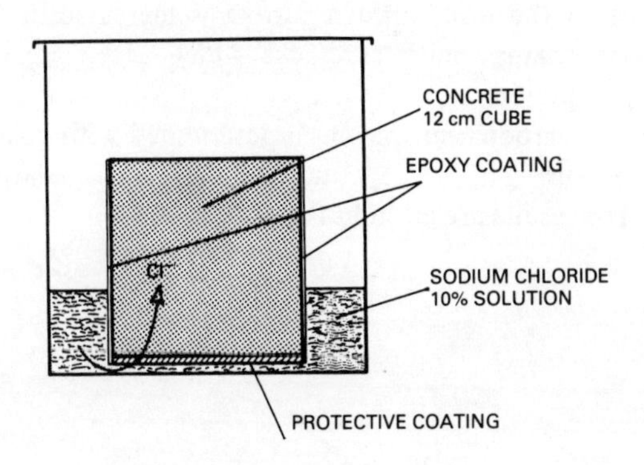

Fig. 5 – Chloride penetration test.

The chloride content was determined for each centimetre between 7 days and 112 days. With no coating the chloride penetrated 9 cm into concrete; through the coated concrete, chloride penetration was only 1 cm.

CARBONATION OF CONCRETE

Fresh concrete has a pH value of 12–13 (equal to a saturated lime solution). By reaction with carbon dioxide calcium carbonate is formed, which reacts neutral (pH 7). This neutralization is a continuous process. Depending on the concrete quality, the depth of carbonation may reach several centimetres after 20–30 years. When the pH value near the reinforcing bars drops to 8.5–9, the iron starts to corrode under formation of different and complex iron oxides.

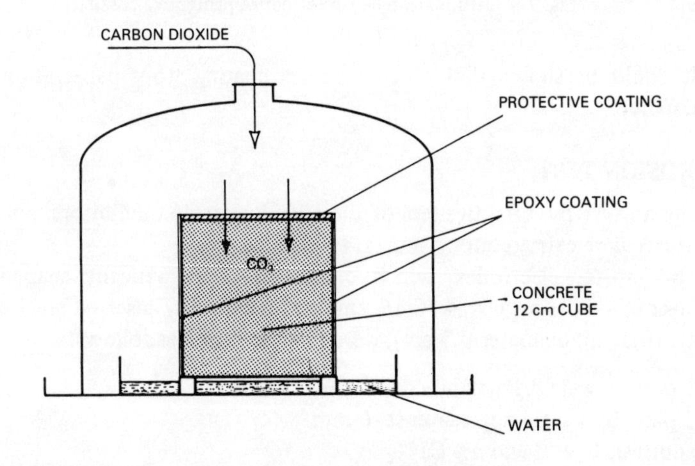

Fig. 6 – Carbon dioxide penetration test.

The volume of the iron through rusting is increased by 2.5. The pressure from this volume change initiates cracks which accelerate further corrosion, especially at high humidity.

The depth of carbonation has been determined with coated and uncoated 12 cm concrete cubes in a CO_2-atmosphere at 100% relative humidity and 20°C (Fig. 6). The results are given in Fig. 7.

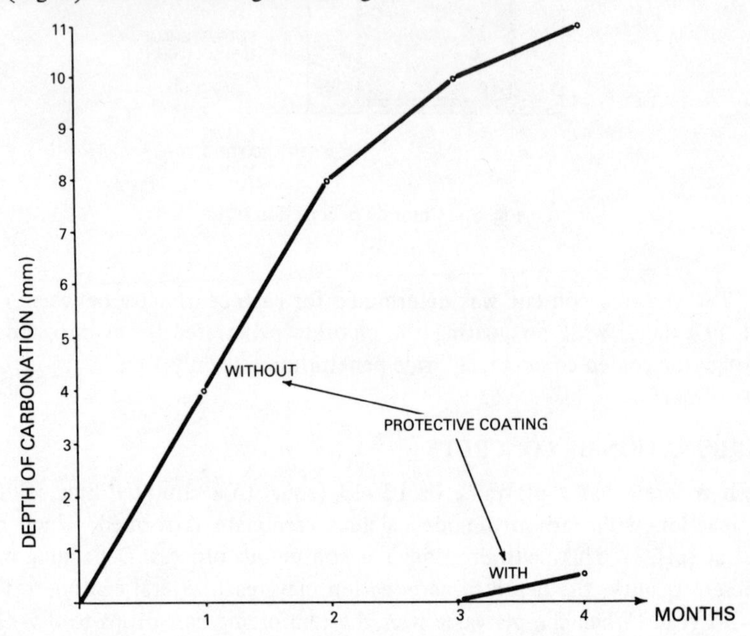

Fig. 7 – Carbonation depth (effect of protective coating).

It could be shown that a cement-silica coating stops penetration of CO_2 into concrete.

CORROSION TEST

In order to test the effectiveness of the added corrosion inhibitors, polarization curve tests were carried out on mortar electrodes.

The mortar electrodes, which consist of a prismatically shaped mortar (dimensions of prism 4 × 4 × 16 cm) and a centrally inserted steel electrode (length 10.5 cm; diameter 0.7 cm), were manufactured as follows:

1 part by weight Portland cement, W/C = 0.5
3 parts by weight aggregates 0–6 mm
additives 0, 1, 2 and 3% $CaCl_2$
corrosion inhibiting agent.

The steel electrode was coated with the protection material; components A:B = 1:4.

Component A: aqueous, anionic polymer dispersion.
Component B: high-quality Portland cement (Type III), fine aggregates and processing aids and corrosion inhibitors.

Immediately prior to the coating, the steel electrodes were degreased by means of a solvent and polished bright by an abrasive paper.

The corrosion measurements were carried out by means of the well-known three-electrode pattern (mortar electrode, platinum-plated auxiliary electrode, reference electrode) (Fig. 8).

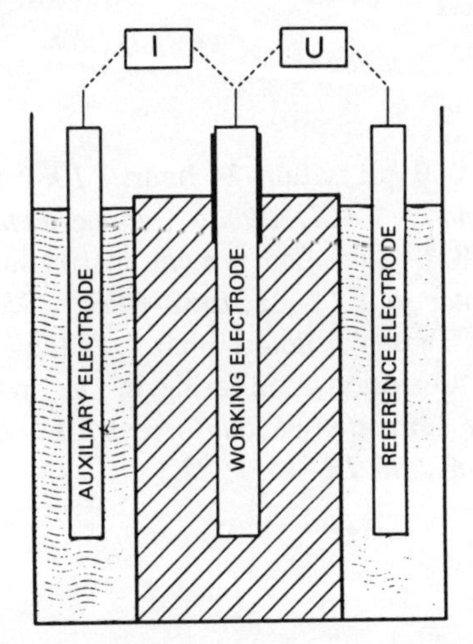

Fig. 8 – Corrosion test (3-electrode pattern).

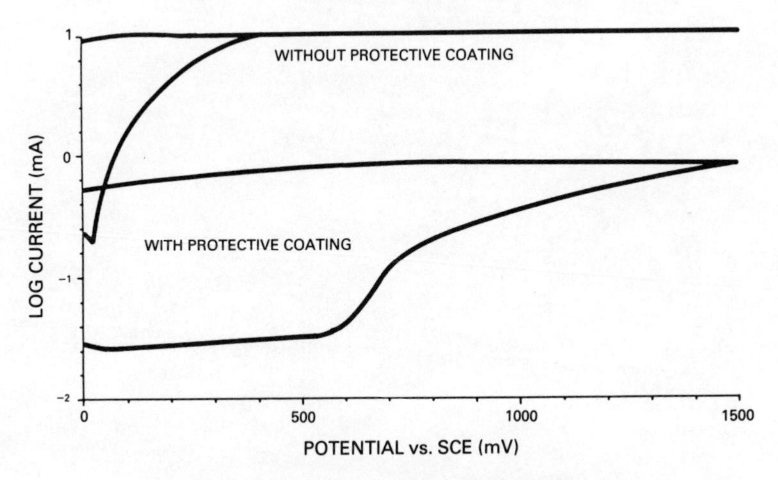

Fig. 9 – Polarization curves of mortar electrodes.

The polarization curves of mortar electrodes according to Bracher [4] clearly showed the corrosion protecting effect of cement-silica coatings (Fig. 9).

CONCLUDING REMARKS

In this paper, a procedure for an effective corrosion protection has been discussed. In order to reach an impermeability for water, chloride and carbon dioxide a coating with a low porosity based on cement, condensed silica fume and a polymer dispersion was formulated. The alkaline reacting cement matrix and added corrosion inhibitors prevents corrosion of the reinforcing bars or other structural members.

REFERENCES

[1] Bürge, Th.A. '14,000 psi within 24 hours'. *1982 Annual Convention on 'High Strength Concrete', ACI, Atlanta, Georgia, January 19–23, 1982.*
[2] Bürge, Th.A. (1982), 'Densified cement matrix improves bond' *Bond in Concrete, Conference June 14–16, 1982, Paisley, Scotland.*
[3] DIN 1048: Prüfverfahren der Beton, Blatt 1, Seiten 1–6.
[4] Bracher, Gustav, 'Corrosion monitoring in the industrial-related research area', Poster at the *'Corrosion of Reinforcements in Concrete Construction', Conference London, June 13–15, 1983.*

Chapter 20

Relation between the alkali content of cements and the corrosion rates of the galvanized reinforcements

C. ANDRADE, A. MOLINA, F. HUETE, J. A. GONZALEZ, Instituto Eduardo Torroja de la Construccion y del Cemento, Madrid, Spain

INTRODUCTION

Galvanized rebars have been used with success in structures which are in contact with marine atmospheres. However, some laboratory tests have shown an incorrect behaviour of the galvanized steel embedded in concrete polluted with chlorides.

In previous papers [1, 2] we have been able to establish the importance of three parameters which are irrelevant for bare steel, but crucial for galvanized steel: (a) the type of cement, that in function mainly of its alkali content generates different pH-values in the aqueous solution of concrete pores, (b) the type of metallographic structure of galvanized coating that governed by the type of base steel and its C and Si content, and (c) the amount of moisture contained in concrete pores.

Very recently [3] we have published experimental results about the influence of pH value on the corrosion rate of galvanized rebars. We used synthetic solutions that try to simulate the concrete aqueous solution. We used polarization resistance as the electrochemical technique of measurement of corrosion rate.

We found, in agreement with the results of Roetheli, Cox and Littreal [4] that the corrosion rate of galvanized rebars exponentially changes, when the pH value increases from 12 to 14. Besides, we found a threshold pH value about 13.00, a few tenths of a unit pH above which, i_{corr} of galvanized steel dramatically increases.

In order to verify if the behaviour tested in solution is reproduced in concrete, in the present paper we study the influence of the total content of alkalis of 11 different cements on the corrosion rate of galvanized rebars.

EXPERIMENTAL METHOD

Reinforcing corrugated 6 mm nominal diameter and 8 cm long steel bars, having a 60–80 μm average coating produced by a hot dip galvanizing at 450°C were used. The galvanized coating is shown in Fig. 1, and the chemical composition of the steel is given in Table 1.

Mortar specimens, as in previous works [1, 2] were 2 × 5.5 × 8 cm in size, the water/cement ratio = 0.5 and the cement/sand ratio = 1/3. They were fabricated with eleven different types of cement:

Table 1 – Chemical composition of the base steel.

% C	% Mn	% Si	% P	% S
0.30	1.24	0.34	0.024	0.027

(1) Normal Portland cement from factory 1 (P_1)
(2) Normal Portland cement from factory 2 (P_2)
(3) Portland cement with low content in C_3A from factory 1 (P-Y_1)
(4) Portland cement with low content in C_3A from factory 2 (P-Y_2)
(5) Pozzolanic cement from factory 1 (PUZ_1)
(6) Pozzolanic cement from factory 2 (PUZ_2)
(7) Slag cement from factory 1 (S_1)
(8) Slag cement from factory 2 (S_2)
(9) Normal Portland cement with high initial strength (P-ARI)
(10) Slag cement from factory 3 (S_3)
(11) Fly-ash cement (PUZ_3)

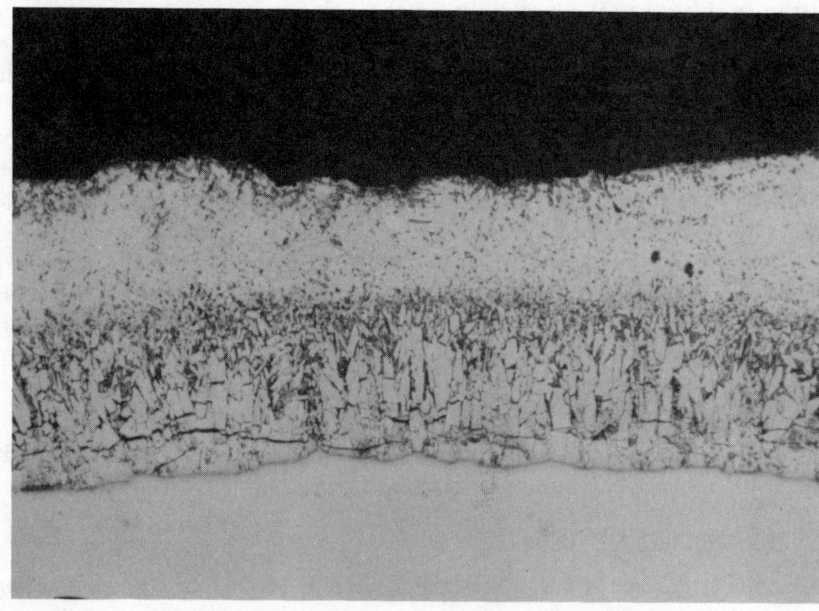

Fig. 1 – Microstructure of galvanized coating (×200).

Table 2 – Chemical analysis of the cements tested and pH of their suspensions.

Cements	$1(P_1)$ P-350	$2(P_2)$ P-450	$3(P\text{-}Y_1)$ P-350-Y	$4(P\text{-}Y_2)$ P-450-Y	$5(PUZ_1)$ PUZ-I-350	$6(PUZ_2)$ PUZ-I-350	$7(S_1)$ S-I-350	$8(S_2)$ S-I-350	9(P-ARI) P-450-ARI	$10(S_3)$ S-II-350	$11(PUZ_3)$ PUZ-II-350
I.L.	2.20	3.60	3.03	0.60	2.74	2.69	–	–	2.90	–	0.80
I.R.	0.15	1.20	2.74	0.40	1.89	5.87	2.99	1.17	1.40	2.10	28.40
SiO_2	20.00	18.50	20.40	20.10	21.28	25.77	19.95	19.97	18.60	30.60	17.60
Al_2O_3	6.02	4.90	4.53	4.30	5.05	6.91	4.67	7.54	4.80	9.20	6.10
Fe_2O_3	3.21	3.00	5.85	6.60	11.14	3.75	11.78	15.88	4.00	4.00	3.60
OCa	64.00	63.30	58.93	64.30	52.39	46.96	52.24	48.14	62.30	43.90	40.20
OMg	1.80	1.20	1.40	1.10	1.88	3.38	2.06	1.36	1.20	6.70	0.80
SO_3	2.86	3.60	2.67	2.40	2.70	2.95	2.56	2.31	3.50	0.80	2.20
free OCa	1.72	0.90	0.98	0.50	2.31	1.62	2.56	0.73	1.10	0.50	0.60
Na_2O	0.18	0.14	0.11	0.10	0.15	0.48	0.14	0.24	0.23	0.23	0.14
K_2O	0.39	0.78	0.30	0.68	0.37	0.84	0.36	0.63	1.00	0.76	0.34
MnO										0.029	
S^{2-}										0.70	
Suspension pH	12.70	12.57	12.62	12.72	12.62	12.75	12.47	12.65	12.84	12.08	12.50

The chemical analysis of the eleven cements tested and their pH suspension values are given in Table 2.

One series of the specimens was kept for 1 year at 100% R.H. and another one was partially immersed (P.S) in distilled water in individual polyethylene bottles for 1 year also.

The cement suspensions were prepared as follows: 25 g of cement were mixed for 15 min with 50 ml distilled water in a closed polyethylene bottle, after which it was filtered in a Büchner funnel. pH of the filtrate being immediately measured.

Corrosion intensity was estimated as in previous works [1, 2] with a potentiostat with IR-electronic compensation, from the polarization resistance, Rp, by the formula of Stein and Geary

$$I_{corr} = \frac{\beta a \times \beta c}{2.3\,(\beta a + \beta c)\,Rp} \doteq \frac{B}{Rp}$$

The pH was measured with a digital pH-meter and combined electrode for the range 0–14 pH.

The weight loss for each bar was also determined gravimetrically.

RESULTS

We show in Fig. 2 the pH value of cement suspensions as a function of the total alkali content, as equivalent to Na_2O [5]. In general, a higher content of

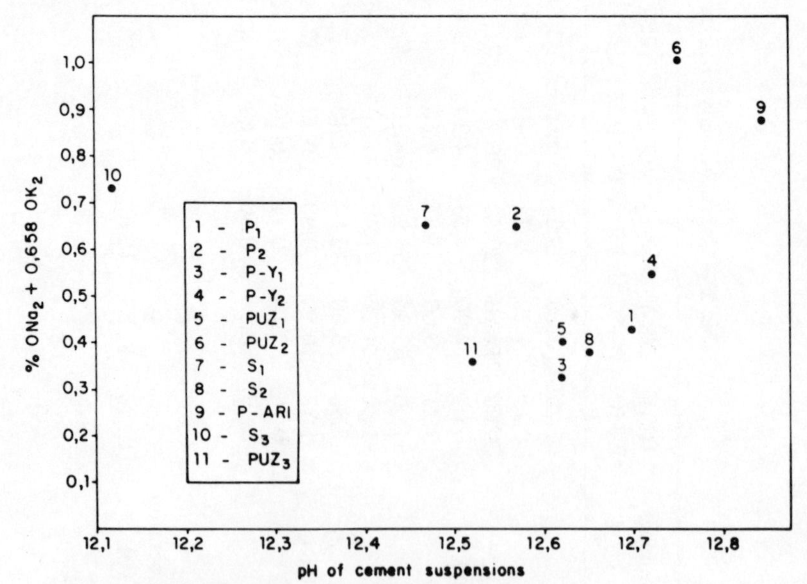

Fig. 2 – pH value of cement suspension versus the total alkali content (equivalent Na_2O) (the number corresponds to those of Table 2).

alkali corresponds to a higher suspension pH value, except for the cements no. 2 (normal Portland cement), nos. 7 and 10 (slag cements), No. 6 (natural pozzolanic cement) and no. 11 (fly-ash cement).

The values of i_{corr} of the galvanized rebars embedded in all mortars at 1, 28, 90 and 365 days are given in Fig. 3, as a function of the pH value of cement suspensions. Twenty-four hours after mixing there is a pH threshold above which the measured i_{corr} is very much higher, such as we were able to see in the synthetic solutions [3]. The slag cements are an exception (nos. 7, 8 and 10).

The results of the 28, 90 and 365 days also show a quasi-logarithmic dependence between the pH value of cement suspensions and the i_{corr} measured. The slag and fly-ash cements present values of their i_{corr} higher than can be deduced from their alkali content. The cements nos. 1, 3 and 5 show a lower i_{corr}.

Another trend of interest is that the behaviour of the galvanized rebars embedded in the mortars kept in 100% R.H. is more random than that of the ones kept in partial immersion. This different behaviour is made clearer in Fig. 4 where we show the evolution with time of i_{corr}. In general, the i_{corr} values are higher in 100% R.H. than in the partial immersion condition. In the latter condition the i_{corr} fall suddenly during the first 20 days and then decrease very gradually.

The thinner coating is in the convex zones of the rebar. There the coating is about 50 μm. We have batched in the figure between 0.07 and 0.1 $\mu A/cm^2$ because this corresponds to a homogeneous attack of about 1.05 to 1.5 μm/year, that is to say the coating would disappear in about 50 years.

Only the cements nos. 4, 6 and 9 kept in partial immersion and nos. 4, 6 and 7 kept in 100% R.H. show at 365 days a i_{corr} greater than 0.1 $\mu A/cm^2$.

We give in Fig. 5 the electrochemical weight loss produced during the 365 days of the test. The trends are similar to those deduced from the previous figures.

The morphology of the attack

By microscopic observation of transverse views of the rebars, we can see the following phenomena:

(a) In the places directly exposed to the mortar when corrosion is produced, the only phase attacked is that of pure Zn (η layer), with perferations as wells and tunnels (Fig. 6). In the concave zones (valleys) of the rebars the dissolution frequently starts in the place nearest the top of the convex zone and it continually excavates all the frontier between the η and ξ layers (Figs. 7 and 8). In no case did we find any attack on the ξ layer; that remains intact.

(b) In the zones guarded by the tape and not exposed to the mortar we can observe some attack, sometimes intense, in the η layer, but only in the mortars kept at 100% R.H.

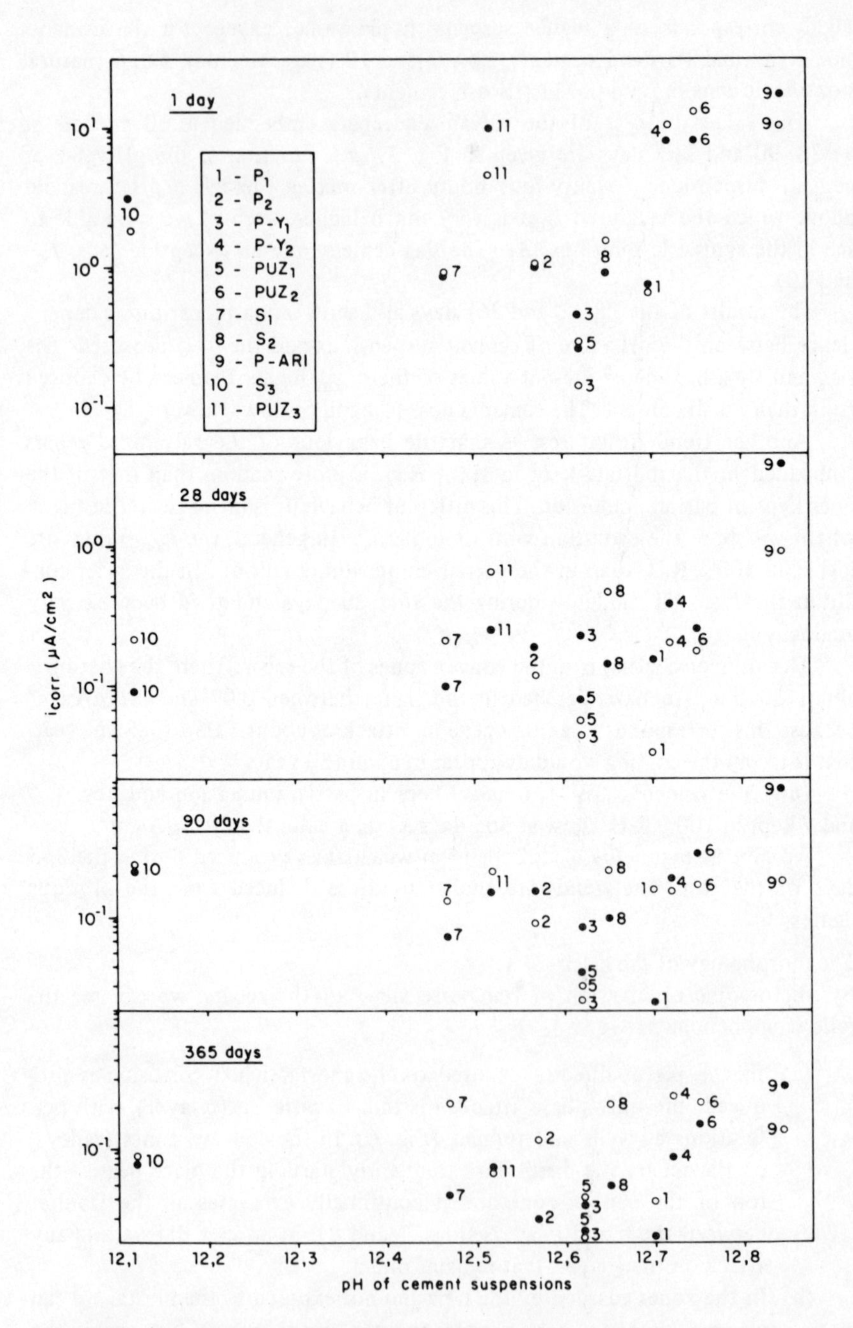

Fig. 3 – Values of i_{corr} at 1, 28, 90 and 365 days as a function of pH of cement suspensions.

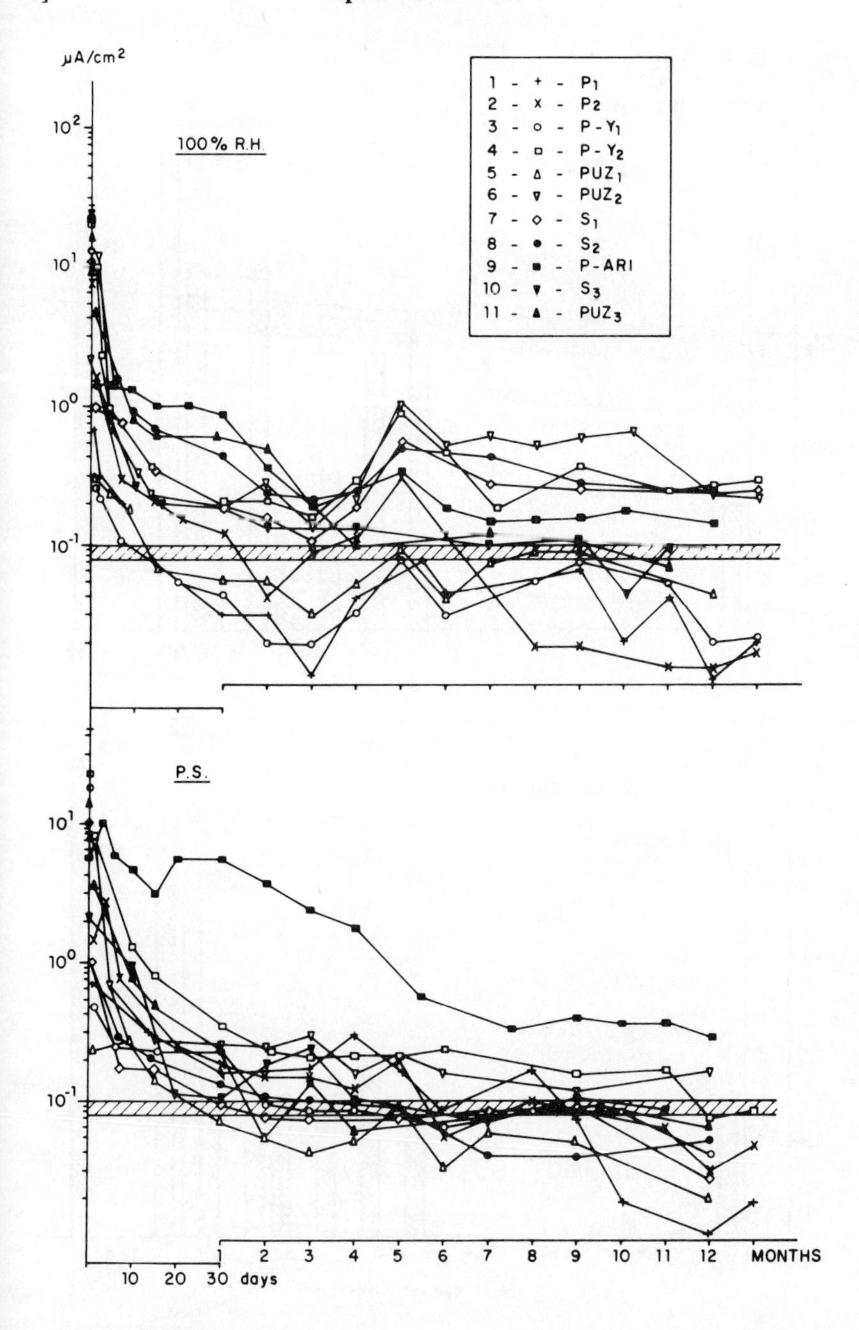

Fig. 4 – Evolution of i_{corr} with time of the galvanized steel embedded in the mortars made with the eleven cements tested.

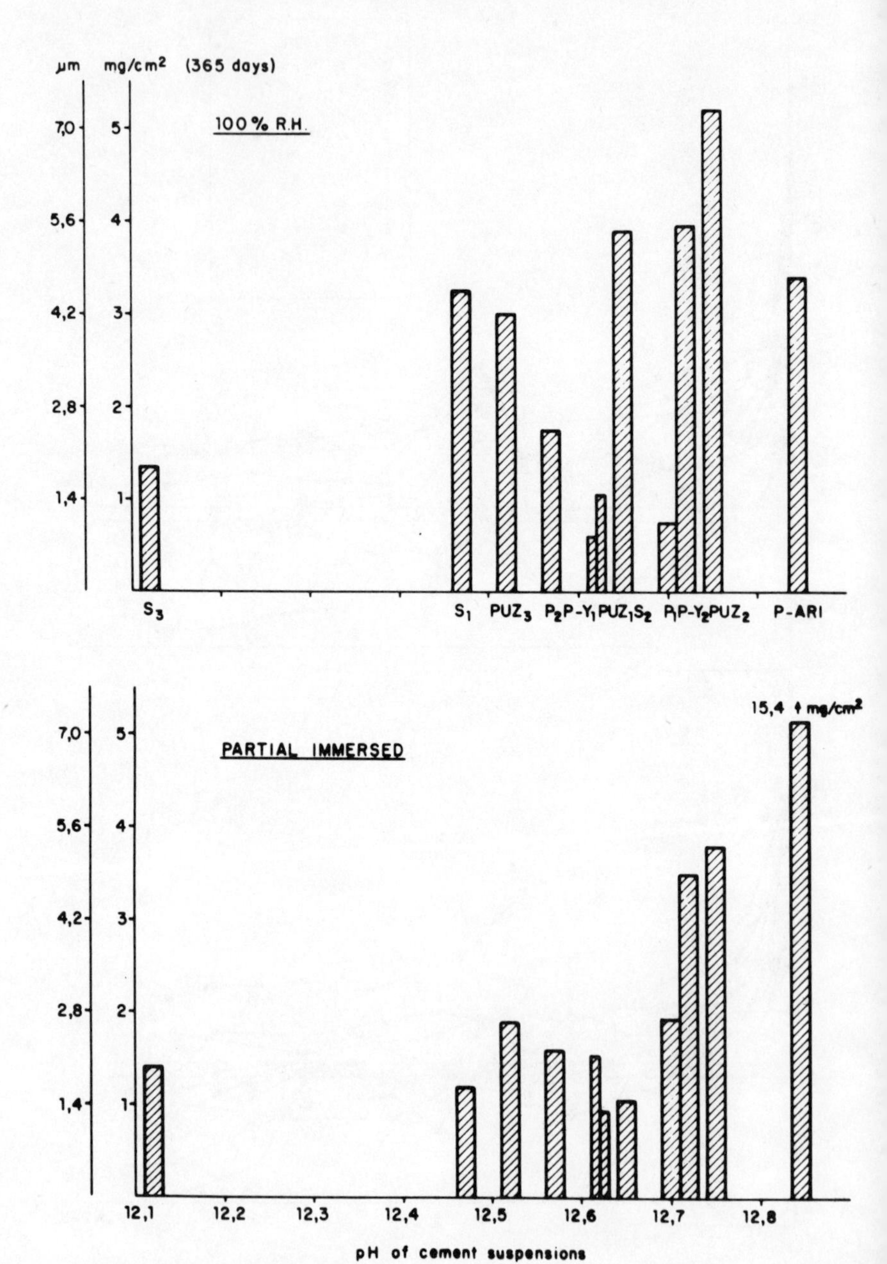

Fig. 5 – Total electrochemical weight loss of the galvanized rebars versus pH of cement suspensions.

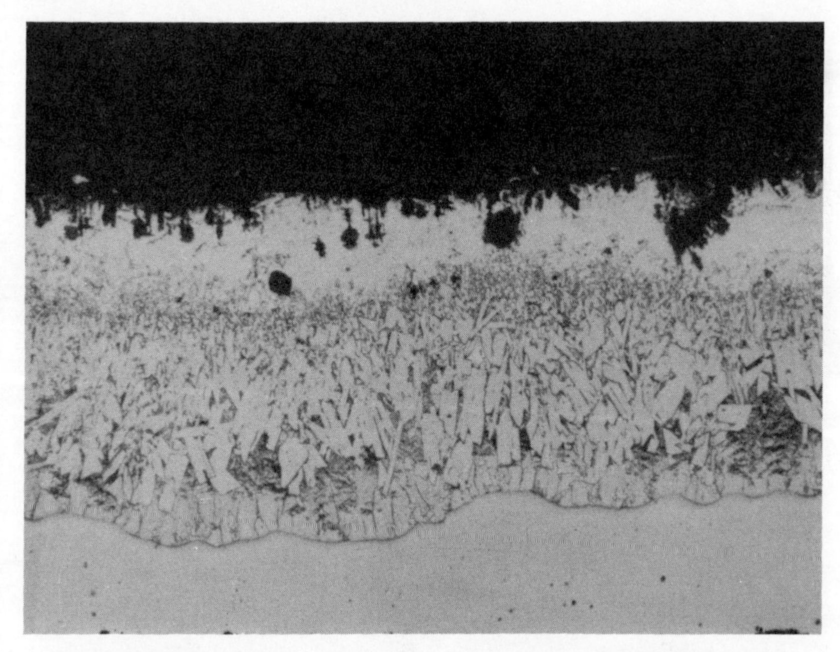

Fig. 6 – Wells and tunnels in the η layer of galvanized coating at the 365 days of the test (×200).

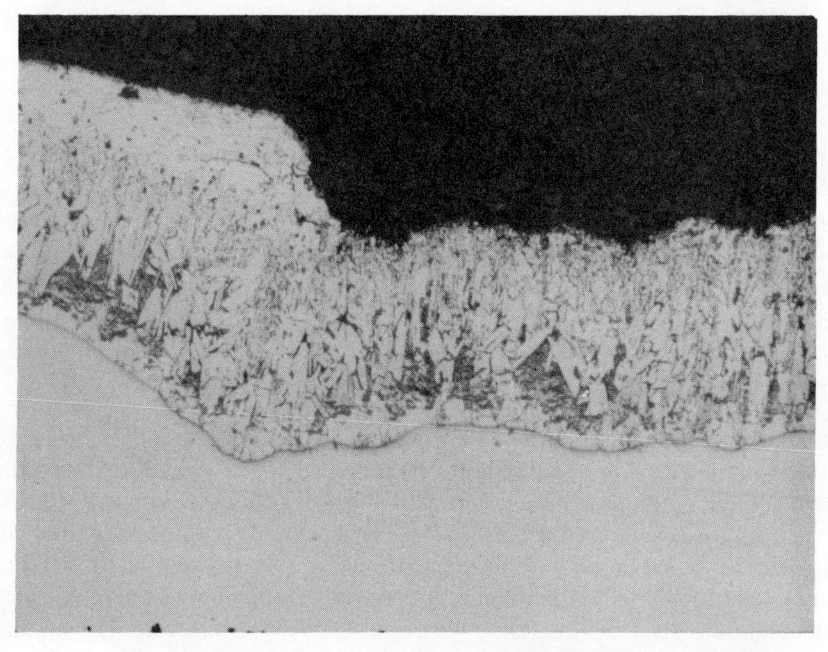

Fig. 7 – Type of dissolution of the η layer of galvanized coating (×200).

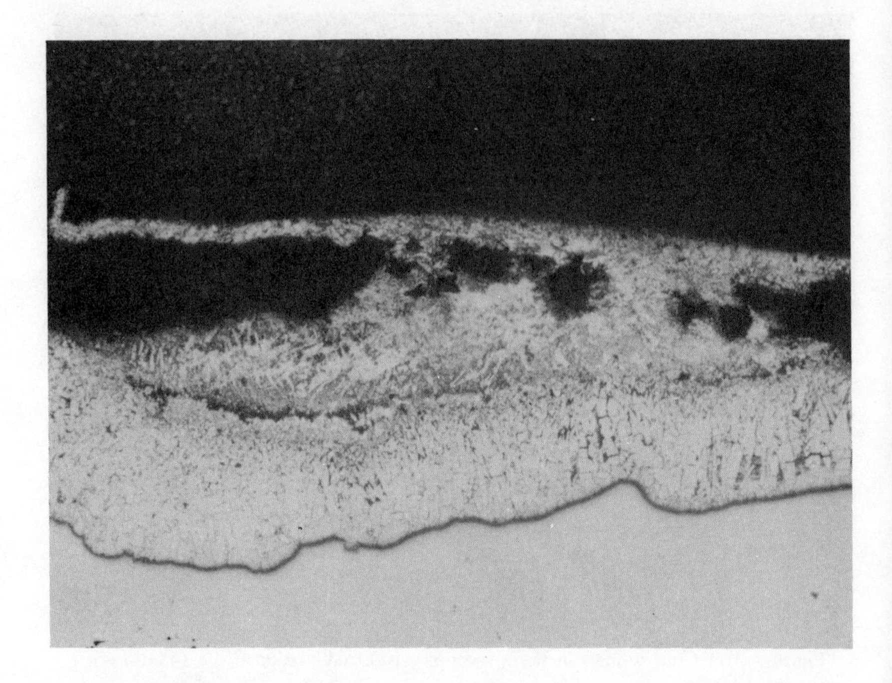

Fig. 8 – Other aspect of the attack suffered by the η layer of the galvanized coating (×100).

We attributed this phenomenon to the penetration of moisture by capillarity between the tape and the rebar. If the electrochemical weight losses were lower than the gravimetric ones, in the mortars kept at 100% R.H., there was no continuity between the area exposed directly to the mortar and the one below the tape.

Comparison between gravimetric and electrochemical results

We show in Fig. 9 the relationship between gravimetric losses and those estimated from *Rp* measurements. The best results are obtained with $B = 26$ mV when i_{corr} is greater than 0.1 $\mu A/cm^2$ and $B = 53$ mV for i_{corr} is less than 0.1 $\mu A/cm^2$. The agreement is good except for some mortars kept at 100% R.H.; that we attribute to the phenomenon previously mentioned of the attack produced below the insulating tape. We could not explain satisfactorily why this phenomenon was not produced in the mortars kept in partial immersion.

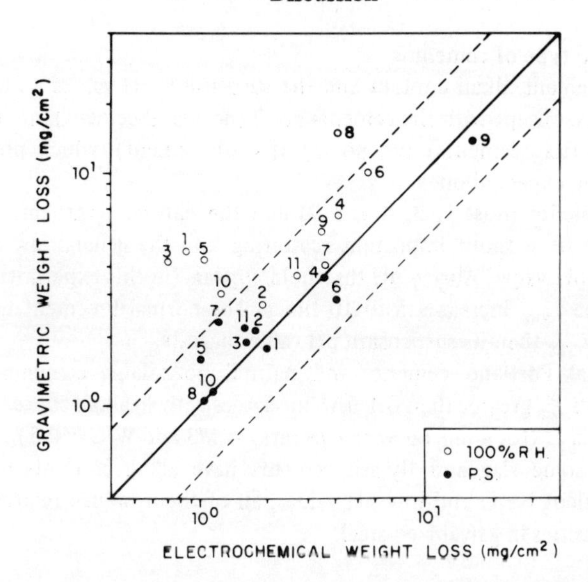

Fig. 9 – Comparison between total gravimetric and electrochemical weight loss of the galvanized bars at the end of the 365 days of the test.

DISCUSSION

The exact determination of the pH values of the concrete pore solution could only be obtained by squeezing the specimen. Through this technique it has been possible to prove that the pH value develops quickly from the moment of mixing to reach values of about 13.5 for normal Portland cements and of about 13.0 for slag cements [6, 7]. The Ca^{2+} ion disappears after a few days and the alkalinity is supported by Na^+ and K^+.

The pH determination of a cement gives only instant values and therefore, does not let us know the pH evolution with time. Also it gives values below the true ones, because the suspensions are more dilute. For lack of a more exact technique it permits us to make comparisons and, in the present paper, it has allowed us to study, in a first approximation, the relation between the pH value and the galvanized corrosion rate.

All the results obtained are more random than the ones obtained in solution. This shows the difficulty of the mortar tests; they are, however, the only true ones.

In spite of this scattering found we have been able to deduce, in an approximate form that, in the first hours after mixing, an increase in the pH value of the cement suspension above a threshold supposes that i_{corr} multiplies by a factor of about ten. But, the contrary is only true for normal Portland cement because values of pH below that threshold do not suppose respectively a low corrosion rate, as slag and fly-ash comments show.

Influence of the type of cement

Between the cement alkali content and the suspension pH values a quasi-lineal relation appears, except with the cements no. 2 (normal P cement), no. 6 (Pozz.) nos. 7 and 10 (slag cements) and no. 11 (fly-ash cement), which present pH values below the expected ones.

The P cements (nos. 1, 3, 4 and 9) and the natural pozzolanic cements (nos. 5 and 6) fit without important scattering into the general law of higher i_{corr} to higher pH value. Also, a pH threshold appears (in this experiment 12.70) from which the i_{corr} increases 5 to 10 times. The normal P cement no. 2 produces a higher i_{corr} than its suspension pH value suggests.

For normal Portland cements and natural pozzolanic cements this pH threshold or an i_{corr} greater than 0.1 $\mu A/cm^2$ agrees with an alkali content of 0.5% of equivalent Na_2O (for a mortar with C/S ratio = 1/3 and W/C = 0.5).

Although some slag and fly-ash cements have alkali contents below the 0.5% of equivalent Na_2O and low pH values, all of them induce relatively risky corrosion intensities in galvanized steel.

Amount and morphology of the attack

We have seen in the metallographic study, that it is the pure Zn (η layer) which is the only material that is attacked in this extremely alkaline medium. The ξ layer remains unaltered in spite of the fact it is directly in contact in some zones with the mortar.

In all the bars, including those attacked more, some zones of η layer remain unaltered at the end (365 days) of the test. Longer tests would be needed to study whether, when the whole η layer disappears, the attack would stop or would continue in the ξ layer.

The lowest measured penetration in the 365 days was of 0.84 μm for mortar kept in 100% R.H. and made with cement no. 3 (P-Y_1). As, at the end of this first year it presents an i_{corr} of about 0.02 $\mu A/cm^2$, say, 0.3 μm/year, the expected homogeneous attack would be about 30 μm in 100 years.

The highest measured penetration in the first year, was 21.56 μm for the mortar kept in partial immersion and made with cement no. 3 (P-ARI). The galvanized bars embedded in it present at the end of the 365 days an i_{corr} of 0.27 $\mu A/cm^2$, say 4 μm/year; therefore, if a homogeneous attack does occur the whole galvanized coating would disappear after 11 years.

Influence of the conservation conditions

In all of our experiments the conservation moisture has been shown to have a high influence on the passivation or non-passivation of the galvanized reinforcement.

The concrete pore moisture determines three characteristics of the concrete (*vis-a-vis* the metallic corrosion); (a) the ohmic resistivity; (b) the pH of the aqueous pore solution; and (c) the ability of the O_2 to reach the metal.

The effects of these three characteristics would sometimes be additive and would sometimes oppose each other. Therefore it is necessary to measure separately each of the characteristics in order to predict the direction of its influence.

In the present tests, the pore moisture has no influence on the 430 °C metallographic structure tested without additives; but it is decisive when Cl^- is added. When Cl^- is present the 100% of relative humidity was always much more damaging than the partial immersion.

CONCLUSIONS

(1) We conclude that, although the scattering is higher than in solution, the i_{corr} of the galvanized rebars embedded in mortar increases with the increase of the pH value.

A pH threshold seems to exist above which the i_{corr} developed implies a risk of dissolution of the galvanized coating during the lifetime of the structure.

(2) The slag and fly-ash cements present suspension pH values below those expected. The i_{corr} were higher than the pH indicated.

(3) The only layer attacked during the 365 days of the test was the η (pure Zn). In spite of the fact that some extensive zones of the ξ layers were in direct contact with the mortar, no attack on it could be detected.

(4) The moisture of the concrete pores has an important influence on the corrosion rate of galvanized steel. Up to now we have not found any convincing explanation for this influence.

REFERENCES

[1] J. A. Gonzalez and C. Andrade, *British Corrosion J.*, **17**, 1 (1982), 21.

[2] J. A. Gonzalez, A. J. Vazquez and C. Andrade, *Matèriaux et Constructions,* **15**, 88 (1982), 271.

[3] C. Andrade and A. Macias, 13th Galvanizing Conference, London, May 1982.

[4] B. Roetheli, G. Cox and W. Littreal, *Metals and Alloys,* **3** (1932), 73.

[5] ASTM – C – 150.

[6] P. Longuet, L. Burglen and Z. Zelwer, *Rev. Matèr. Const. et Traveux Publics,* **676** (1973), 35.

[7] P. Delmas, *Ciments, Bètons, Platres, Chaux,* **3** (1979), 167.

Chapter 21

Long-term corrosion resistance of epoxy-coated reinforcing bars

JIRO SATAKE, MASASHI KAMAKURA, KIYOSHI SHIRAKAWA, NAOTO MIKAMI,
Simitomo Metal Industries Ltd., Japan
NARYAAN SWAMY, University of Sheffield, Sheffield, UK

SUMMARY

This paper presents a detailed study on the long-term resistance to corrosion of epoxy-coated reinforcing bars, and these results are compared with plain uncoated and galvanized bars. The tests were carried out on centrally reinforced concrete prisms with variable cover and pre-formed cracks of maximum width of 0.10 to 0.25 mm. The specimens under stress were subjected to accelerated corrosion tests and marine exposure tests in a tidal zone up to two years. It is shown that whilst plain bars undergo extensive corrosion and pitting, galvanized bars show superior performance but cannot provide full protection against corrosion or pitting. A macrocell model is presented to demonstrate that plain and galvanized bars without an insulation film from alkali concentration cells leading to corrosion. The test results show that epoxy coating can provide an effective long-term protection to steel against chloride attack and corrosion.

INTRODUCTION

Corrosion of reinforcing steel in concrete structures leading to cracking and subsequent spalling is now recognized as a major problem confronting the construction industry in almost all parts of the world [1]. The major causes of breakdown of the passivating layer surrounding reinforcements embedded in concrete are

(i) carbonation;
(ii) chloride attack, where the presence of the chloride ions may be due to absorption from external agencies such as de-icing salts and marine environment or from within the concrete itself from contaminated aggregates or water or accelerating admixtures containing chlorides.
(iii) Cracking which permits ingress of water, air and deleterious chemicals.

The breakdown of passivity can be greatly accelerated by poor workmanship and quality control affecting the depth of cover and the quality of the concrete as well as by the presence of a high chloride content.

The build-up of corrosive conditions at the steel surface is a time-dependent phenomenon so that there is no certainty that structures not yet affected by corrosion will never have such problems in the future. Indeed the wider use of marine aggregates and the increasing number and size of offshore structures is likely to create more problems of concrete durability. It is now known that the top steel in a bridge deck is highly susceptible to corrosion and that serious deterioration of bridge decks has already occurred in some countries, and that the cost of repairing and upgrading such structures will be enormous [2–4].

The need to incorporate during design and construction various protective methods to prevent corrosion is universally recognized [5–7]. Apart from the permeability of the concrete and the depth of cover provided over the reinforcement, cracking of the concrete reinforcement is also recognized as an important factor influencing the degree of protection of the reinforcement [3, 8]. Even when the effects of cracking are localized as happens when the cracks are transverse to the reinforcing steel, and corrosion is confined to the immediate vicinity of the crack, there is some concern that the static and fatigue strengths of the reinforcement might be jeopardized by such local effects [9].

The various construction practices designed to prevent corrosion, particularly in bridge decks, have been extensively studied and discussed [2, 3, 7, 10]. Although it is now generally recognized that there is probably no single method of corrosion protection that can be presumed to be effective throughout the life of a structure, the use of epoxy coating of the reinforcing bars has been a favoured solution in many countries [3, 5, 6, 7, 11]. Wills [9] has made a critical examination of the use of epoxy-coated reinforcement in bridge decks, and has suggested that such a coating together with a suitable water-proofing system can provide an economic and effective long-term protection of steel reinforcement.

The authors have been involved in the development of an epoxy-coated bar having adequate bendability and bonding properties. This paper presents the results of a detailed study on the long-term corrosion resistance of epoxy-coated reinforcing bars, and these are compared to the performance of untreated and galvanized bars. The test specimens had varying depths of concrete cover, and stressed initially to crack the concrete. The specimens were then subjected to a 24-month accelerated corrosion test or a 36-month exposure test at a tidal zone, with the bars stressed to 2000 kg/cm^2 throughout the test period.

EXPERIMENTAL PROGRAMME

Reinforcement details

Three different thicknesses of epoxy coatings varying from 100 to 300 μm were used in this study (Table 1). To obtain a comparative evaluation, the control tests consisted of plain uncoated (as rolled) bars and galvanized (chromated) bars.

Table 1 – Reinforcement details.

Bars	Coating thickness μm
Uncoated (as rolled)	–
Galvanized† (chromated)	–
Epoxy-coated	100
Epoxy-coated	200
Epoxy-coated	300

†Zn layer thickness is about 70 μm.

All the reinforcing bars were hot rolled and deformed having a nominal diameter of 19 mm and the configuration shown in Fig. 1. The size and unit weight of the bar as well as the surface shape were in accordance with the appropriate Japanese, ASTM and DIN standards. The 19 mm bars had typically minimum yield and tensile strengths of 3500 kg/cm^2 and 5000 kg/cm^2 respectively and a minimum elongation of 20%.

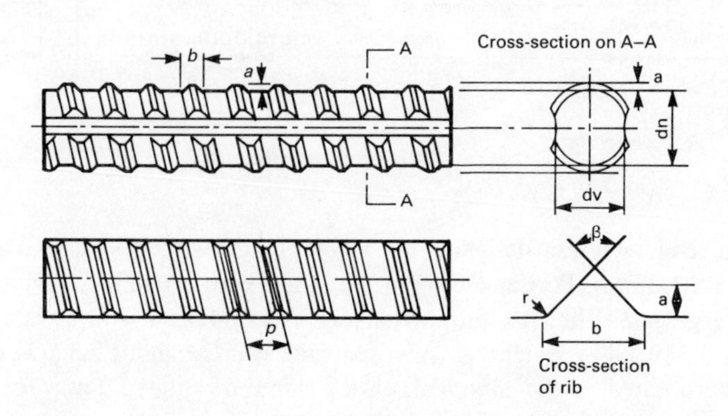

Fig. 1 – Configuration of steel reinforcement (Sumineji bar).

The epoxy-coated bars were produced by the process shown in Fig. 2. The epoxy powder coating material used was a composition containing an epoxy resin of bisphenol A type, a curing agent (acid anhydride), a softener and an inhibitive pigment which was shown from preliminary tests to have satisfactory sea water and alkali resistance.

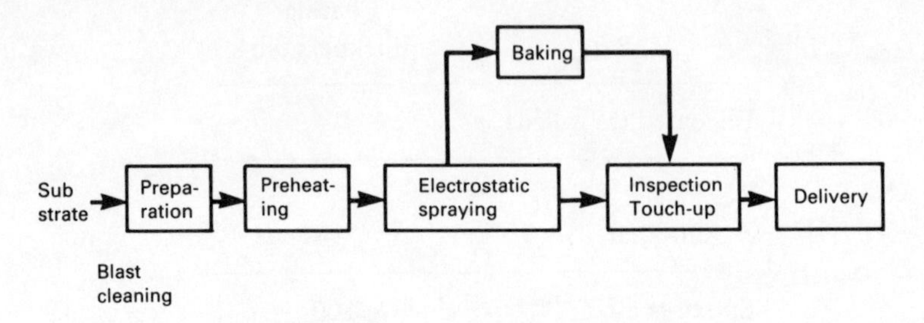

Fig. 2 – Production process of epoxy-coated bar.

PREPARATION OF TEST SPECIMENS

The test specimens consisted of square concrete prisms with a central reinforcing bar projecting at both ends and having an embedded length of 40 metres. Three sizes of prisms were used (Fig. 3), the cover and the type of bar being the main test variables.

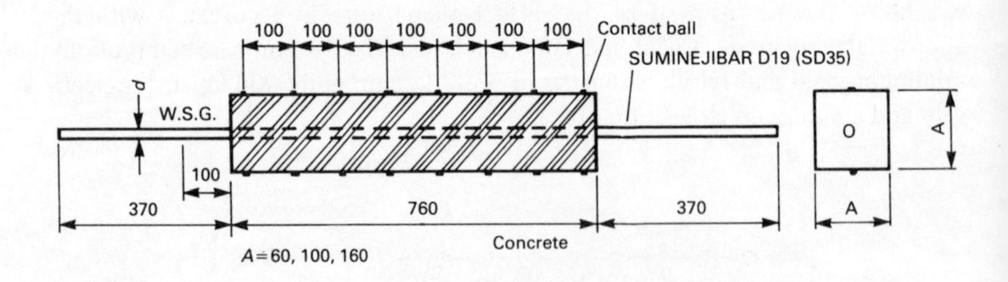

Fig. 3 – Details of test specimens.

The concrete was designed for a compressive strength of 240 kg/cm^2 at 28 days. Ordinary Portland cement was used with 25 mm maximum size of coarse aggregate. The mix proportions were 1:2.84:4.54 with a water–cement ratio of 0.549, all by weight. A water-reducing agent at about 2.5 g/kg of cement was incorporated in the mix. This gave a slump of about 80 mm for the fresh concrete. At 28 days, the compressive and tensile splitting strengths averaged 246 and 25.2 kg/cm^2 respectively.

The test specimens were cast horizontally, and the concrete well compacted with a vibrator. The specimens were cured in water at 20°C for 28 days, and then subjected to a static tensile test to generate cracks in the concrete prior to to the exposure tests. The specimens were loaded gradually until the stress in the bar reached a value of 3000 kg/cm^2. Both the number and the width of cracks were measured at predetermined stress levels with a contact gauge. Six test specimens prepared under identical conditions were tested for each variable. The number and width of cracks are both influenced by concrete cover and the type of coating on the bar; the larger the cover, the fewer the number of cracks and the larger the maximum crack width. The average values of the number of cracks and the maximum crack width on the surface of the concrete of the six test specimens are shown in Figs. 4 and 5.

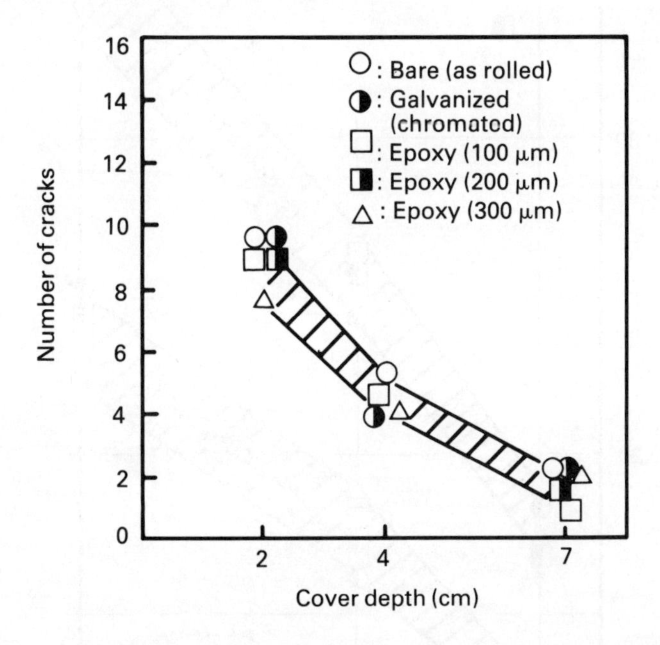

Fig. 4 – Influence of depth of cover on cracking of the test specimens.

EXPOSURE TEST CONDITIONS

The exposure tests were carried out on test specimens in a permanently loaded state, precracked, and then subjected to a constant stress of 2000 kg/cm^2 in the reinforcement bars. To keep the bar stress constant throughout the exposure tests, a special reaction frame was designed, each to hold nine test specimens. Each specimen was mounted on the frame designed to take up the reaction force by utilizing the threads of the bar, and a tensile force was applied to the bar by tightening the associated nuts. The stress in the bars was then adjusted to a value of 2000 kg/cm².

Figure 5 shows the maximum crack widths (average of 6) on the surface of the concrete at the test stress of 2000 kg/cm² in the reinforcing steel. As shown in the figure, the maximum crack widths at the beginning of the exposure tests, varied from 0.11 to 0.12 mm, from 1.16 to 0.19 mm and from 0.22 to 0.25 mm at the specified covers of 20, 40 and 70 mm.

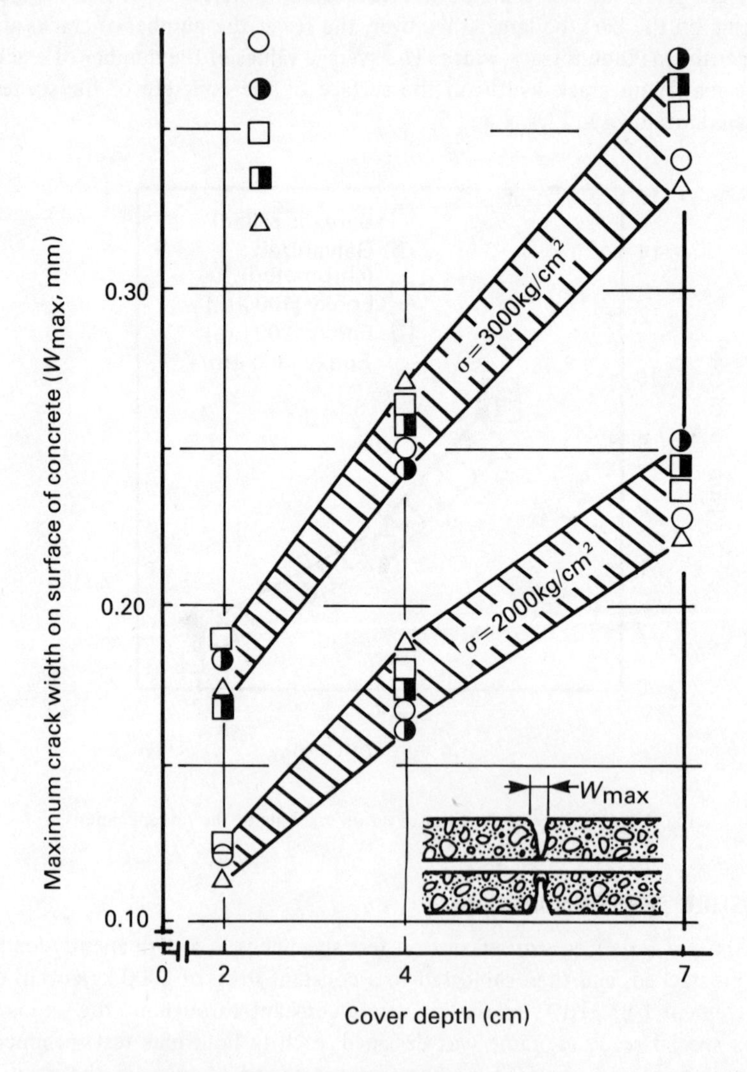

Fig. 5 – Variation of maximum crack width with cover at steel stresses of 3000 kg/cm² and 2000 kg/cm².

EXPOSURE TESTS

Two types of exposure tests, namely accelerated corrosion tests and marine exposure tests, were carried out on the loaded specimens.

Accelerated corrosion test conditions

With the bar subjected to a tensile stress of 2000 kg/cm^2, each specimen was immersed in sea water at 60°C for six hours and, then, allowed to dry in the atmosphere for six hours. This wet-and-dry cycle was repeated for six, twelve and twenty-four consecutive months. The wet-and-dry cycle of six hours was chosen to simulate the high and low tides.

Conditions of exposure test at tidal zone

With the bar subjected to a tensile stress of 2000 kg/cm^2, the test specimens have been exposed in the tidal zone which is known to be the most corrosive, in accordance with the arrangement shown in Fig. 6. The longest scheduled duration of exposure is sixty months and the results reported in this paper are those after twelve and twenty-four months.

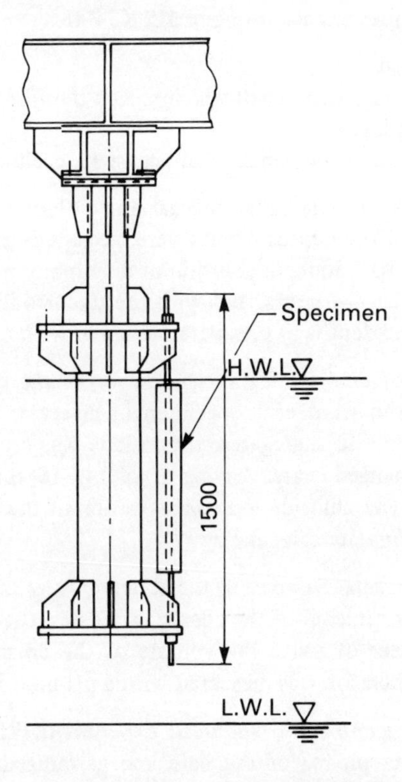

Fig. 6 – Arrangement of exposure test at tidal zone (Kashima Harbour).

EVALUATION OF EXPOSURE TEST RESULTS

Each specimen was detached from the reaction force frame and photographed to record its appearance and the pattern of cracks. The concrete cover was then removed and both the bar and the concrete were evaluated according to the following schedules.

(a) *Appearance inspection.* Each specimen was investigated for red rust covering ratio of state of corrosion. The red rust covering ratio is defined as follows.

$$\text{Red rust covering ratio (\%)} = \frac{\text{Red rust formation area}}{\text{Surface area of bar covered with concrete}} \times 100$$

(b) *Adhesive strength of the epoxy film.* The film was slit with a knife and the peeling resistance of the film was measured at the slit to evaluate its adhesive strength.

(c) *Hardness of the epoxy film.* The hardness of the epoxy film was measured by the pencil test method according to JIS K5400.

- Scoring hardness
  The hardness of the pencil which scored the film.
- Peeling hardness
  The hardness of the pencil which caused a peeling of the film.

(d) *Pitting depth.* The uncoated and galvanized bars which developed red rust were de-rusted and the depths of pits were measured. Each bar was de-rusted by immersing it in a 10% aqueous solution of diammonium hydrogen citrate (room temperature) for 15–18 hours, and with the uncorroded surface of the bar as a reference plane, the depths of pits were measured with a 15° point micrometer.

(e) *Distribution of chloride ion concentration in the concrete.* Samples of the concrete were taken from each specimen in layers at intervals of 5 mm and, after removal of coarse aggregates, the residue was crushed and shaken for six hours in ion-exchanged water having a volume 10 times that of the crushed concrete sample. The chloride ion concentration of the solution was determined by mercuric chloranylate calorimetry.

(f) *pH of the concrete.* Samples of the concrete were taken from each specimen in layers and after removal of the coarse aggregates, the residue was crushed and ion-exchanged water of twice the volume of the crushed concrete was added. The pH of the suspension was measured with a pH indicator.

(g) *The alkali concentration macrocell experiment.* The accelerated corrosion test revealed severe pitting of the bare and galvanized bars. The cause of this condition is thought to be due to the existence of an alkali concentration macrocell formed between the exposed surface of the bar at the crack after neutralization and the surface of the bar adjacent to the zone alkalinized due to penetration of sea water (Fig. 7). Based on this postulate, the experimental set-up shown in Fig. 8 was carried out. In this test, one of a pair of glass vessels is filled with artificial sea water (ASTM D-1141-52, neutral) and the other with a supernatant of artificial sea water containing an excess of cement (alkaline pH), and a test piece was immersed in each of the baths. The lead wires from test pieces

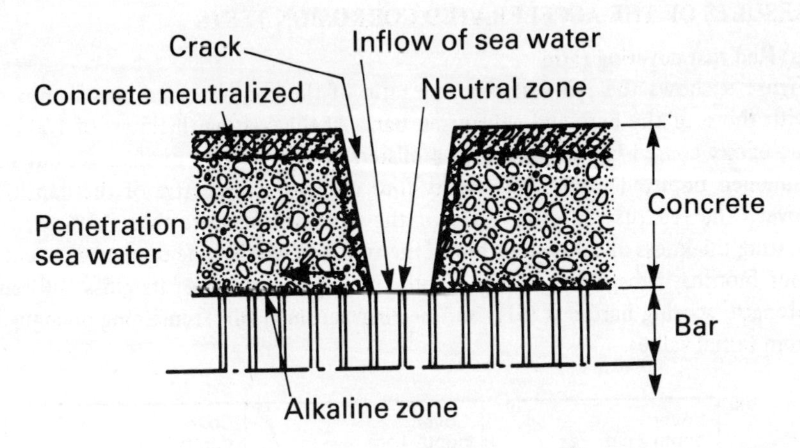

Fig. 7 – Alkali concentration macrocell model.

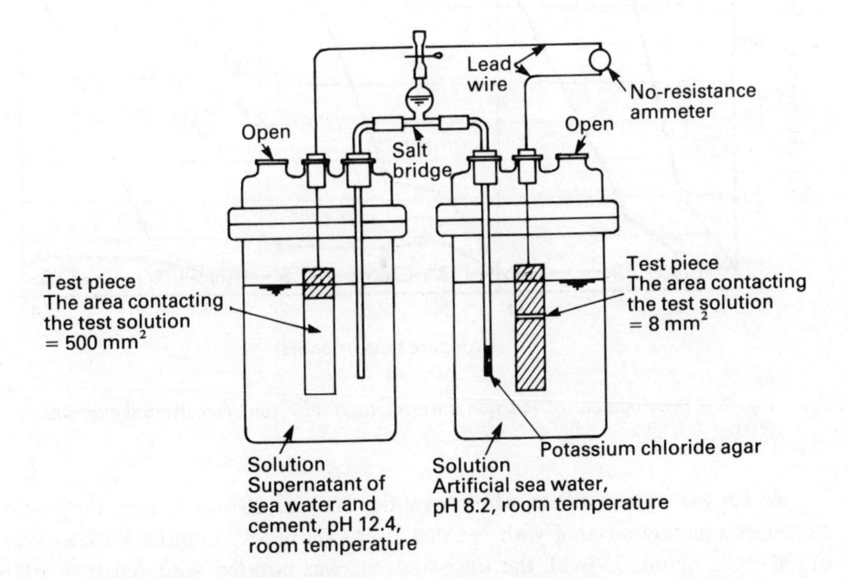

Fig. 8 – Test set-up for alkali concentration macrocell corrosion experiment.

were connected to a no-resistance ammeter and the vessels were interconnected with a salt bridge to form an electric circuit. In this set-up, corrosion current and corrosion potential were measured. The test piece in contact with artificial sea water in this set up corresponded to the bar exposed to the natural zone in Fig. 7, while the test piece in contact with the supernatant of artificial sea water and cement corresponded to the bar adjacent to the alkaline zone of Fig. 7.

RESULTS OF THE ACCELERATED CORROSION TESTS

(a) Red rust covering ratio

Figure 9 shows the red rust covering ratio of the epoxy-coated bars compared with those of the bare and galvanized bars. At the coating thickness of 100 μm, the epoxy-coated bar showed 2 to 4 blisters less than 3 mm in diameter but still remained unpitted even after twenty-four months, irrespective of the depth of cover. The red rust covering ratio of the bar was not more than 0.1%. At the coating thickness of 200 μm or more, the film was wholesome even after twenty-four months, irrespective of the depth of cover, with all of its gloss, adhesive strength, scoring hardness (5H) and peeling hardness (6H) remaining unchanged from initial values.

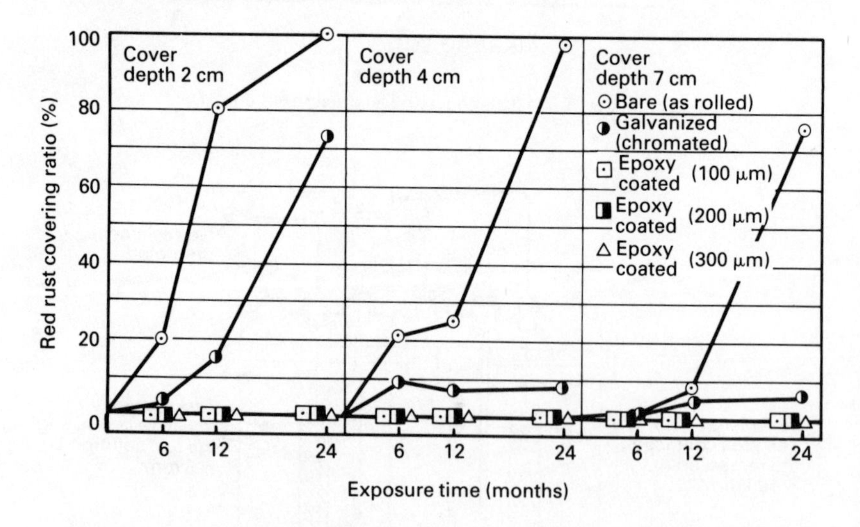

Fig. 9 – Development of red rust covering ratio with time (accelerated exposure test).

As for the uncoated bars, when cover depths were 20 and 40 mm, they were almost completely covered with red rust after twenty-four months. With a cover of 70 mm, about 75% of the uncoated bar was covered with red rust after twenty-four months. The galvanized bar after twenty-four months showed a

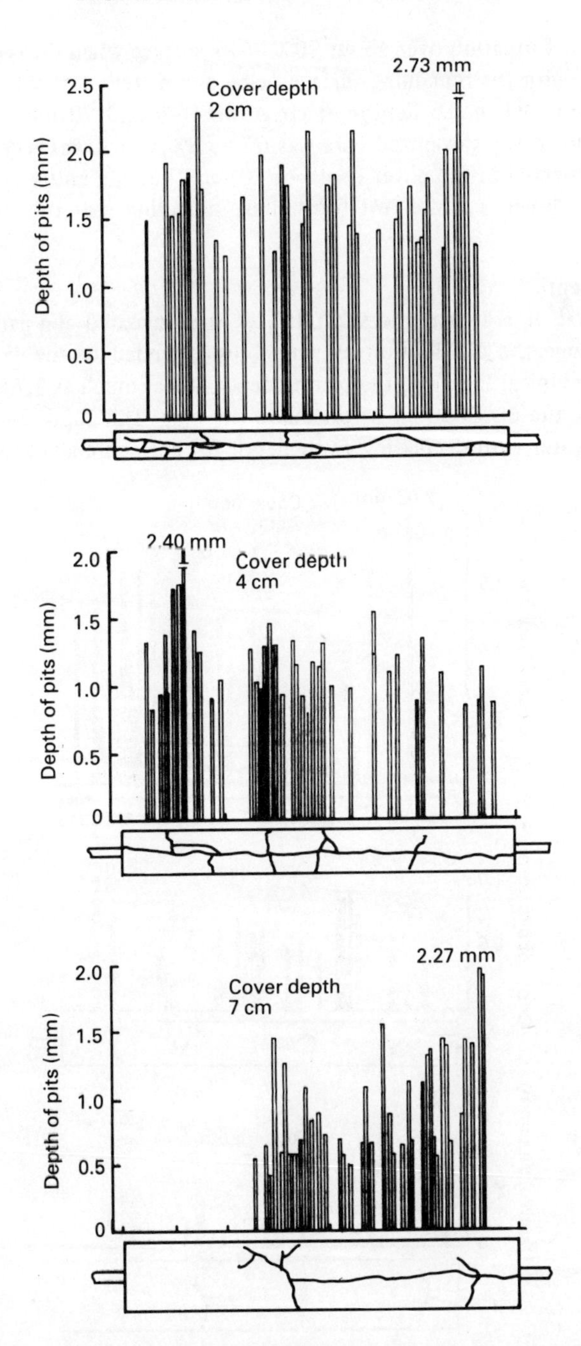

Fig. 10 – Correlation between concrete cracking and distribution and depth of pitting in uncoated bars (after twenty-four months accelerated corrosion test).

thick red rust formation over about 70% of its surface when the concrete cover was 20 mm, with the remaining surface being completely covered with a white rust formation. When the depths of cover were 40 and 70 mm, the red rust covering ratio of the galvanized bars was 6% to 8%, which was very much lower than that observed at the cover depth of 20 mm, but the entire remaining surface had developed a white rust formation, indicating that the zinc layer was corroded.

(b) Pitting depth

After removal of red rust, the substrates of the uncoated and galvanized bars showed pitting, the distribution of which corresponded to the distribution of cracks. At a typical location, the pitting depth was as much as 2.7 mm. Figs. 10 and 11 show the distributions of pitting in the bars. The depth and number of pits were greater with decreasing cover depth for both bare and galvanized bars.

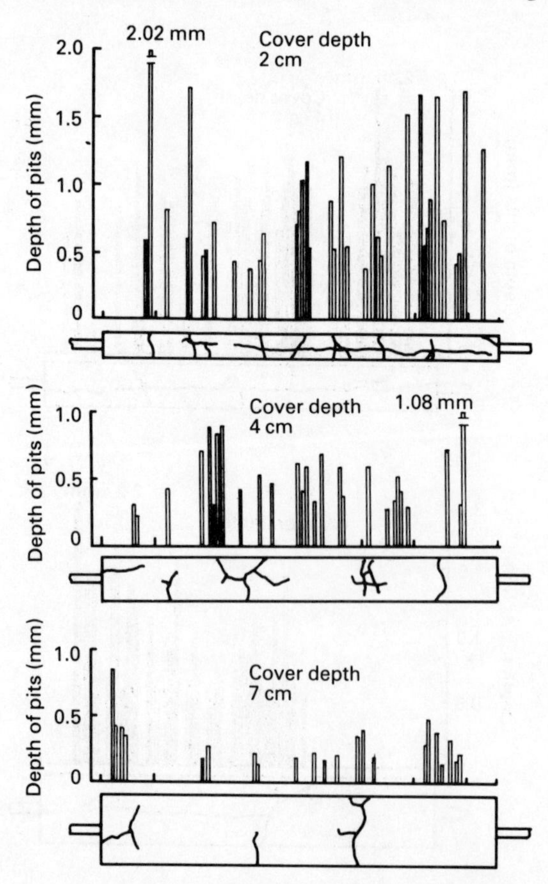

Fig. 11 – Correlation between concrete cracking and distribution and depth of pitting in galvanized bars (after twenty-four months accelerated corrosion test).

These results show that the galvanized bars were superior to the untreated bars, but that the galvanizing did not provide complete protection against pitting.

RESULTS OF THE EXPOSURE TEST AT TIDAL ZONE

(a) Red rust covering ratio

Figure 12 shows the red rust covering ratio of the epoxy-coated bars compared with those of the bare and galvanized bars. It is clear that, irrespective of the depth of concrete cover and film thickness, the epoxy-coated bars remained totally unaffected by rust. The bars not only retained the intial values of gloss, adhesive strength and hardness, but were found to be completely wholesome.

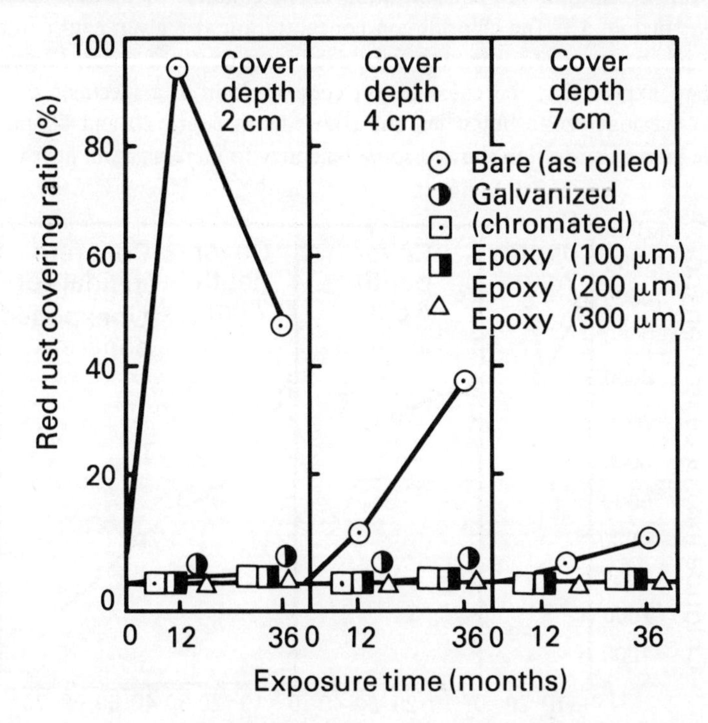

Fig. 12 – Development of red rust covering ratio with time (exposure test at tidal zone).

Compared with the other bars, the uncoated bars showed extensive and deep red rust formation. At the cover depth of 20 mm, about 50% of its surface was covered with red rust after thirty-six months, the rust being distributed throughout the length. Generally, red rust covering ratio increased with time. However, the smaller the cover depth, the greater was the variation of red rust covering ratio and, in some instances, the ratio after twelve months was larger than that after thirty-six months when the concrete cover depth was 20 mm.

When the cover depths were 40 and 70 mm, the red rust covering ratios were about 40% and about 10% respectively, after thirty-six months, the red rust formation was generally greater on the lower part of the tidal zone.

The red rust covering ratio of the galvanized bars was as low as 1.4% at the maximum even after thirty-six months of exposure but showed some localized white rust formation, suggesting a progress of corrosion in the zinc layer. The rust formation of the galvanized bars was greater than the upper part of the tidal zone, contrary to the phenomenon observed with the uncoated bars.

(b) Distribution of chloride ion concentration in concrete

The profile of chloride ion concentration in the concrete of the test specimens is shown in Fig. 13. The chloride ion concentration at a given depth from the surface of the specimen is greater with decreasing depth of cover. It would be natural to expect that the chloride ion concentration of a specimen decreases towards the inner zone but when the cover depths were 20 and 40 mm, the chloride ion concentration showed some tendency to increase again near the bar.

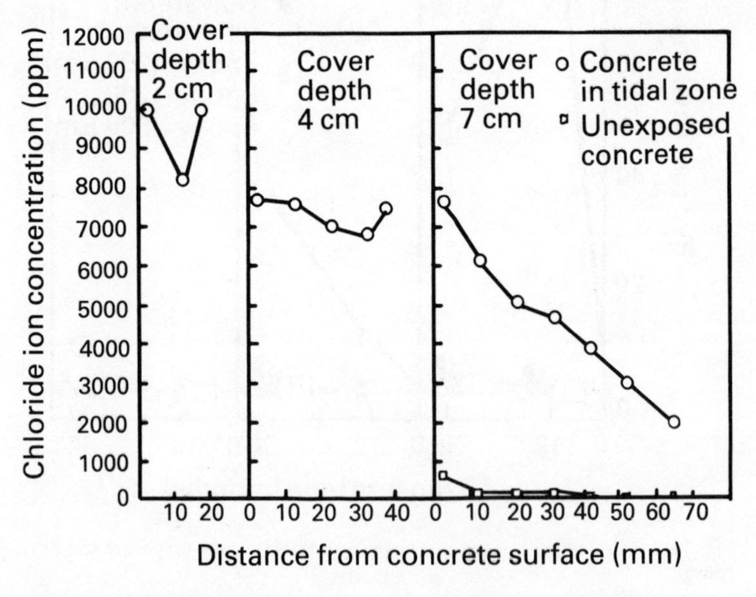

Fig. 13 – Development of chloride ion concentration in concrete (after 36 months of marine exposure).

(c) pH of concrete

The measured pH values of concrete in the test specimens are shown in Fig. 14. The value on the exposed surface or crack surface of the specimen is the average value for a 1 mm-thick layer perpendicular to the surface. Accordingly, it is considered that the pH value of the exposed surface is lower than the indicated value and the pH value was substantially low in the neutral region. The pH within the concrete was more than 12 at 10 mm from the surface and for all cover depths, the pH of concrete in the neighbourhood of the bar remained in the range 12.6 to 12.8.

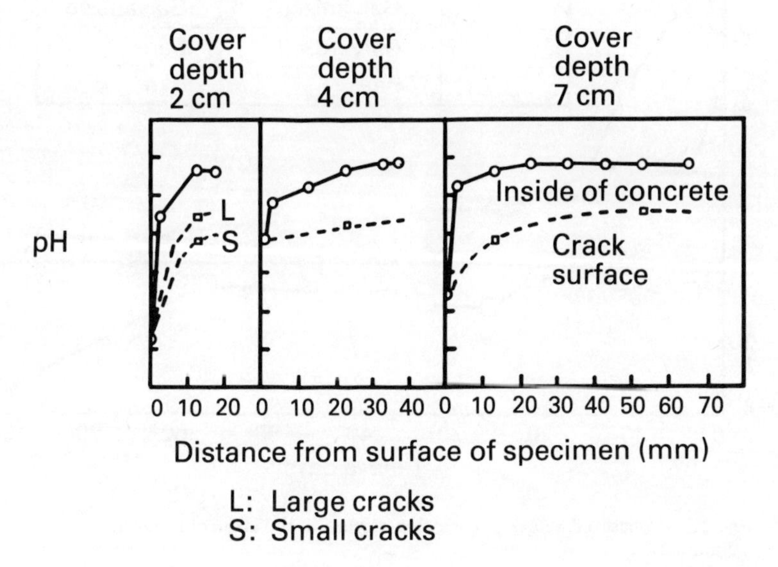

Fig. 14 – Variation of pH in concrete (after 36 months of marine exposure).

(d) Results of the alkali concentration macrocell experiment

The variation with time of the macrocell corrosion current flowing in the model of Fig. 8 is shown in Fig. 15. The direction of current flowing from the test piece immersed in the supernatant of artificial sea water and cement (alkaline solution) though the lead wire to the test piece immersed in the artificial sea water (neutral solution) was taken to be positive. The open circuit potential of the test piece is shown in Fig. 16.

In the case of macrocell corrosion with a pair of galvanized bars, the zinc layer of the test piece in the neutral solution functions as an anode and is corroded, with the zinc layer of the test piece in the alkaline solution acting as a cathode. A corrosion current of 40 μA or more flows in the initial period of the experiment but this decreases with time. In a pair of galvanized bar and bare bar, a negative current is generated. Thus, the test piece on the neutral solution side acts as a cathode and the test piece on the alkaline solution side acts as an anode and is corroded. In this case, the formation of a galvanic cell owing to contact of zinc and steel is considered to be the governing factor in the progress of corrosion of the zinc layer. The high corrosion between zinc and steel and the consequent increase of electromotive forces. In a pair of bare bars, the bar on

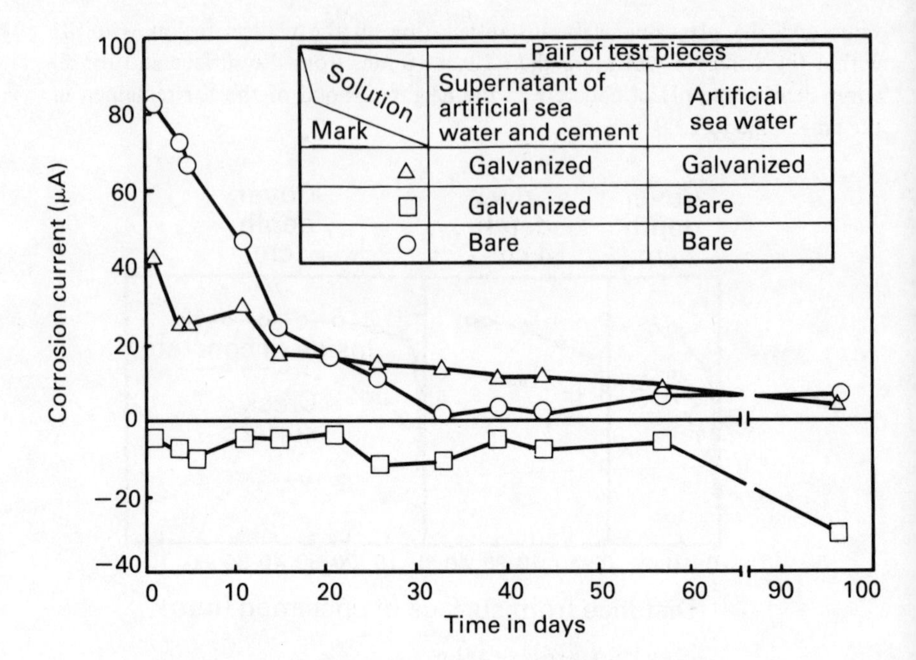

Fig. 15 – Measured values of corrosion current in the alkali concentration macro-cell model.

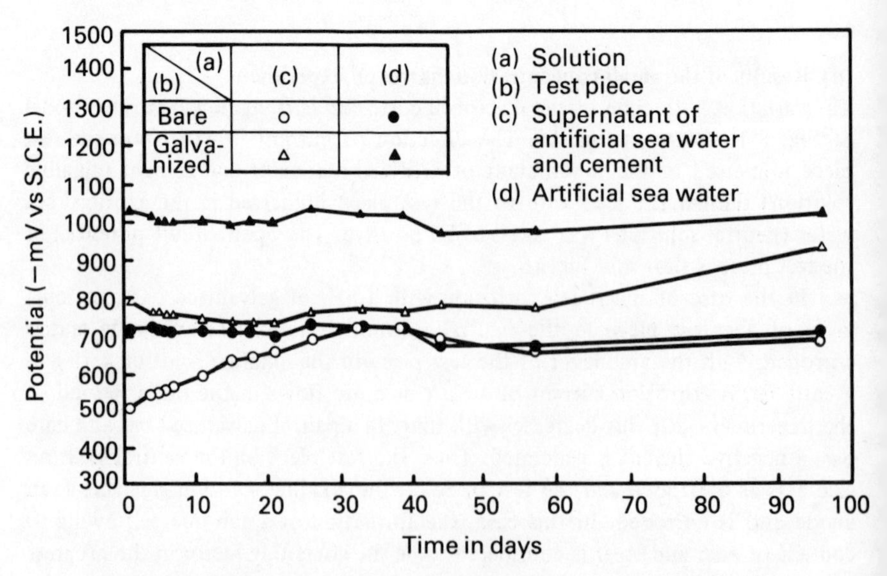

Fig. 16 – Open circuit potential in the artificial sea water and supernatant of artificial sea water and cement.

the neutral solution side acts as an anode and is corroded, with the bar on the alkaline side acting as a cathode. While a corrosion current of 80 μA or more flows in the initial period of the experiment, it decreases to a few milliamperes as the anode surface is covered with corrosion products.

DISCUSSION OF TEST RESULTS

(a) Corrosion tests with epoxy-coated bar

In both the accelerated corrosion test and the exposure test at the tidal zone, epoxy coating protected the reinforcing bars and any abnormality noted in the epoxy film was a limited degree of blistering which occurred only when the epoxy coating thickness was small (100 μm). The epoxy-coated bar used in this study was selected as it had shown satisfactory results in all of the 500-hour salt spray test (JIS Z2371), the 2-week artificial sea water immersion test (sea water saturated with air at room temperature, ASTM D1141-52), the 2-week cement supernatant immersion test (pH 12.6, room temperature) and the 2-week NaOH (3M) solution immersion test (room temperature). It was found that the film giving satisfactory results in such laboratory tests would also display desirable corrosion resistance in long-term corrosion tests.

(b) Influence of concrete cover depth on the corrosion of reinforcing bar

From the test results it appears that there is no reasonable correlation between Cl^- ion concentration and red rust covering ratio in the cracked condition.

If a crack is produced in a reinforced concrete structure, water, air and other corrosive elements find their way into the structure in an amount corresponding to the width of the crack. Duffaut *et al.* conducted an exposure test in a marine environment and showed that cracking is a dominant cause for the corrosion of bar and that corrosion progresses considerably when the width of cracks reaches 0.15 to 0.5 mm [12].

In the specimens used in the present study, the number of cracks produced by applying a given tensile stress to a bar was dependent on the depth of concrete cover, and as shown in Fig. 4, the larger the cover depth, the smaller was the number of cracks produced. Since the sea water entering into cracks corrodes the bar, red rust covering ratio is governed by concrete cover depth and although its magnitude varies somewhat with the presence or absence of an inhibitive film, the greater the concrete cover depth, the smaller is the red rust covering ratio.

(c) Influence of environment on the corrosion of reinforcing bar

The main cause of high alkalinity in the liquid phase of Portland cement concrete is that a large amount of $Ca(OH)_2$ corresponding to about 30% by weight of the cement used is produced as a result of pozzolanic reaction [13].

At a pH of 9.5 or more, a film is produced on the bar which is due to a decrease of solubility of the ferrous hydroxide produced in an initial stage of corrosion and as it is oxidized by oxygen, the hydroxide is converted to iron oxide and, then, the film is passivated. As this occurs, the rate of diffusion of oxygen decreases and the rate of corrosion is accordingly reduced. However, as carbon dioxide gas enters and diffuses into the concrete to lower its pH, the effect of the protective layer is reduced and the rate of corrosion increases as a result [14].

The neutralization of concrete begins on the surface and progresses into the concrete with time. Regarding the degree of influence of environmental conditions on the depth of neutralization in a given time period, Kishitani [13] has suggested that with the influence of general outdoor conditions (concentration of CO_2 = 0.03%) being taken as unity, that under general indoor conditions (CO_2 = 0.1%) is 1.5 to 3, and that if the carbon dioxide concentration and humidity be held constant and the influence at 20°C be taken as unity, the influence at 40°C will be 2. He suggested, also, that if carbon dioxide concentration and temperature be held constant and the influence at a relative humidity of 80% be taken as unity, the influence of 40% R.H. will be 1.8.

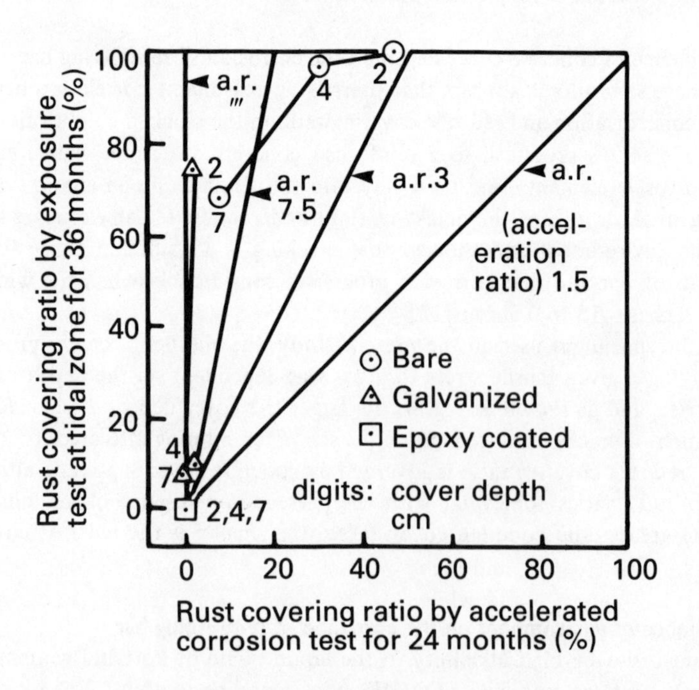

Fig. 17 – Acceleration ratio of accelerated corrosion test to exposure test.

As will be seen from Fig. 17, there was a significant difference in red rust covering ratio between accelerated corrosion and corrosion at tidal zone and despite the fact that the duration of exposure at tidal zone was 1.5 times as long as that of accelerated corrosion, the red rust covering ratio in the latter case was smaller than that under accelerated conditions. Based on the test data referred to above, this difference may be accounted for as follows.

- Accelerated corrosion
  The neutralization reaction of concrete in the presence of a crack deep enough to reach the bar, i.e.

$$Ca(OH)_2 + CO_2 \rightarrow CaCO_3 + H_2O,$$

  progresses vigorously under accelerated conditions (60°C, indoors) and the bar is rusted without protection by alkali.
- Corrosion at tidal zone
  In corrosion at tidal zone (atmospheric temperature, outdoors), the neutralization reaction proceeds only at a low rate and the sea water reaching the bar through a crack is alkalinized by the previously formed calcium hydroxide with the result that the bar is rusted only locally.

(d) Mechanism of pitting

Based on the results on the alkali concentration macrocell experiment, the pits in the galvanized bar used in the accelerated corrosion test were considered to have progressed in the following three stages.

First stage: The stage where the zinc layer exposed to a crack is corroded.

As the neutralization of concrete progresses to the extent that the zinc layer exposed to the crack is surrounded by a neutral liquid, an alkali concentration macrocell is formed between it and the adjoining zinc layer in contact with the alkaline liquid with the result that the zinc layer exposed to the crack acts as an anode and is corroded.

Second stage: The stage where the corrosion of the zinc layer spreads to the adjoining layer.

As the zinc layer exposed to the crack is consumed, a galvanic cell is formed between the bar surface and the zinc layer so that the latter layer is corroded.

Third stage: The stage in which the bar is pitted.

As the zinc layer around the crack is consumed, some of the bar surface is exposed to a neutral liquid while the other surface is exposed to an alkaline liquid. In this stage, the bar surface exposed to the neutral liquid acts as an anode and is corroded.

In the case of a bare bar, pitting commences with the third stage, bypassing the first and second stages.

CONCLUSIONS

From the results of the tests presented here, the following conclusions can be drawn.

(1) After twenty-four months of accelerated corrosion tests, the uncoated bar showed extensive corrosion even with a cover of 70 mm. With 40 to 70 mm cover, the galvanized bars showed signs of corrosion. With an epoxy coating of 200–300 μm, the bars remained unaffected and retained all the original properties of the coating even with as small a cover as 20 mm.

(2) Although galvanized bars were superior to the untreated bars after twenty-four months of accelerated tests, the galvanizing did not provide complete protection against pitting.

(3) The marine exposure tests showed extensive and deep rusting in plain bars. Although the rust formation in the galvanized bars was low, there was localized rusting, suggesting a progress of corrosion in the zinc layer.

(4) The penetration of chloride ions under marine exposure decreased with distance from the concrete surface, but with covers up to 40 mm, there was some indication of increased chloride ion concentration near the bar surface.

(5) Under marine exposure, the pH within the concrete was more than 12 at 10 mm from the surface whilst the pH in the vicinity of the bar remained between 12.6 and 12.8.

(6) The acceleration ratio for the untreated bars in the accelerated corrosion test was more than five. For galvanized bars also, the conditions were rigorous and the acceleration rate high.

(7) The suspected mechanism of pitting appears to be that as the crack surface is neutralized, an alkali concentration macrocell is formed between the steel or zinc surface exposed to the cracks and the steel or zinc surface adjacent to the high-alkalinity zone.

(8) With a coating thickness of 200 μm, epoxy coating protected the bar substrate well against chloride attack and corrosion, and the film itself remained intact, irrespective of depth of concrete cover. It is suggested that epoxy coating can provide effective protection against corrosion for a long time even when chloride ions directly reach the metal surface of the bar.

REFERENCES

[1] Hidejima, S., *Report on the Corrosion and Corrosion Inhibition Committee of the Japanese Society of Industrial Materials,* Vol. 14, No. 74, Part 1, Jan. 27, 1976.

[2] Cleer, K. C., Time-to-corrosion of reinforcing steel in concrete slabs, Vol. 3, Performance after 830 daily salt applications, *Interim Report, FHWA-RD-76-70,* Federal Highway Administration, Washington DC, April 1976.

[3] Transportation Research Board, Durability of concrete bridge decks, *National cooperative highway research program, synthesis of highway practice 57,* Transportation Research Board, National Research Council, Washington DC, May 1979.

[4] Anon., Salt corrosion confirmed in Cumbrian Bridges, *New Civil Engineer,* 26 March 1981.

[5] Pike, R. G., Hay, R. E., Clifton, J. R. and Beeghly, H. F., Nonmetallic protective coatings for concrete reinforcing steel, corrosion and corrosion protection, *Transportation Research Record 500,* Transportation Research Board, National Research Council, Washington DC, 1974.

[6] Clifton, J. R., Beeghley, H. F. and Mathey, R. G., Nonmetallic coatings for concrete reinforcing bars, *Building Science Series 65,* US Department of Commerce, National Bureau of Standards, August 1975.

[7] Brown, M. G., Control and prevention of deterioration of concrete bridge decks, *Report FHWA-R-1034,* Testing and Research Division, Michigan Department of State Highway and Transportation, Michigan, December 1976.

[8] Beeby, A. W., Corrosion of reinforcing steel in concrete and its relation to cracking, *The Structural Engineer,* **56A** (3), March 1978, pp. 77–81.

[9] Wills, J., Epoxy coated reinforcement in bridge decks, *Supplementary Report 667,* Transport and Road Research Laboratory of Transport, 1982, pp. 20.

[10] *Corrosion of reinforcement and prestressing tendons, state-of-the-art report,* RILEM Technical Committee 12-CRC Materiaux et Constructions, Vol. 9, No. 51.

[11] Federal Highway Administration, Concrete bridge decks, US Department of Transportation, *Federal-Aid Highway Program Manual,* April 5, 1976.

[12] Duffaut, P., *et al., The Road,* October 1973.

[13] Kishitani, K., *Corrosion Engineering (Boshoku Gijutsu),* **24** (3), 1975.

[14] Okada, K., *Report of the Corrosion and Corrosion Inhibition Committee, Japanese Society of Industrial Materials,* Vol. 14, No. 74, Part I, Jan. 27 1976.

Chapter 22

The influence of constructional and manufacturing conditions on the corrosion behaviour of prestressed wires before grout injection

B. ISECKE, Bundesanstalt für Materialprüfung, 1000 Berlin 45, Federal Republic of Germany

ABSTRACT

This report describes the environmental conditions for prestressed steels in un-injected ducts which are important for corrosion. Endurance tests simulating practical conditions provide information about the resistance of different types of prestressed steels to hydrogen-induced stress corrosion cracking after pre-corrosion of the steels.

INTRODUCTION

In recent years there have been many failures of prestressed concrete constructions due to hydrogen-induced stress corrosion cracking [1], some of these failures occurring before the grout was injected into the ducts. The following report is aimed at finding out what conditions cause corrosion in the ducts of prestressed concrete buildings during construction. The environmental conditions to which the steels were subjected before injection were also simulated in order to analyse their influence on the corrosion behaviour of different steels.

EXAMINATIONS AT THE BUILDING SITES

To the present day no study of the conditions causing corrosion in non-injected ducts has been made. The essential parameters for corrosion attack on prestressed steel in non-injected ducts are the relative humidity, the temperature and the oxygen content of the atmosphere of the duct. Furthermore, the ducts were examined to see if they contained water and, if so, of what composition.

TEMPERATURE AND RELATIVE HUMIDITY IN THE DUCTS

Combined measuring probes, which allowed measurement of temperature and relative humidity in the ducts between the casting of concrete and the injection of the ducts, were inserted from above through the air ducts and the results recorded continuously.

Figure 1 shows the variation in the duct of relative humidity with time after concreting. Measuring probes were introduced at two points along the duct, point 1 at the beginning of the duct; point 2 in the middle. Another measuring point 3 was situated outside the duct in the external atmosphere for comparison with the inner atmosphere. It was found that only 1 hour after concreting the relative humidity at both measuring points in the duct was significantly higher than in the outer atmosphere. This led to the assumption that water had penetrated the duct during concreting and was confirmed by blowing compressed air through the duct before grout injection. The heat of reaction during concreting drove the temperature during measurement to 60°C. At first the relative humidity in the duct dropped a little, since time was needed to reach equilibrium of relative humidity, temperature and water in the duct. 24 hours after concreting, both measuring points in the duct reached a steady state of 100% relative humidity and stayed unchanged until injection. Simultaneously the temperature in the duct dropped steadily over about 180 hours from 60°C to ambient values. Other stressing systems yielded similar results. Therefore we may assume a relative humidity of 100% in non-injected ducts.

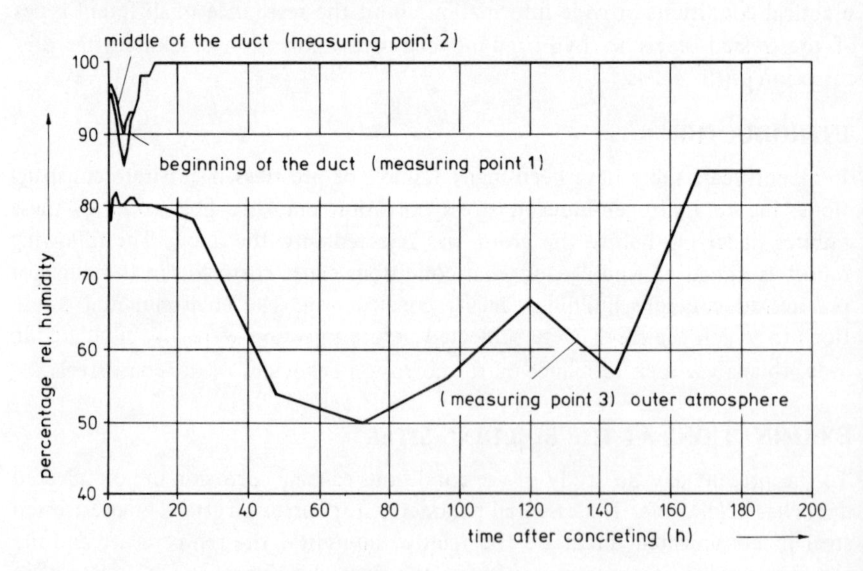

Fig. 1 – Percentage relative humidity in the duct after concreting.

OXYGEN CONTENT

Oxygen content in non-injected ducts was determined by inserting oxygen sensors into the ducts at various points and the oxygen content registered until grout injection. The results varied for different techniques. In longer tension elements, which after concreting had a good seal against the outer atmosphere, an oxygen gradient was set up over the length of the duct. Whereas near the anchoring (about 5 m distance) the oxygen content dropped to about 16%, oxygen contents below 10% could be measured at greater distances (about 30 m).

In other tension techniques with less efficient sealing of the ducts against the outer atmosphere there was no variation of oxygen content from that of air.

COMPOSITION OF WATER IN THE DUCTS

Another important aspect of the corrosion of steel was whether or not water could gather in the ducts before and after concreting and what its composition was. Therefore at different sites the ducts were blown through with compressed air after concreting. The outflowing water was collected and analysed in the laboratory.

The compositions of a few samples are listed in Table 1. Nearly all samples had very high pH-values and their concentrations of chlorides and sulphates lay above the composition of tap water. Probes 7 and 14 show extraordinarily high concentrations of chlorides and sulphates which are of importance in corrosion. The concentrations indicate that the water was a product of concreting. These conditions were simulated in the laboratory and their effect on prestressed steels observed.

Table 1 – Water samples taken from ducts.

Number	pH-value at 20°C	Chlorides mg/l	Sulphates mg/l
1	11.4	81	869
2	10.9	58	512
3	11.4	115	475
4	10.3	82	235
5	11.5	32	100
6	11.3	145	–
7	11.8	220	2830
8	11.9	301	4304
9	7.7	337	4327
10	11.1	284	4881
11	9.0	277	4007
12	7.1	291	3969
13	7.1	267	3933
14	9.7	250	3667

LABORATORY TESTS

Three different commercial steels were taken for study, their compositions are shown in Table 2. The first part of the code number indicates the yield stress, the second the maximum stress.

Table 2 – Composition of the prestressed steels.

Steel	Percentage							
	C	Si	Mn	P	S	Al	Cr	Cu
St 1080/1230 hot rolled $d = 26$ mm	0.71	0.81	1.41	0.012	0.021	0.032	0.04	0.15
St 1420/1570 quenched and tempered $d = 12.2$ mm	0.50	1.80	0.62	0.012	0.018	0.036	0.40	0.05
St 1375/1570 cold drawn and tempered $d = 12.2$ mm	0.83	0.26	0.65	0.011	0.016	0.060	0.03	0.02

PRECORROSION

Firstly, the steel is subjected to a humid atmosphere in the unstressed condition. To achieve this, a precorrosion system was developed, whose construction is shown schematically in Fig. 2.

The tensile steel is placed in a corrugated PVC-duct. Dilute solutions (250 to 500 mg/l) of chlorides, sulphates, nitrates or sulphides are fed under control into the duct as electrolytes. The bar remains immersed in solution for 1 hour, then the solution is drained from the duct. To simulate the effect of gaps, the bottom of the bar rests on the corrugations, which trap the solution. The steel is left for two hours in this condition, then the duct is flooded once more and the whole process repeated continuously for three weeks. The effect of each type of ion is tested separately.

After removal from the precorrosion apparatus, the bars are first given a tensile test to ascertain whether or not precorrosion alone leads to mechanical embrittlement. It was found that in spite of heavy surface corrosion, none of the precorrosion media affected the mechanical properties of the steels.

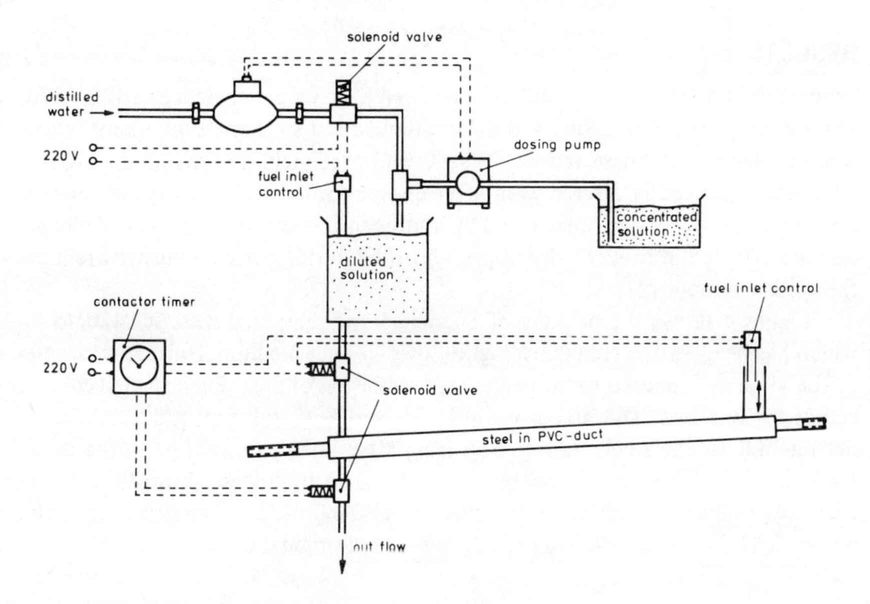

Fig. 2 – Precorrosion system.

ENDURANCE TESTS AFTER PRECORROSION

To examine the effect of mechanical stressing after precorrosion, the bars are then loaded to 80% maximum stress in an endurance test, shown schematically in Fig. 3. The precorroded bar is stressed against the frame and is surrounded by a duct, divided into two sections. This duct is filled either with tap water, or with a duct medium such as would be found on a building site (e.g. containing $Ca(OH)_2$ with additions of sulphates and chlorides), so that only the bottom of the duct is open to air, the other is airtight to study the effect of decreasing oxygen concentration. Premature failures are examined by the usual methods. If failure does not occur in 1000 hours a tensile test is made and the results compared with untreated steels from the same melt.

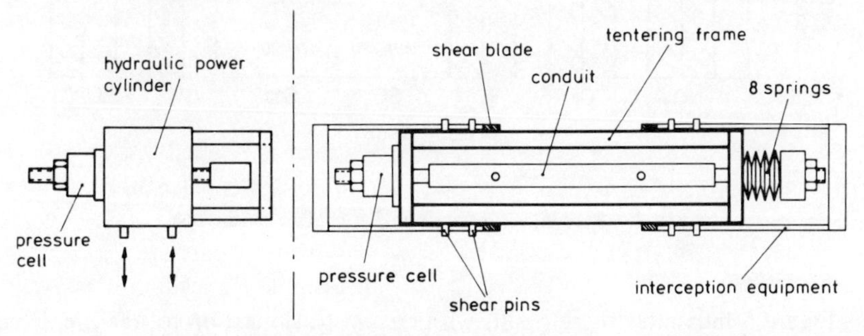

Fig. 3 – Endurance test system.

RESULTS

Premature failures occurred only in quenched and tempered steel St 1420/1570. The majority of samples survived the endurance test so that a subsequent tensile test was necessary. These tensile tests showed that yield and maximum stresses were not significantly altered even after endurance loading. This is not unusual since these properties are unaffected by hydrogen absorption. Necking and elongation are affected, however – hydrogen absorbed during corrosion usually reduces these considerably [2].

Figure 4 shows the necking of quenched and tempered steel St 1420/1570, which has been tensile tested after endurance in duct medium. Untreated samples of the same melt had 40 to 50 percentage reduction of area. Each shaded column stands for one test. The precorrosion in S^{2-} and SO_3^{2-} solution proved the most detrimental to the steels, failure occurring after only a short time in the endurance test. This behaviour can be traced back to hydrogen absorption during corrosion. Chloride and sulphate solutions also reduced necking considerably, while nitrate precorrosion had little effect on the original values.

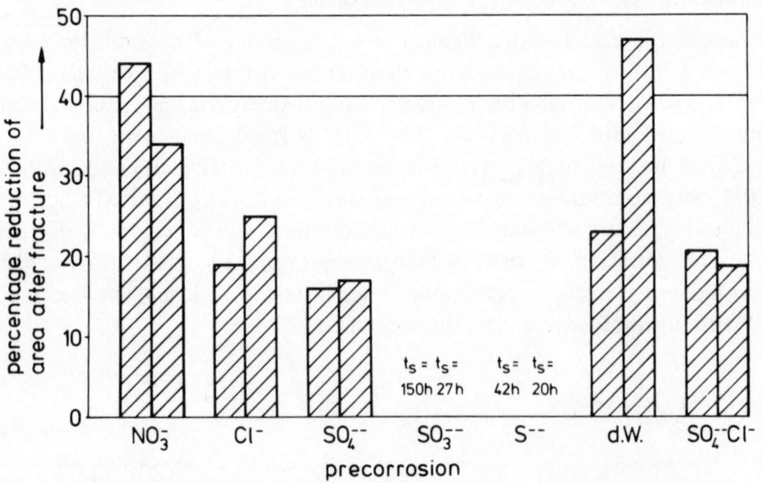

Fig. 4 – Percentage reduction of area after tensile test.

Figure 5 illustrates these results with respect to elongation to fracture. Even these values lie well below those of the untreated specimens.

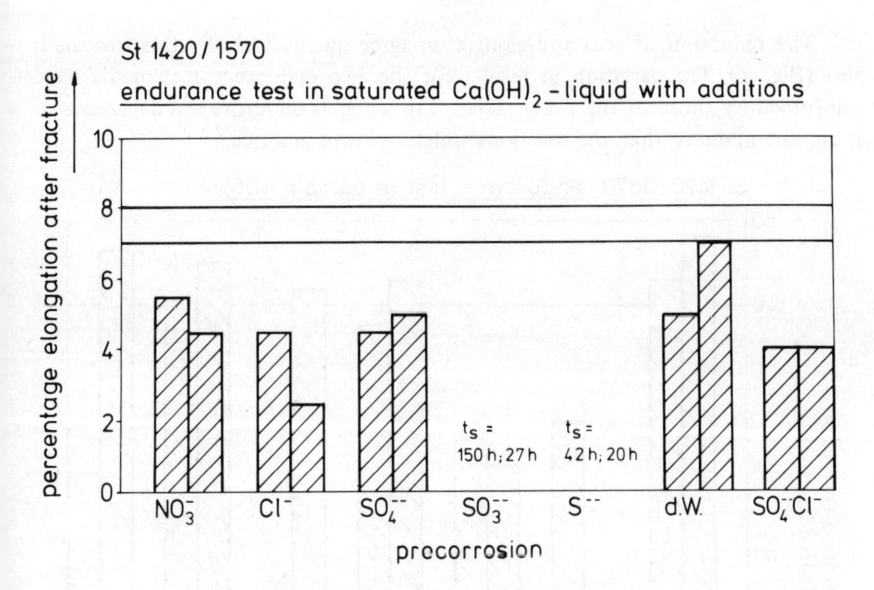

Fig. 5 – Percentage elongation after tensile test.

Substitution of tap water for duct medium produces other, different results. Figure 6 shows how in this case only two failures occurred SO_3^{2-} precorrosion had no effect and Cl^- and SO_4^{2-} precorrosion produced similar effects as for duct medium.

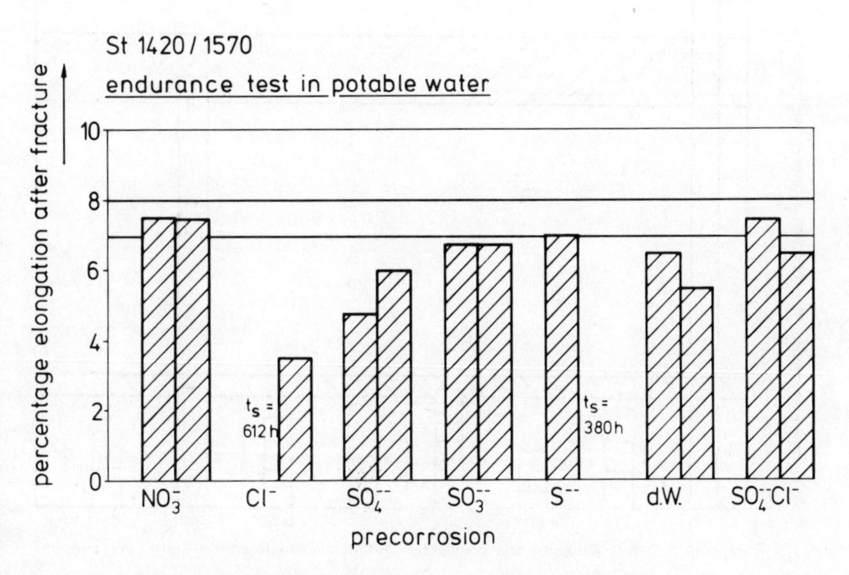

Fig. 6 – Percentage reduction of area after tensile test.

The reduction of area and elongation agree qualitatively by this treatment too (Fig. 7). The variation in results for the two endurance test media were reinforced by those of the other steels. Tap water is therefore less dangerous, if it appears in ducts, than the solutions which occur in practice.

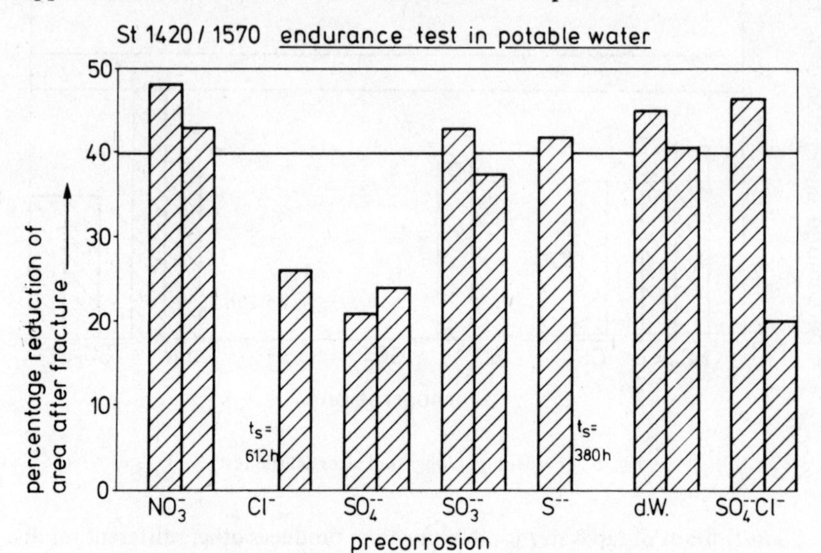

Fig. 7 – Percentage elongation after tensile test.

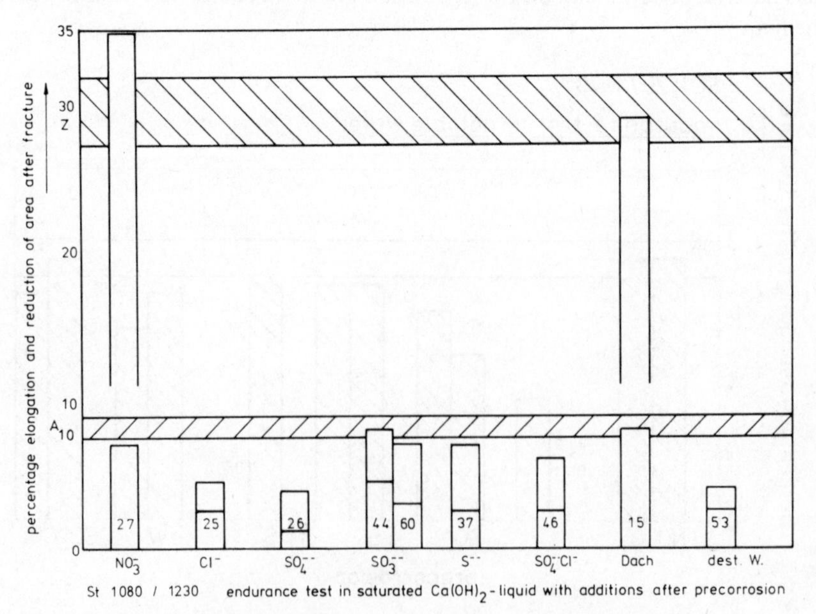

Fig. 8 – Percentage elongation and reduction of area after tensile test.

Figure 8 shows the elongation (lower section) and necking of perlitic tensile steel St 1080/1230 after endurance testing in duct medium as a function of precorrosion. No failures occurred during endurance testing. With the exception of nitrate precorrosion and a previous exposure to the atmosphere, all precorrosion types lead to distinct reduction in the two properties.

The greatest resistance to hydrogen embrittlement was shown by cold drawn steel St 1375/1570. Only in the case of sulphide precorrosion were the properties altered to an extent (Figs. 9 and 10) where hydrogen embrittlement could be considered the cause.

St 1375 / 1570 endurance test in saturated $Ca(OH)_2$ - liquid with additions

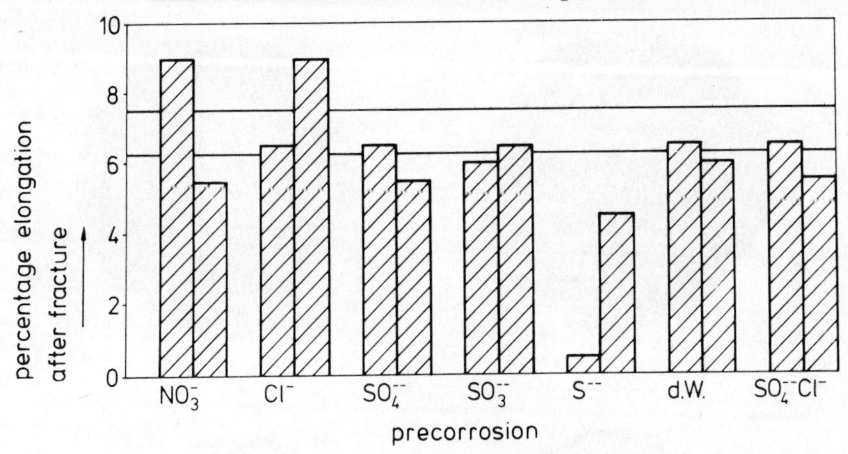

Fig. 9 – Percentage elongation after tensile test.

St 1375 / 1570 endurance test in saturated $Ca(OH)_2$ - liquid with additions

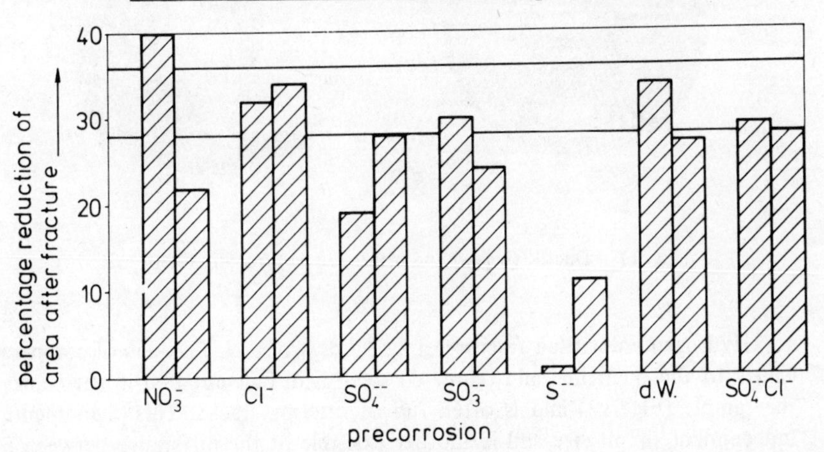

Fig. 10 – Percentage reduction of area after tensile test.

EVALUATION OF FRACTURES

All fractures occurring in endurance and tensile test were studied and described. Macroscopically, the brittle and ductile fractures are distinguishable by the disappearance of necking. Figure 11 illustrates this by comparison with the ductile fracture of an untreated quenched and tempered steel (lower pair). Additionally, it may be seen that in the embrittled steel fracture does not initiate at the centre but on the surface, usually at a corrosion pit. Smaller pits, invisible to the naked eye, tend to have a more serious effect than those with large areas.

Fig. 11 – Ductile (below) and brittle (above) fracture (St 1420/1570).

Hydrogen embrittled fracture of a cold drawn steel, although of comparable values to the quenched and tempered steel, is in fact initiated in the centre of the sample (Fig. 12) and is often full of interior cracks. This phenomenon is independent of pit size and is another example of the difference between SCC and hydrogen embrittled failure.

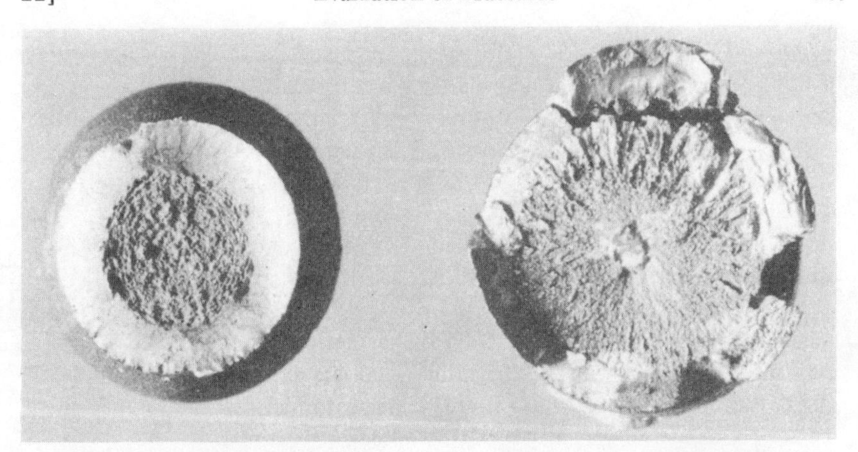

Fig. 12 – Ductile (left) and brittle (right) fracture (St 1375/1570).

This difference in fracture behaviour due to hydrogen absorption through corrosion in cold drawn and quenched and tempered steel is also visible on a microscopic scale. The initial fracture in the quenched and tempered steel is characterized by dissolved Austenite grain boundaries (Fig. 13), the remainder of the fracture surface being of the dimple type. A macrospically brittle fracture can therefore be shown to be microscopically ductile.

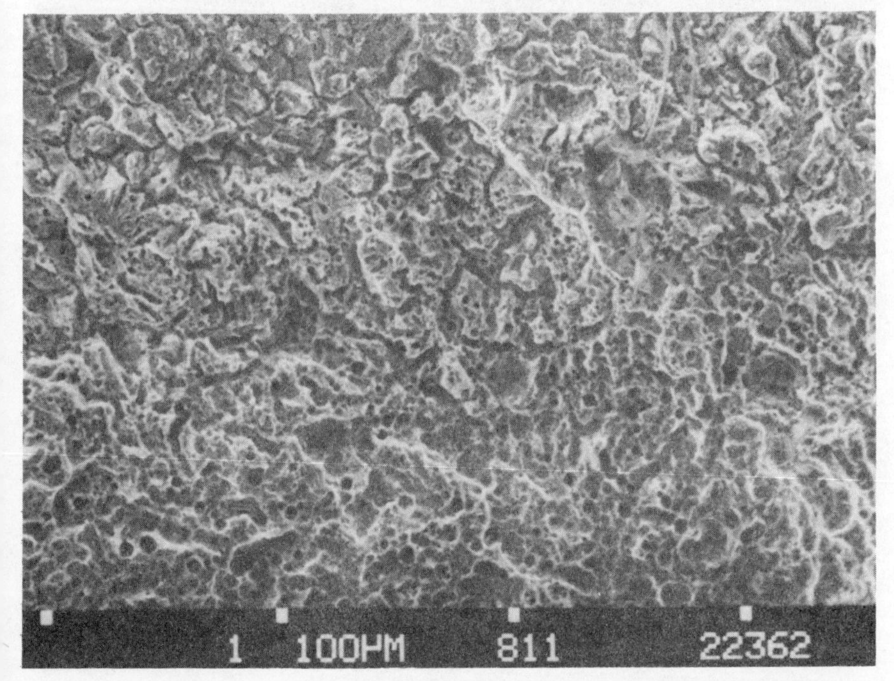

Fig. 13 – Dissolved areas along former Austenite grain boundaries (St 1420/1570)

Fig. 14 – Brittle fracture of a cold drawn and tempered prestressed steel.

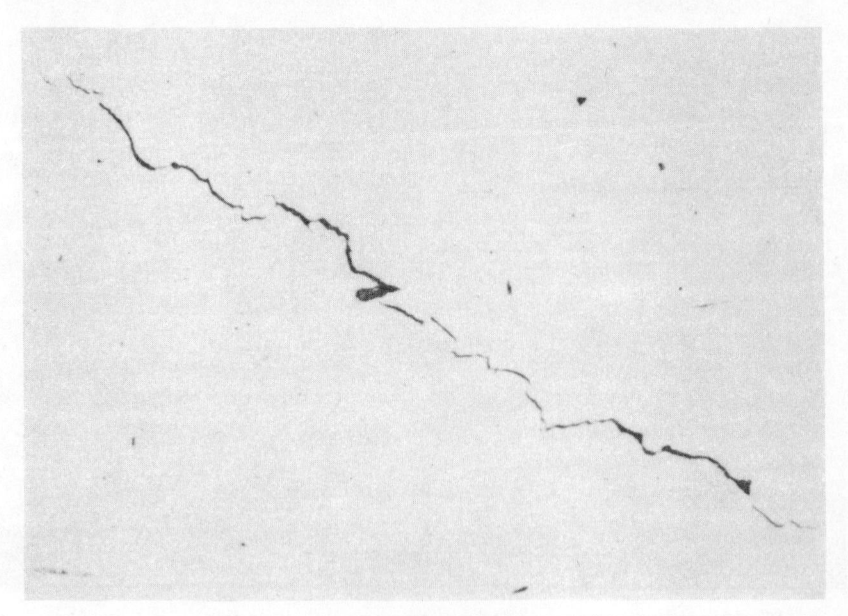

Fig. 15 Internal cracks.

After hydrogen absorption the cold drawn steel shows dimple rupture with islands of brittle cleavage rupture (Fig. 14). Internal tearing also occurs (Fig. 15), particularly at the boundaries of the grains elongating by working.

CONCLUSION

As a conclusion one can say that there is a line-up in the susceptibility of hydrogen embrittlement of prestressing steels under practical conditions. The most susceptible steel is the quenched and tempered, followed by the hot rolled perlitic, whereas the highest resistance against hydrogen-induced failure was found in cold drawn and tempered steel.

ACKNOWLEDGEMENT

I would like to express my thanks to the EGKS and the Institut für Bautechnik for their financial support of this research.

REFERENCES

[1] Nürnberger, U., Analysis and evaluation of failures of prestressed steels. Bundesminister für Verkehr, *Bulletin 308,* p. 1, 1980.

[2] Hofmann, W. and Rauls, W., *Arch. Eisenhüttenwesen,* **34,** p. 925 (1963).

Chapter 23

The influence of cement type on the electrochemical behaviour of steel in concrete

C. M. PREECE, F. O. GRØNVOLD, T. FRØLUND,
Danish Corrosion Centre, Glostrup, Denmark

ABSTRACT

All good quality cements are generally considered to protect steel reinforcement from corrosion because of their high alkalinity. However, even under ideal conditions, the ability of concrete to passivate steel varies significantly with cement type, giving rise to variations in corrosion rates of several orders of magnitude. Moreover, the degree and extent of passivation in a particular concrete cannot readily be predicted from the cement properties such as electrical resistance, chloride ion diffusion rates, permeability, etc. Thus, a cement which may prove superior to other cements in salt water may be inferior in fresh water. In order to understand some of these factors, the behaviour of steel in mortars made from different cement types is being investigated electrochemically. Results are presented for the first stage of this investigation in which steel embedded in ordinary Portland cement, fly-ash cement, blast furnace slag cement, sulphate resistant Portland cement and dense silica cement has been tested in deaerated $Ca(OH)_2$ solution and subsequently in NaCl solution.

INTRODUCTION

The use of blended cements containing Pozzolanic waste materials, such as fly-ash, blast furnace slag or condensed silica fume ('micro silica') is becoming increasingly common but, as yet, there are few quantitative data on the effect of these additions on the corrosion of steel reinforcement. There are both potentially detrimental and potentially beneficial effects of the additives. On the detrimental side, there are two major concerns. The first is that the Pozzolanic reaction will proceed to a degree at which the pH level of the cement is reduced to below the level necessary for steel to passivate. The second concern is that, even if the Pozzolanic reaction is not complete, the reduced OH content of the cement will result in more rapid and deeper levels of carbonation than in other cements which, again will cause depassivation of the steel. While there is some evidence for the latter [1], depassivation of the steel by the Pozzolanic reaction alone has not been reported. The beneficial influence of Pozzolans on the corrosion of embedded steel include increased electrical resistivity of the cement [2], and decreased pore size [3,4]. Unfortunately, it is not yet possible to predict the net result of the beneficial physical effects and potentially detrimental chemical effects of waste material additions on the corrosion of embedded steel. The purpose of this paper, therefore, is to present the first results of a study of actual corrosion rates of steel embedded in mortars based on five different cements and to describe the differences in terms of what is known about the structure and physical properties of the cements.

In order to investigate the fundamental differences in the ability of the different cement pastes to protect steel from corroding, we have chosen to keep the samples under water throughout the investigation to avoid any possibility of carbonation and to hold the samples at 35°C to accelerate any effects due to the Pozzolanic reaction. Parallel experiments, however, have also been conducted at room temperature on a range of fly-ash cements. We have also chosen to study 'good quality' laboratory samples to avoid the complicating factors of engineering flaws. In such samples, the corrosion rate in the passive state or in the low potential active state is extremely low and conventional methods, such as visual inspection or weight loss measurements of the steel after a given period, are not suitable. The electrochemical potentiostatic polarization method, on the other hand, is capable of measuring corrosion rates as low as 0.01 μm/year and has, therefore, been employed in order to detect the onset of active corrosion at an early stage.

EXPERIMENTAL PROCEDURE

(a) Sample Preparation

The mortars used in this study were based on ordinary Portland cement (OPC), sulphate-resistant Portland cement (SRPC), pulverized fly-ash cement (PFA), blast furnace slag cement (BFSC) and a highly dense microsilica cement (DSP).

Table 1

Composition of cements

	OPC	SRPC	BFSC
CaO	63.7	67.6	47.5
SiO_2	19.8	24.6	28.5
Al_2O_3	4.3	1.9	7.8
Fe_2O_3	2.0	1.9	1.4
MgO	3.6	0.7	6.9
K_2O	1.4	0.11	1.1
Na_2O	0.3	0.20	0.6
SO_3	3.5	1.5	1.7 (total S)
Loss on ignition	1.9	0.8	0.4

Table 2

Proportions of the mortar mixes tested at 35°C

	OPC	SRPC	BFSC	DSP
Cement	1.00 (OPC)	1.00 (SRPC)	1.00 (65% slag + 35 OPC)	1.00 (SRPC)
Microsilica†	–	–	–	0.24
Superplasticizer	–	–	–	0.03
Water	0.50	0.30	0.50	0.18
Sand	3.00	3.00	3.00	1.80
w/c	0.50	0.30	0.50	w/c + $SiO_2 = 0.15$

†Microsilica is approx. 96% SiO_2 obtained as a by-product during the production of silicon

Proportions of concrete mixes tested at 22°C

	OPC	OPC-15% PFA	OPC-30% PFA	OPC-45% PFA
Cement	1.00	1.00	1.00	1.00
Fly-ash (PFA)	–	0.176	0.43	0.82
Water	0.53	0.56	0.60	0.66
Sand	3.00	3.00	3.00	3.00
Stone	2.00	2.00	2.00	2.00
w/c + PFA	0.53	0.53	0.53	0.53

The composition of the cements are given in Table 1. The mix proportions of the mortars are given in Table 2.

Rectangular 4 X 4 X 16 cm^3 specimens were cast with an embedded steel electrode as illustrated in Fig. 1. The rod had press-fit plastic end pieces giving an exposed area of 30 cm^2 and plastic end-plates were glued to the cement using epoxy resin. After casting, the specimens were compacted by vibration and were kept at 100% R.H. for 24–48 hours before demoulding. They were then kept in $Ca(OH)_2$-saturated water at ambient temperature for 13 weeks prior to electrochemical testing. At least five samples of each mortar type were tested.

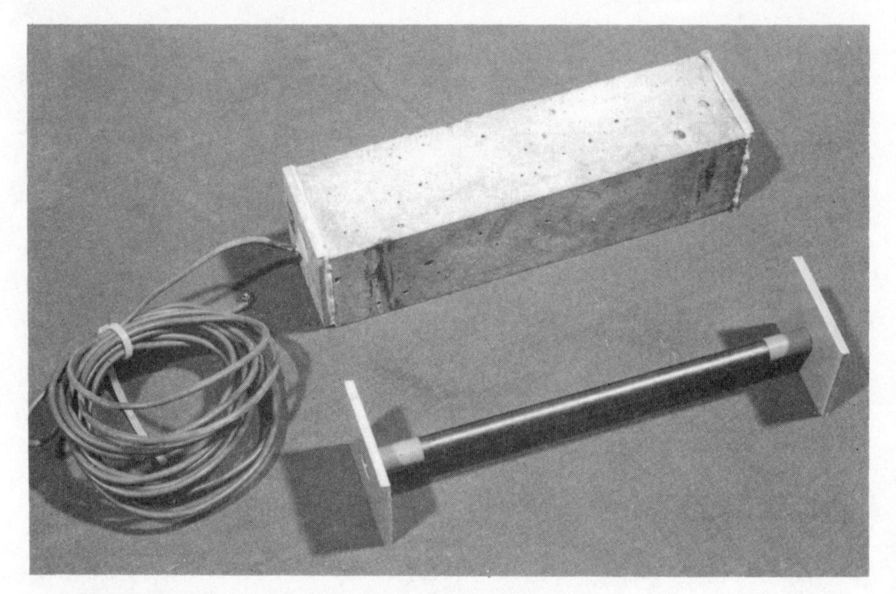

Fig. 1 – Sample used in electrochemical experiments. An 8 mm dia. mild steel rod as shown, is cast into a 4 X 4 X 16 cm^3 prism.

(b) Electrochemical Measurement

The electrochemical reaction of corrosion of steel in high pH solutions can be considered of consisting of two half cell reactions. The anodic reaction is the oxidation of iron and the cathodic reaction is generally the reduction of oxygen. These can be represented by the reactions:

$$Fe \rightarrow Fe^{2+} + 2e^-$$

$$Fe^{2+} \rightarrow Fe^{3+} + e^-$$

$$2\,H_2O + O_2 + 4e^- \rightarrow 4\,OH^-$$

The anodic reactions, in fact, result in the formation of a passive film on the iron surface which may be represented by the overall reaction:

$$3\,Fe + 4\,H_2O \rightarrow Fe_3O_4 + 8\,H^+ + 8e^-$$

During actual corrosion, the rate of emission of electrons by the iron is exactly equal to the rate of consumption of electrons by the oxygen and, therefore, there is no measurable net current. In order to obtain an estimate of the current, it is necessary to shift the potential away from the equilibrium, measure the resultant net current and extrapolate the data to the equilibrium potential.

Experimentally, this is accomplished as illustrated schematically in Fig. 2. The mortar sample containing a steel electrode is immersed in water and the steel is held at a constant electrochemical potential with respect to a reference electrode. The current flowing between the embedded steel (the working electrode) and an external steel plate (the counter electrode) is then measured. The process is then repeated for different values of applied potential.

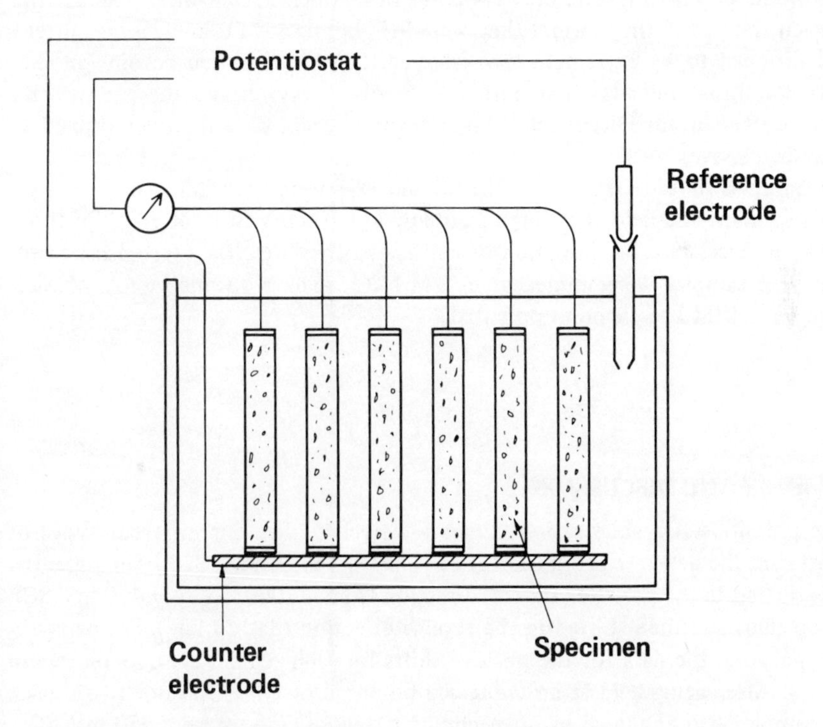

Fig. 2 – Schematic representation of the potentiostatic polarization experiment. Samples are connected in parallel and their embedded steel electrodes are held at a specific potential with respect to the reference electrode. The current flowing between individual samples and a counter electrode is then measured.

Because the current takes a long time to stabilize after the potential is changed, the experiment takes several weeks and it is not practicable to measure a complete polarization curve on one sample at a time. Therefore, for the 35°C experiments, all specimens were held in parallel at the same constant potential as shown in Fig. 2 and, after 24 hours, the current for each specimen was measured and the potential shifted. Although the currents were not completely stable after 24 hours they were sufficiently so to allow reasonable comparisons between cement types. For the room temperature study of PFA, seven specimens of each type were used. Each specimen was held at a specific constant potential over a period of at least two months. The long-term steady-state value of the current for each potential was used to construct the polarization curve.

Steel embedded in cement or mortar and exposed to air or to aerated water is passivated and its free electrochemical potential can assume any value in the passive range (i.e. +200 → −644 mV on the saturated calomel (SCE) scale). Therefore, in order to obtain the same starting conditions for the different samples, all the samples were kept in N_2 saturated $Ca(OH)_2$ solution and held at a cathodic potential (−900 or −1000 mV SCE) to electrochemically reduce the oxygen trapped in the mortar. They were held in this condition until the current had dropped to an extremely low value ($<10^{-4}$ A/m^2) or had become anodic. This condition indicates that both the dissolved oxygen and the passive film have been reduced. Potentiostatic polarization curves were then determined as described above.

Samples of the OPC, SRPC, BFSC and DSP were then immersed in a 1 M NaCl solution and held at an anodic potential (+100 mV SCE) at 35°C for three weeks to accelerate the initiation of corrosion. They were then tested as before. The PFA samples were immersed in 2 M NaCl, again with one sample of each type being held at a different potential.

RESULTS AND DISCUSSION

The potentiostatic polarization curves for steel in the four different types of mortar in the deaerated condition at 35°C are shown in Figs. 3–6. The potential was shifted in the positive (anodic) direction from −1000 mV to +450 mV SCE (solid line) and then shifted in the reverse direction (dashed line). For comparison purpose, the data for the positive shifts for each cement type are plotted in Fig. 7. After being held at an impressed potential of −900 mV for two weeks, all samples had assumed an open circuit potential, E_R, of ~ −950 mV SCE. There was no significant effect of mortar type in the low potential region of the plot. The corrosion current density at the open circuit potential was of the order of 10^{-4} A/m^2 for all samples. It should be noted that this is the corrosion rate in the active state after depassivation by cathodic polarization but it is, in fact, of the same order of magnitude as the corrosion rate of the passive steel in mortar with access to air [5]. The corrosion rate in this case must be limited by the cathodic reaction.

$$2\,H^+ + 2e^- \rightarrow H_2$$

The limitations are both the availability of H^+ ions and thermodynamical maximum pressure of H_2 which can develop before the reaction ceases or, alternatively, the rate at which H_2 can escape from the steel surface. At an electrochemical potential of −900 mV SCE, the maximum hydrogen pressure that can develop in mortar with a pH 13.5 is only 10^{-5} atm.

It is apparent that, for all the cement types, there was a small active peak in the region of −850 → −900 mV SCE and that the onset of passivity occurred between −600 and −700 mV SCE. This is in good agreement with the experiments of Pourbaix [6] who found the passive film on iron in aqueous solution of pH 12.4–13.5 to be stable only above approximately −650 mV SCE.

In the passive state, on the other hand, there was a marked effect of cement type. The steel in the blast furnace slag cement has a passive corrosion rate approximately 50 times greater than that of OPC or SRPC. However, this rate corresponds to only 10 μm/year and should not be regarded as a high corrosion rate, but does indicate that the passive film is less protective. At the higher potentials, the DSP samples also exhibited a higher passive current density than that of OPC and SRPC.

After the potentiostatic creation of the passive film, the value of E_R was shifted to higher values: ~ −350 mV SCE for the OPC and SRPC and ~ −550 mV for the BFSC and DSP as indicated by the dashed curves in Figs. 3–6.

In deaerated cement, one of the rate determining reactions in the passive film formation is:

$$Fe^{2+} + 2OH^- \rightarrow Fe(OH)_2$$

It is possible that the above reaction is restricted in BFSC cement by the availability of hydroxyl ions. This could be due to a reduction in the concentration of OH^- ions in the pore water by slag reaction with the alkalis and with $Ca(OH)_2$. Longuet *et al.* [7] have, in fact, shown a reduction in OH^- concentration in slag cements relative to that in OPC but the available OH^- was more than adequate for passivation. However, the samples considered here were kept at elevated temperatures for long periods and the reaction rate of the slag may have been increased with a correspondingly greater decrease in OH^- content. An alternative explanation, however, is that the higher temperature and greater 'free' water content in the BFSC allows a thicker but less protective film to form.

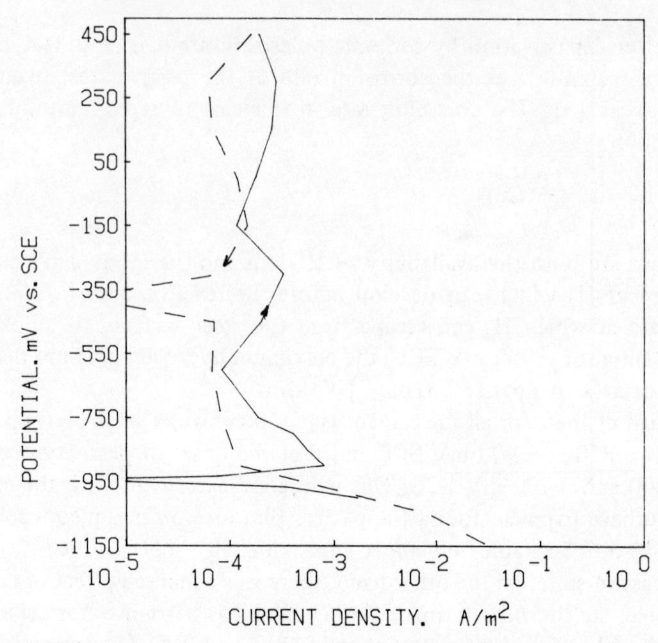

Fig. 3 – Potentiostatic polarization curve for OPC samples in deaerated $Ca(OH)_2$ solution at 35°C.

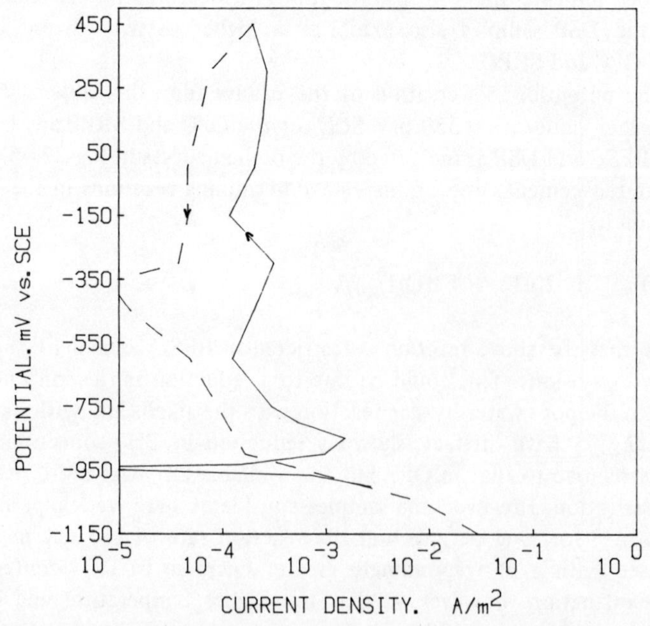

Fig. 4 – Potentiostatic polarization curve for SRPC samples in deaerated $Ca(OH)_2$ solution at 35°C.

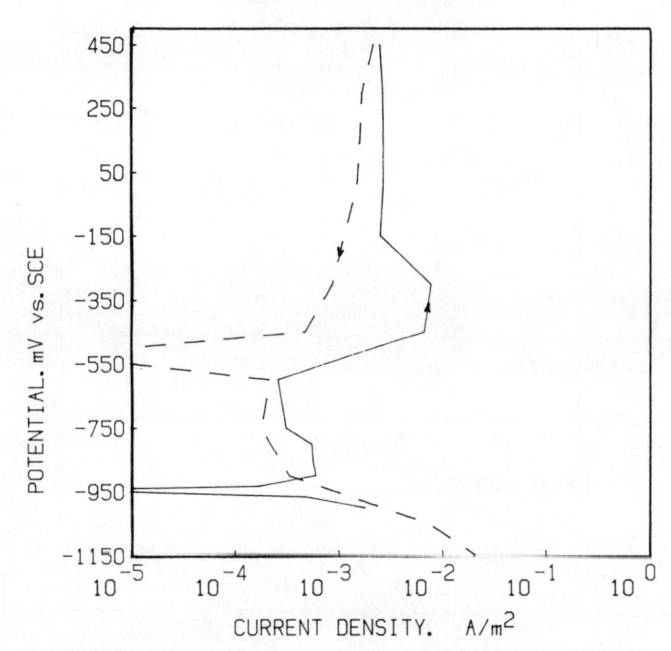

Fig. 5 – Potentiostatic polarization curve for BFSC samples in deaerated $Ca(OH)_2$ solution at 35°C.

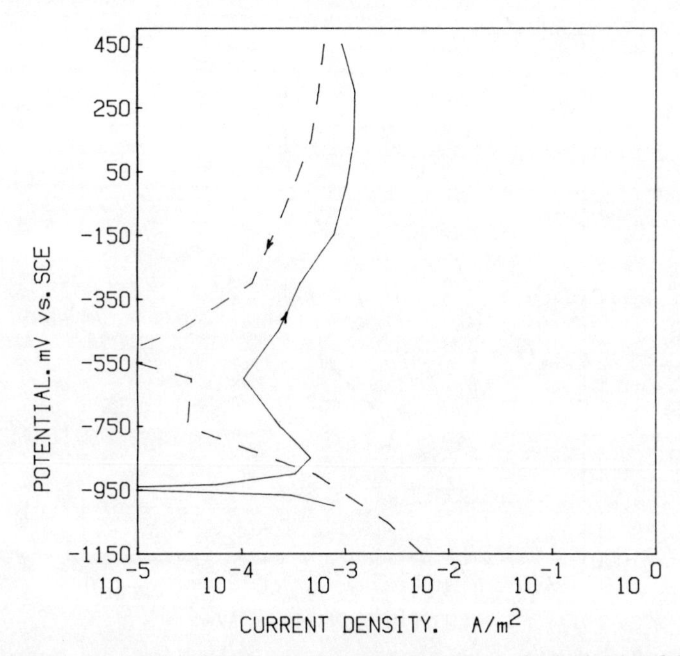

Fig. 6 – Potentiostatic polarization curve for DSP samples in deaerated $Ca(OH)_2$ solution at 35°C.

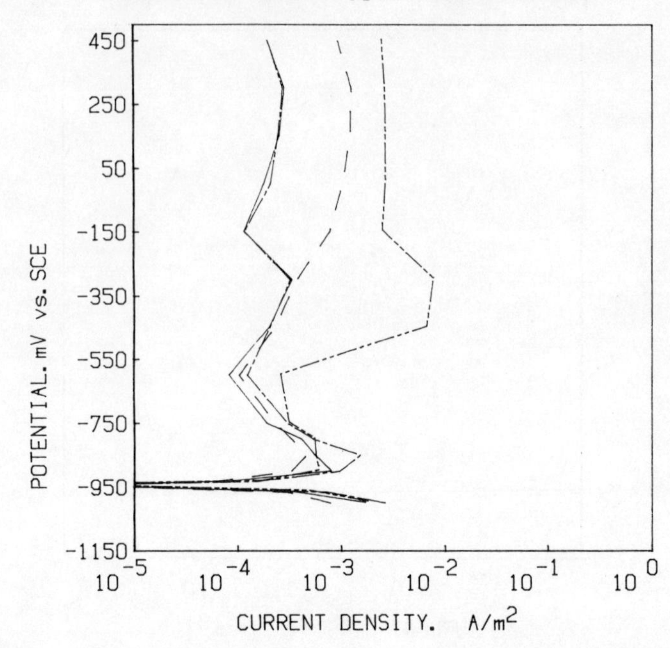

Fig. 7 – For comparison purposes, the data from solid curves of Figs. 4–6 are plotted together: ——— OPC, – ′ – ′ – - - SRPC, – – – – DSP, ----------- BFSC.

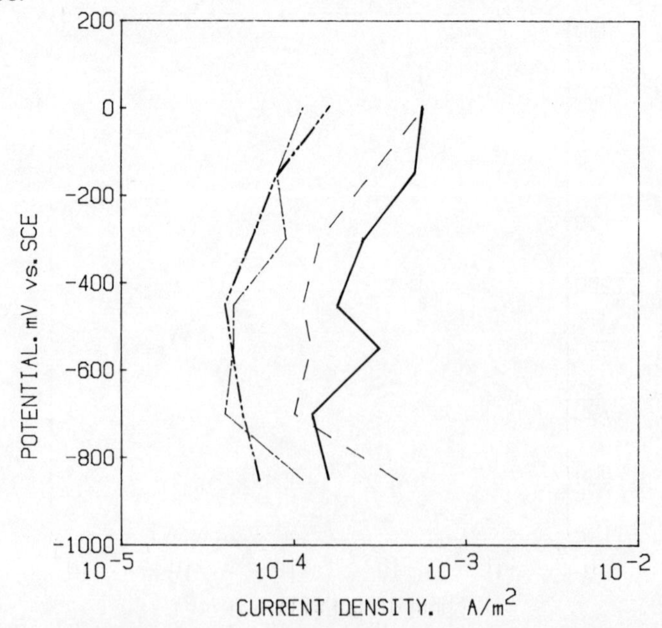

Fig. 8 – Potentiostatic curves for PFA samples tested in deaerated $Ca(OH)_2$ solution at 22°C: ——— OPC, – – – – OPC + 15% PFA, –·–·– OPC + 30% PFA, ----------- OPC + 45% PFA.

The effect of fly-ash on the passive state was the reverse of that exhibited by BFSC and DSP. Fig. 8 shows that increasing the fly-ash content resulted in a decrease in passive current density, the values of which were found to be approximately in inverse proportion to the resistivity of the mortars [8].

The polarization curves of samples tested at 35°C in NaCl solutions are given in Figs. 9 and 10. The chloride attack of the steel in SRPC was very marked whereas the samples containing blast furnace slag or silica were totally unaffected by the chloride at this stage. The steel in OPC had just begun to corrode at the time of the measurements and so there was considerable scatter in the data as shows in Fig. 10. After 33 weeks at room temperature in 2 M NaCl (without any applied potential) the PFA samples were unaffected whereas corrosion of the OPC samples began after only 2 weeks exposure to the salt solution.

These data agree well with chloride diffusion measurements and porosity measurements on the different cement pastes [9, 10]. The effective diffusion coefficients for chloride through the different cement types are given in Table 3. The very low diffusion coefficients for Pozzolanic cements can be attributed to the fineness of the pore structure in these cements despite the fact, mentioned above, that BFSC has a greater amount of free (evaporable) water than has OPC with the same water:cement (w/c) ratio. The higher free water content may be attributed to a lower water requirement for the Pozzolanic reactions of the slag than that for Portland cement clinker hydration reaction. The same is probably also true for the PFA cements. Nevertheless, the water in BFSC is present in a much finer pore distribution than that of OPC. The DSP material has an even finer structure [3]: it is essentially free of capillary pores and its evaporable water is present in pores $\leqslant$ 20 Å.

Table 3

Effective diffusion coefficient for Cl^- ion through cement pastes

Cement	w/c	T°C	D,10^{13} m^2/s
OPC	0.50	22	45
		40	73
SRPC	0.26	22	22
		40	35
BFSC	0.50	22	1.7
		40	10
DSP	0.15†	22	–
		40	–
PFA	0.50‡	25	15

†w/(c + SiO_2).
‡w/(c + PFA).

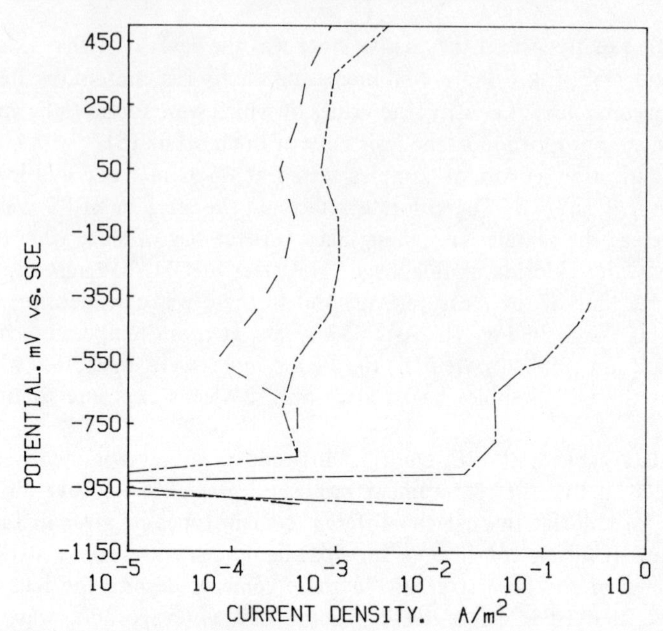

Fig. 9 – Potentiostatic polarization curves of samples in 1 M NaCl at 35°C after having been held at a potential of +100 mV SCE for 3 weeks, ---- DSP, --------- BFSC, ----- SRPC.

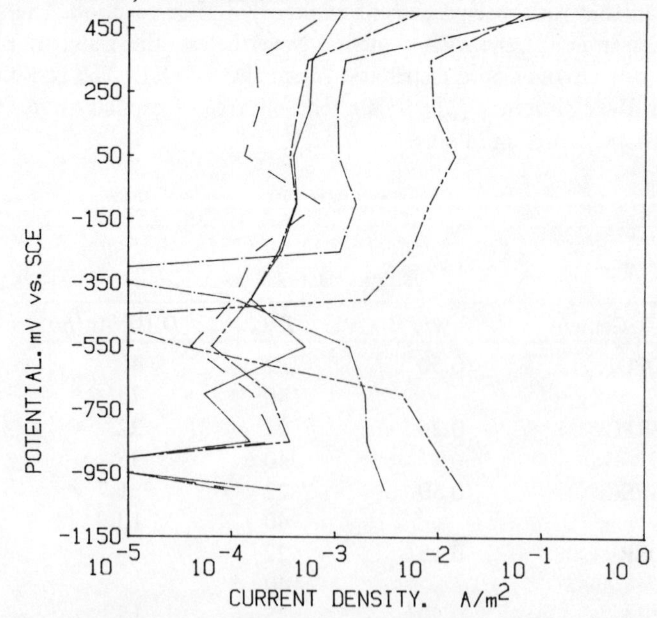

Fig. 10 – Potentiostatic polarization curves of five individual samples of OPC tested in 1 M NaCl at 35°C after having been held at a potential of +100 mV SCE for 3 weeks. The scatter in the data is due to the fact that some samples had begun to corrode, whereas others had not.

SUMMARY AND CONCLUSIONS

(1) Within the age limit of the samples tested in this project (up to 34 weeks), there was no evidence of the breakdown of passivity due to the Pozzolanic reaction, even after samples had been held at 35°C for 23 weeks.

(2) Additions of fly-ash decrease the room temperature passive current density of steel in concrete relative to OPC. In contrast, at 35°C, highly dense silica cement causes a slight increase and blast furnace slag results in a considerable increase in passive current density. The reasons for this behaviour are not certain at this time but are being investigated further.

(3) Additions of the different Pozzolanic materials provide excellent protection of steel reinforcement from chloride attack. It is probable that chloride-initiated corrosion in the PFA and BFSC is simply delayed and the experiments are being continued to determine if and when corrosion will eventually occur. Other experiments [9] on chloride penetration of the DSP, on the other hand, suggest that chloride attack will not occur in this material because it has no interconnected pore network.

ACKNOWLEDGEMENTS

The authors would like to express their appreciation of the advice and encouragement given by Hans Arup throughout this programme. Furthermore the financial support for this project provided by the Aalborg Portland Cement-Fabrik, The Swedish Nuclear Fuel Supply Co., Project KBS, the Danish Council for Technology and the Danish Council for Scientific and Industrial Research is gratefully acknowledged.

REFERENCES

[1] R. Tsukayama, H. Abe and S. Nagataki, *Proc. of the Congress on Chemistry of Cement, Paris 1980,* Vol. IV, p. 30.

[2] I. L. H. Hansson and C. M. Preece, to be published.

[3] C. M. Preece, T. Frølund and D. H. Bager, in *Condensed Silica Fume in Concrete* (eds. O. E. Gjørv and K. E. Løland), Norwegian Institute of Technology, 1982, p. 51.

[4] C. M. Preece, Project Report to the Swedish Nuclear Supply Co., Project KBS, 1982.

[5] J. A. Gonzales, S. Algaba and C. Andrade, *Br. Corr. J.,* **15,** 135 (1980).

[6] M. Pourbaix, *Corr. Sci.,* **14,** 25 (1974).

[7] P. Longuet, L. Burglen and A. Zelver, *Revue de Materiaux de Construction,* **676,** 35 (1973).

[8] F. O. Grønvold, to be published.

[9] C. M. Preece and T. Frølund, to be published.

[10] C. L. Page, N. R. Short and A. El Tarras, *Cement and Concrete Research,* **11,** 395 (1981).

Chapter 24

Sea water corrosion attack on concrete blocks embedding zinc galvanized steel rebars

HARUO SHIMADA, Nippon Steel Corporation, Kawasaki 211, Japan
SEIYA NISHI, Onoda Cement Co. Japan

INTRODUCTION

It has been hitherto reported that the zinc coating on the surface of steel rebars is effective for the prevention of the deterioration of concrete exposed to chloride ion attack [1–3]. However, its effect is not clear in case of exposing concrete embedding steel rebars covered with zinc film to such a high chloride ion environment as offshore, seaside for long periods. The purpose of this study is to examine its effect by preparing the concrete with various qualities, embedding zinc galvanized steel rebars and hot rolled steel rebars, exposing them to the most corrosive offshore environment for long periods and observing the concrete cleavage caused by the corrosion of the embedded rebars.

EXPERIMENT

Steel rebars preparation

In order to examine the effect of zinc coating, we prepared the hot rolled steel rebars and the zinc galvanized steel rebars with zinc film of 40 μm thickness. For this purpose, we used the ordinary commercial grade round bars with 13 mm diameter whose chemical composition is 0.20% C, 0.17% Si, 0.45% Mn, 0.02% P, 0.02% S. Zinc galvanized steel rebars were dipped into chromate solution with the aim of covering the zinc film surface with very thin chromate film in order to delay rust formation.

Concrete preparation

In order to examine the effect of the zinc coating in detail, we tried to observe the concrete deterioration behaviour by embedding the prepared rebars in concretes of various qualities and exposing them to severe chloride ion attack for long periods. For this purpose, we prepared the higher density concrete with artificial crack and the lower density concrete with no crack. In the higher density concrete, we increased the covering depth and decreased the water/cement ratio in order to emphasize the sea water penetration to the embedded rebars through the crack. In the lower density concrete, we decreased the covering depth and increased the water/cement ratio in order to emphasize the sea water penetration through concrete wall. In any case, the value of the covering depth was set much lower and the value of the water/cement ratio was much higher than that of the practical civil concrete construction in order to

Table 1 – Mix proportions and the mechanical strength of concrete.

Concrete quality	Mix proportions, unit weight (kg/m^3)				Air entrained agents (weight % in cement)	Compressive strength (kg/mm^2 28 days)
	Portland cement	Water	Fine aggregate	Coarse aggregate		
Higher density concrete	300	170	733	1072	0.01	337
Lower density concrete	300	195	803	944	0.01	284

Table 2 – Outlines of the experiment.

Item		Specifications
Steel bar specimens	Specimens	SR-24 ordinary steel round rebar whose diameter is 13 mm
	Surface state	Hot rolled steel bar covered with mill scale, zinc galvanized steel bar with thin chromate film
Concrete quality	Higher density concrete	Covering depth. 40 mm Water cement ratio; 0.56 Crack width; about 0.2 mm and 0.02 mm
	Lower density concrete	Covering depth. 20 mm Water/cement ratio; 0.60 No crack
Exposure test	Exposure environment	At the splash and submerged zone in Tokyo Bay
	Exposure periods	Two years; five years

accelerate the corrosion rate of the embedded rebars. In addition, the amounts of NaCl(%) in fine aggregate were set to 0, 0.2 and 0.4% by mixing concrete materials with fresh water and sea water, in order to check the effect of initial salt contents [4] in fine aggregate on the acceleration of the corrosion rate of the embedded rebars. The detail of the concrete quality is shown in Table 1.

Exposure test

The cured concrete blocks embedding rebars, as shown in Fig. 1, whose size is 100 mm width and 1000 mm length were attached to the buoy-type exposure equipment in order to expose them to the splash zone of sea water which is most corrosive [5] in offshore environment (Fig. 2). By using this equipment, 850 mm of concrete length was steadily exposed at the splash zone of sea water and 150 mm of concrete length was steadily exposed at the submerged zone of sea water [6–8]. The aim of this experimental technique is to simulate the exposure conditions of actual offshore civil constructions. The outline of the experimental procedures are summarized in Table 2.

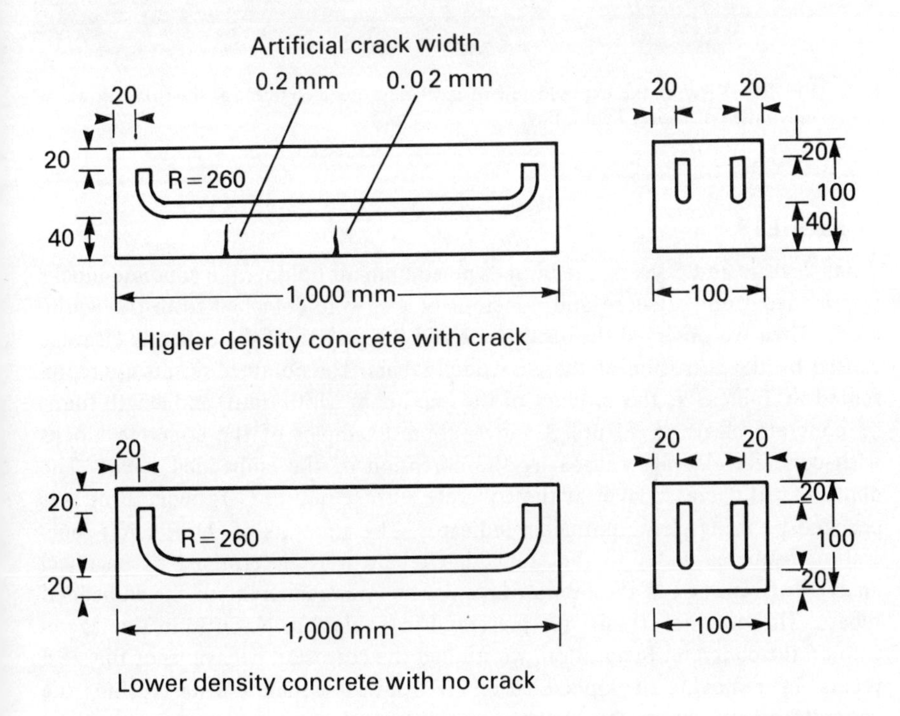

Fig. 1 – The arrangements of the rebars embedded in the concrete blocks prepared for this study.

Fig. 2 – View of the exposed test of steel reinforced concrete at the splash zone in Kimitsu offshore, Tokyo Bay.

2. RESULTS

After 2 years and 5 years, the buoy-type equipment holding the concrete blocks was landed from offshore and concrete blocks were detached from the equipment. Then we observed the occurrence and the growth of the concrete cleavage caused by the corrosion of the embedded rebars. The obtained results are represented in Table 3 as the product of the maximum width (mm) and length (mm) of concrete cleavage. Figure 3 shows some examples of the concrete blocks with concrete cleavage caused by the corrosion of the embedded rebars. The depth of pH decrease layer at the concrete surface due to CO_2 penetration was checked by using phenolphthalein indicator. The amounts of chloride ion penetration from sea water to the embedded rebars were determined by chemical analysis of NaCl(%) in the interior layer of concrete adherent to the embedded rebars. The obtained results are presented in Table 4 as NaCl(%) in fine aggregate of the concrete. In addition, we studied the corrosion behaviour of the steel rebars, by removing the embedded rebars from the concrete after crushing the concrete blocks. From the observation of the corrosion behaviour of the steel rebars, it was found that the corrosion attack is detected only at the splash zone of sea water in any case. The corrosion behaviour was studied by determining

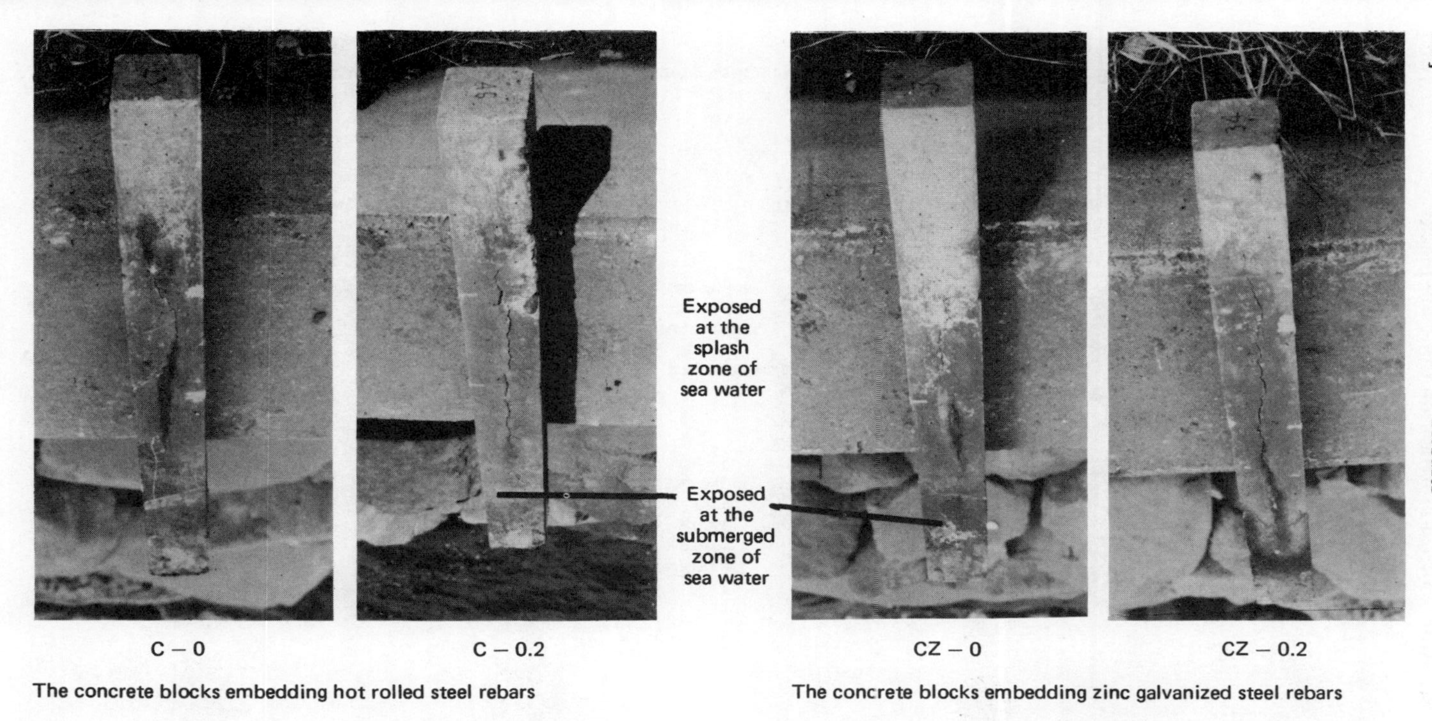

Fig. 3.1 – The cleavage of the higher density steel reinforced concrete with artificial crack caused by the corrosion of the embedded rebars during 5 years' exposure test in an offshore environment.

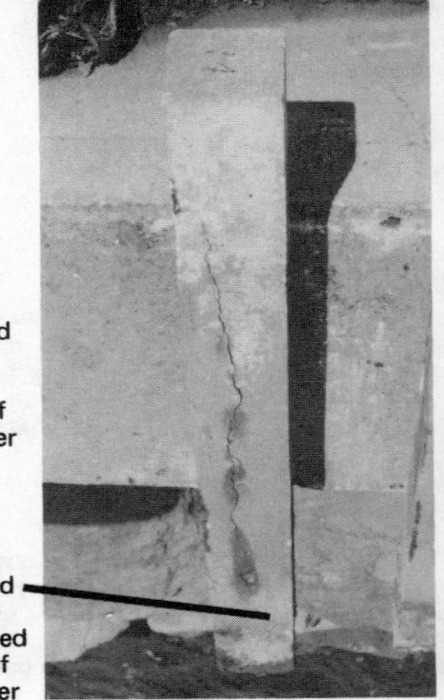

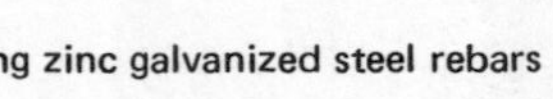
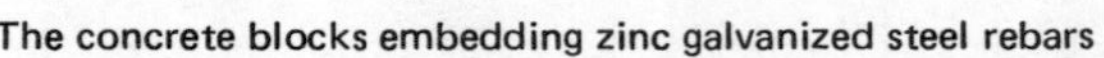

Fig. 3.2 – The cleavage of the lower density steel reinforced concrete with no crack caused by the corrosion of the embedded rebars during 5 years' exposure test in an offshore environment.

Table 3 – Concrete cleavage in the concrete blocks exposed at the splash zone of sea water.

Concrete blocks embedding rebars		Maximum width (mm) X length (mm) of the cleavage	
Embedded rebars	Mark†	After 2 years' exposure	After 5 years' exposure
Hot rolled steel rebars	C – 0	0 X 0	1.0 X 100
	C – 0.2	0 X 0	2.5 X 200
	A – 0	0 X 0	2.0 X 300
	A – 0.2	0 X 0	2.5 X 220
	A – 0.4	–	–
Zinc galvanized steel rebars	CZ – 0	0 X 0	1.0 X 150
	CZ – 0.2	0 X 0	2.0 X 250
	AZ – 0	–	–
	AZ – 0.2	0 X 0	3.0 X 400
	AZ – 0.4	0 X 0	2.0 X 400

†C, CZ: higher density concrete blocks. A, AZ: lower density concrete blocks.
0, 0.2, 0.4: initial salt contents in fine aggregate of the concrete prepared for this study.

Table 4 – The amounts of NaCl(%) in the concretes exposed to sea water penetration (represented as NaCl(%) in fine aggregate).

Concrete blocks† embedding rebars	In the layer adherent to the rebars embedded in the concrete exposed at the submerged zone of sea water		In the layer adherent to the rebars embedded in the concrete exposed at the splash zone of sea water	
	After 2 years	After 5 years	After 2 years	After 5 years
C – 0 , CZ – 0	0.7	1.9	0.3	1.6
C – 0.2, CZ – 0.2	0.6	2.4	0.6	1.5
A – 0	1.4	2.0	0.5	1.1
A – 0.2, AZ – 0.2	1.0	2.2	0.9	1.7
AZ – 0.4	1.2	2.4	1.2	1.4

†C, A: concrete blocks embedding hot rolled steel rebars.
Cz, AZ: Concrete blocks embedding zinc galvanized steel rebars.

the rust formation area (%) and the maximum pit depth (mm). The maximum pit depth caused by the localized corrosion due to chloride ion attack was determined by means of a point micrometer. The results obtained are shown in Table 5 and Fig. 4. Figure 4.1 shows the view that hot rolled rebars whose corrosion products were eliminated by immersing them in acidic solution with inhibitor.

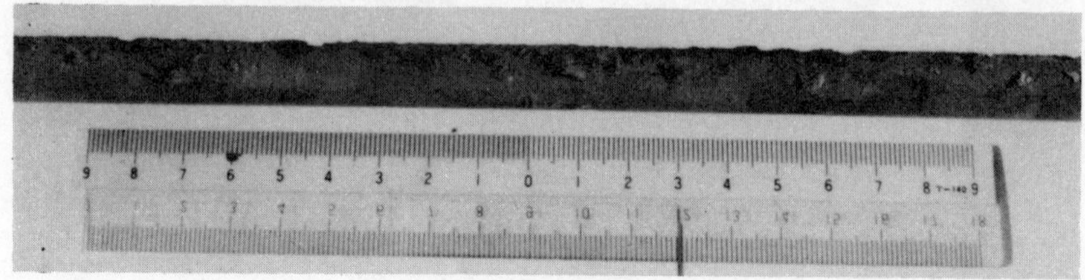

C – 0 The rebars embedded in the higher density concrete with artificial crack

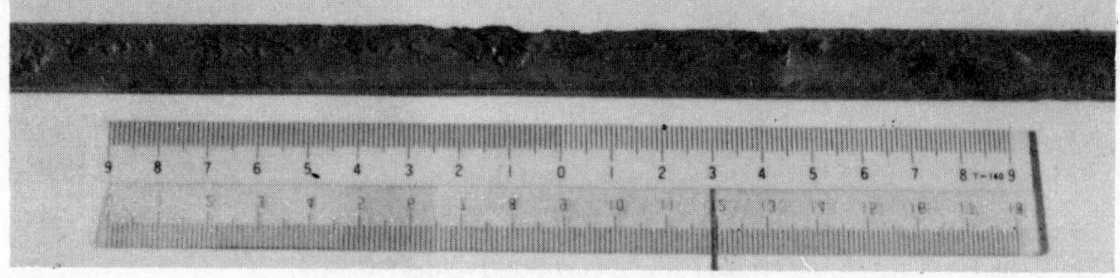

A – 0 The rebars embedded in the lower density concrete with no crack

Fig. 4.1 – The corrosion behaviour of the hot rolled steel rebars embedded in the concrete blocks exposed in an offshore environment for 5 years.

Table 5 – Corrosion behaviour of the rebars embedded in the concrete exposed at the splash zone of sea water.

Concrete blocks embedding bars		Rust formation area (%)		Maximum pit depth (mm)	
Embedded rebars	Mark†	After 2 years	After 5 years	After 2 years	After 5 years
Hot rolled steel rebars	C – 0	8.4	42.1	0.47	3.0
	C – 0.2	11.5	55.1	1.20	2.7
	A – 0	9.7	66.3	0.71	2.6
	A – 0.2	17.6	62.1	1.17	3.3
	A – 0.4	–	–	–	–
Zinc galvanized steel rebars	CZ – 0	1.6	57.4	0.87	1.0
	CZ – 0.2	0.6	69.5	0.27	1.1
	AZ – 0	–	–	–	–
	AZ – 0.2	6.5	70.4	0.44	1.8
	AZ – 0.4	3.4	74.8	0.40	1.9

†C. CZ: higher density concrete. A, AZ: lower density concrete.
0, 0.2, 0.4: initial salt contents in fine aggregate of the concretes prepared for this study.

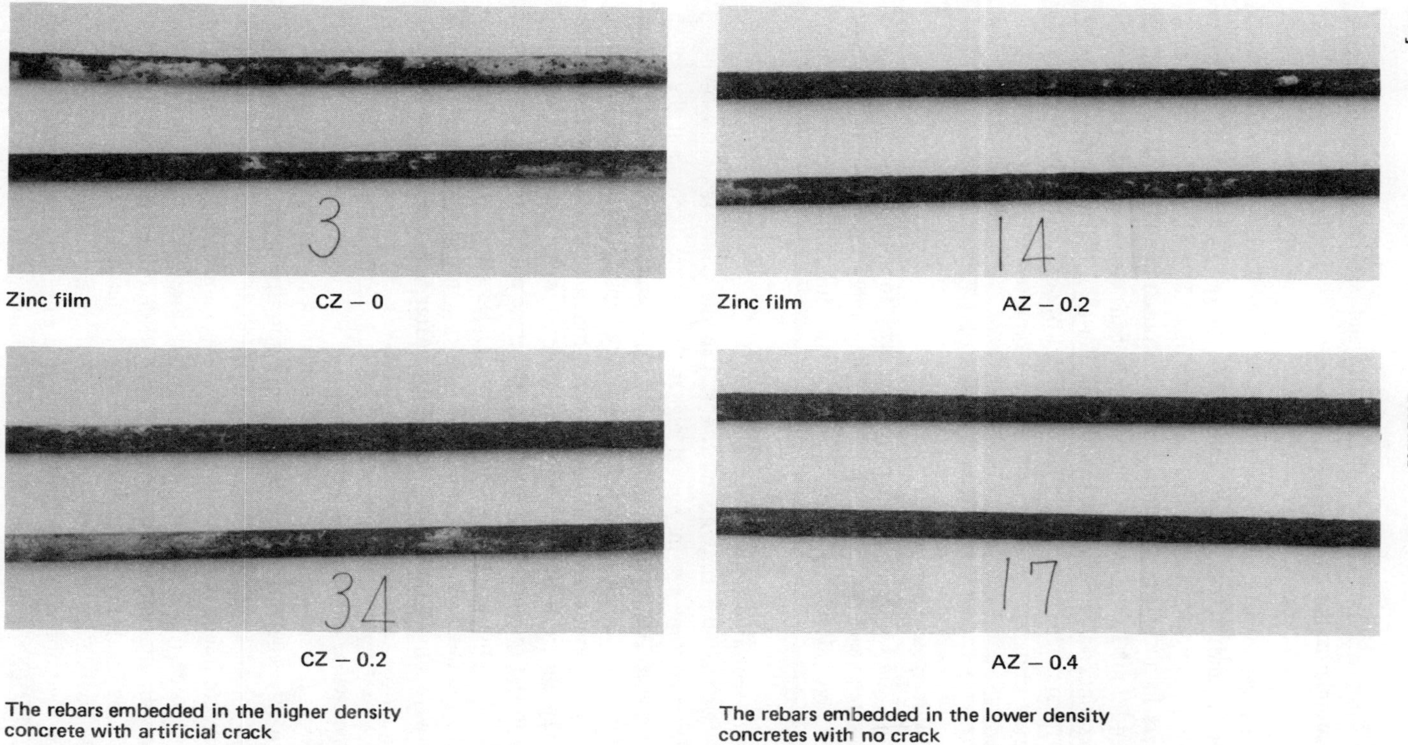

Fig. 4.2 – The corrosion behaviour of zinc galvanized steel rebards embedded in concrete blocks exposed in an offshore environment for 5 years.

Figure 4.2 shows the co-existence of zinc film and corroded steel rebar. These results are summarized as below.

Concrete cleavage

From Table 3 and Fig. 3, it was found that the occurrence of the concrete cleavage is not detected in any concrete blocks after 2 years but detected clearly after 5 years in all of the concrete portions exposed at the splash zone of sea water. In addition, Table 3 shows that the size of the cleavage caused in the concrete blocks embedding zinc steel rebars is higher than that in those embedding hot rolled rebars. This represents that the effect of zinc coating on delaying the occurrence of concrete cleavage can not be expected after such a long-term exposure.

Depth of pH decrease layer

After 2 years, the depth of pH decrease layer was from 1 to 3 mm but after 5 years, that value was from 2 to 5 mm. The remarkable increase of the depth of pH decrease layer after 5 years exposure might be due to the occurrence of the concrete cleavage.

Chloride ion penetrated from sea water

Table 4 shows that the amounts of NaCl(%) in the concrete portion exposed at the submerged zone are in general higher than those exposed at the splash zone of sea water. At the submerged zone of sea water, numerous chloride ions of sea water reached the embedded rebars very rapidly through the concrete wall, and their amounts in the higher density concrete were lower than in the lower density concrete after 2 years; but after 5 years they were almost the same. Also in the splash zone of sea water, the same phenomena were observed. In this case, the amounts of chloride ion determined after 2 years' exposure increased with the increase of the initial salt contents in fine aggregate but after 5 years they were almost the same, owing to the remarkable increase of sea water penetration.

Corrosion behaviour of the embedded rebars

From Table 5, it was found that the rust formation of zinc galvanized steel rebars is negligibly small after 2 years' exposure but after 5 years the zinc film of the zinc galvanized steel rebars are damaged rapidly in any concrete blocks with various concrete qualities and both the bared steel and the zinc film are simultaneously exposed to chloride ion attack. In the case where bared steel and the zinc film are simultaneously exposed to the chloride ion environment, the local cell tends to be created very easily between both metals and to cause hydrogen gas liberation which accelerates the occurrence of concrete cleavage. Table 5 and Fig. 4 show that the corrosion rate of the zinc galvanized steel rebar is much lower than that of the hot rolled steel rebars but the tendency causing cleavage of the concrete blocks is almost the same. This might be due to this hydrogen gas liberation [3] whose accumulation tends to cause concrete cleavage. As shown in Table 5, the initial salt contents in fine aggregate added by mixing concrete materials with fresh water and sea water have not influenced the behaviour of zinc film damage of the zinc galvanized steel rebars embedded in concrete. This might be due to the remarkable increase of chloride ion supplied by sea water penetration of the embedded rebars during long-term exposure.

3. SUMMARY

By exposing the concrete blocks embedding zinc galvanized and hot rolled ordinary steel rebars simultaneously, both in the splash and submerged zone of sea water, we examined the effect of zinc film on the prevention of the concrete deterioration.

The results obtained are described as follows.

(1) During 2 years' exposure, we could recognize the effect of zinc coating but during long-term exposure the zinc film was damaged by the attack of the chloride ion supplied from sea water penetration. And after 5 years the concrete blocks were all deteriorated by the occurrence and the growth of the cleavage at the concrete portions exposed in the splash zone of sea water.

(2) The results of the 5 years' exposure test showed that the corrosion rate of zinc galvanized steel rebars is much lower than that of the hot rolled steel rebars, but the tendency to cause cleavage in the concrete block is almost the same. Therefore, the effect of zinc coating on delaying the occurrence of concrete cleavage could not be expected after such a long-term exposure in the most corrosive offshore environment.

ACKNOWLEDGEMENTS

The authors would like to express their thanks to Mr Shigejirō Mori, Representative Director of Onoda Cement Company, and Dr. Hideya Okada, Director of Nippon Steel Corporation.

REFERENCES

[1] C. E. Bird, *Corrosion Prevention & Control,* July (1964), 17–20.
[2] C. E. Bird and F. J. Strauss, *Materials Protection,* July (1967), 48–52.
[3] C. E. Bird, *Nature,* May 26 (1962), **194,** 798.
[4] Kōichi Kishitani, *Study of the corrosion behaviour of the steel bars embedded in concrete,* No. 1, May 1968; No. 3, June 1969; published by the Architectural Institute of Japan.

[5] Howard F. Finlay, *Corrosion,* **17** (3), (1961), 104t–108t.
[6] Haruo Shimada, Hideya Okada and Seiya Nishi, *OTC,* No. 3194, May 1978, at Houston, USA.
[7] Haruo Shimada, Hideya Okada and Seiya Nishi, *The proceedings of the 5th International Congress on Marine Corrosion and Fouling (1980) at Barcelona, Spain,* pp. 238–253.
[8] Haruo Shimada, Yoshiaki Sakakibara and Hiroshi Hirotani, *The proceedings of the First International Conference on Marine Technology,* Paper No. 42 P1741 *at Goteborg, Sweden,* paper no. 42, pp. 1–41.

Chapter 25

The promising concrete rebars for construction in off-shore, seaside and desert environments

KŌICHI KISHITANI, Tokyo University, Japan
ISAO FUKUSHI, Housing and Urban Development Corporation, Japan
HARUO SHIMADA, Nippon Steel Corporation, Kawasaki 211, Japan

INTRODUCTION

The report entitled 'Sea water corrosion attack on concrete blocks embedding zinc galvanized steel rebars' (see this volume) made it clear that zinc coating is incapable of delaying the occurrence and the growth of the concrete cleavage in the case of concrete construction being exposed to high chloride ion attack for long periods. The purpose of this study is to examine promising hot rolled rebars coated with alloying elements which are capable of delaying the occurrence and the growth of the concrete cleavage caused by the corrosion of the embedded rebars.

1. EXPERIMENTS

According to the report of Kōichi Kishitani [1], the addition of WO_4^{2-} ion inhibitor to the concrete aggregate is effective for the improvement of corrosion resistance of steel rebars in an environment that includes chloride ions. In addition, Cu is very effective for improvement of corrosion resistance in various atmospheric environments, including the marine atmosphere. Therefore, expecting the continuous yielding of WO_4^{2-} ion inhibitor from the steel bar itself in the concrete and the improvement of the resistance against atmospheric corrosion, we picked the Cu-W-bearing steel rebars [2].

As the first step, we prepared the concrete blocks embedding this Cu-W-bearing hot rolled steel rebars and those embedding the ordinary hot rolled steel rebars, exposed them to a very corrosive environment [3] such as the splash zone of sea water for long term periods [2] and checked the concrete deterioration caused by the occurrence and the growth of the cleavage due to the corrosion of the embedded rebars. As the next step, we developed [4, 5] a new laboratory technique capable of forecasting the results of this long-term field test of the steel rebars embedded in concrete, and tried to find out the much more promising concrete rebars by using this technique [5]. And as the final step, we developed the new laboratory-accelerated testing method to create the concrete cleavage within a short period. And by using this method, we examined the resistance of the promising rebars against the occurrence and the growth of the concrete cleavage caused by the expansion force of the corrosion products of the embedded rebars.

2. RESULTS

Field test carried out in the most corrosive ocean environment

In the same way as mentioned in the report entitled 'Sea water corrosion attack on concrete blocks embedding zinc galvanized steel rebars' we prepared the concrete blocks embedding the hot rolled steel rebars whose chemical composition is shown in Table 1 and exposed them simultaneously both at the splash- and submerged zone of sea water for 5 years. After 5 years exposure tests, we observed the occurrence and the growth of the cleavage caused by the corrosion of the embedded rebars due to chloride ion attack. From the results obtained, it was found that the higher density concrete block with thicker covering depth and artificial crack embedding the Cu-W-bearing steel rebars shows much higher resistance against the concrete damage than those embedding the ordinary hot rolled steel- and zinc galvanized steel rebars. Fig. 1.1 shows this comparison and Fig. 1.2 shows the comparison of the total cleavage length among these three concrete blocks. After crushing the concrete blocks, the embedded rebars were removed from the concrete and we examined the rust formation area (%) and determined the maximum pit depth (mm) by means of point micrometer. The results obtained are shown in Fig. 2. From Fig. 2, it was found that the maximum corrosion rate of the Cu-W-bearing hot rolled rebars, was much lower than that of the ordinary hot rolled rebars [5]. In addition, it was found that the maximum corrosion rate of the zinc galvanized steel rebars is lowest but the concrete damage is highest. This might be due to the hydrogen gas liberation which tends to accelerate the occurrence of the concrete cleavage. In any way, the Cu-W-bearing hot rolled steel rebars tend to decrease the amounts of the corrosion products and are therefore to show high resistance against the occurrence and the growth of the concrete caused by the expansion force of the corrosion products of the embedded rebars [5].

Fig. 1.1 – Comparison of the concrete cleavage of the concrete blocks with covering depth of 40 mm and with artificial crack of 0.2 mm width, which were exposed in an offshore environment for 5 years.

Table 1 – The chemical composition of the steel rebars prepared for 5 years field tests

Steel rebars	Surface state	Chemical composition (%)							
		C	Si	Mn	P	S	Cu	W	Al
SR-24 Ordinary steel	Hot rolled rebars with mill scale Zinc galvanized rebars	0.20	0.17	0.45	0.02	0.02	–	–	0.01
Cu-W-bearing steel	Hot rolled rebars with mill scale	0.10	0.20	1.00	0.02	0.02	0.23	0.10	0.02

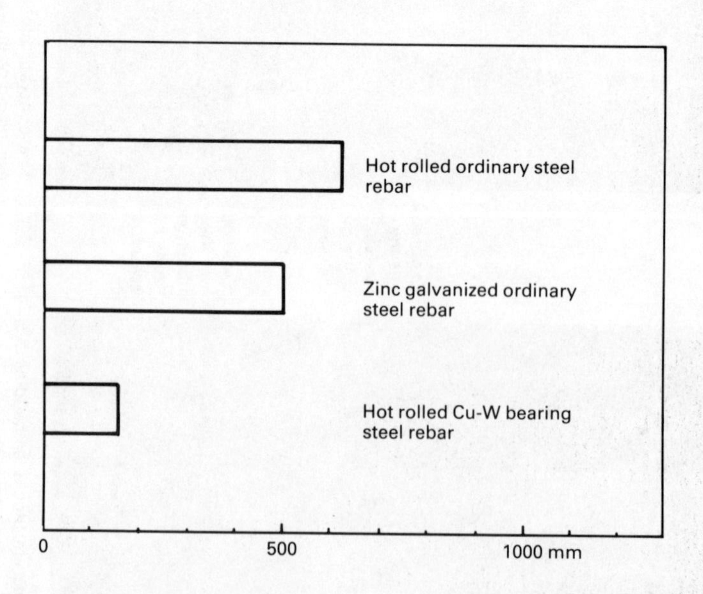

Fig. 1.2 – Total cleavage length (mm) of the concrete blocks caused by the corrosion of the embedded rebars during 5 years' exposure test in an offshore environment.

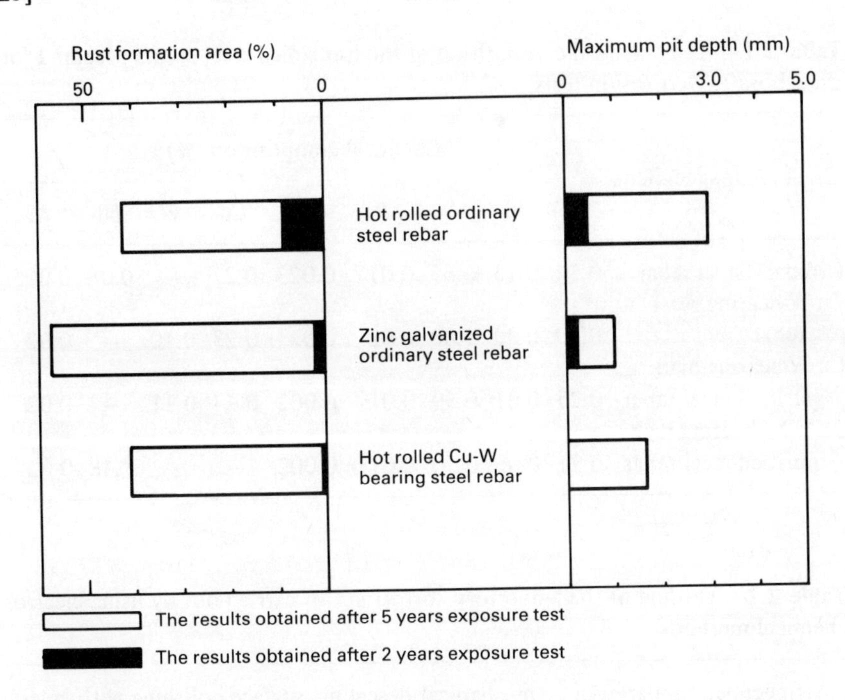

Fig. 2 – The corrosion behaviour of the rebars embedded in concrete blocks exposed in offshore environment for long periods.

Laboratory corrosion test to forecast the results of the long-term field test [5]

In the field test we need as long as 5 years' exposure period, even if we select a very corrosive environment such as the splash zone of sea water and reduce the covering depth and water/cement ratio as low as possible simulating civil engineering construction concrete. However, we do like to obtain many data of the long-term field test. Therefore, we have developed a new laboratory test capable of forecasting such long-term field test data as soon as possible. And by using this technique, we could discover which were the more promising rebars, and the most promising rebar to be used for low temperature conditions such as Arctic ocean or in an LNG storage concrete tank. The brief outline of these laboratory tests is shown in Table 2 [5, 6]. The results obtained are shown in Fig. 3 and Table 3. Fig. 3 shows that the potentials of the Cu-W-bearing high purified steel rebar and 3.5% Ni-bearing high purified steel rebar are much higher than that of the above-mentioned Cu-W-bearing steel rebar and ordinary steel rebar [5]. This means that they show higher resistance to chloride ion attack in such a high pH region as concrete mortar. Table 3 shows the same tendency. From these results, we could recommend the Cu-W-bearing

Table 2.1 – The chemical composition of the hot rolled steel rebars prepared for the laboratory corrosion tests.

Steel rebar specimens	Chemical composition (%)								
	C	Si	Mn	P	S	Cu	W	Ni	Al
Ordinary steel rebar	0.14	0.13	0.65	0.017	0.023	0.27	–	0.08	0.005
Cu-W-bearing steel rebar	0.10	0.30	1.00	0.02	0.02	0.23	0.10	–	0.02
Cu-W-bearing high purified steel rebar	0.25	0.01	0.59	0.01	0.002	0.23	0.11	–	0.02
3.5% Ni-bearing high purified steel rebar	0.21	0.05	0.30	0.011	0.002	–	–	3.48	0.02

Table 2.2 – Outline of the laboratory corrosion test carried out by using electrochemical methods.

(1) Specimen preparation – mechanical descaling, surface polishing with emery paper, degreasing, covering with silicon resin sealant, except the area of measurement
(2) Specimen size – 20 mm X 40 mm X 2 mm
(3) Specimen area for measurement – 0.5 cm^2
(4) Immersion solution – $Ca(OH)_2$ aqueous solution of pH 12 including NaCl
(5) Time required for measurement by Wenking-type potentiostat – 1.5 hours

Table 2.3 – Outline of the immersion test in the $Ca(OH)_2$ aqueous solution of pH 12 including NaCl.

(1) Specimen preparation – mechanical descaling, mechanical grinding, degreasing, covering the reverse and corner sides with silicon resin sealant
(2) Specimen size – 8 mm X 50 mm X 2 mm
(3) Immersion solution – $Ca(OH)_2$ aqueous solution of pH 12 including NaCl
(4) Liquid paraffin on the solution surface prevents the CO_2 penetration from the air atmosphere. The solution was renewed to maintain the high dissolved oxygen content every 3 days during 20 days' immersion test period

Table 3 – The corrosion behaviour of various steel specimens in the $Ca(OH)_2$ aqueous solution with pH 12 including NaCl.

Steel specimens	NaCl(%) in solution	NaCl(%) corresponding to the amounts in fine aggregate of concretes	Maximum pit depth (mm/2 days)		Rust formation area (%)	
Ordinary steel rebar	0.20	0.05	0.23	0.27	3.3	3.0
	0.80	0.20	0.32	0.32	2.7	2.4
Cu-W-bearing steel rebar	0.20	0.05	0.18	0.22	1.7	0.5
	0.80	0.20	–	–	–	–
Cu-W-bearing high purified steel rebar	0.20	0.05	0	0	0	0
	0.80	0.20	0.20	0	1.5	0
3.5% Ni-bearing high purified steel rebar	0.20	0.05	0	0	0	0
	0.80	0.20	0	0	0	0

high purified steel rebar (low Si, Low P, ultra low S) as much more promising for concrete exposed to high chloride ion attack, and 3.5% Ni-bearing steel rebar (low Si, low P, ultra low S) as the more promising for concrete exposed to chloride ion attack and to low-temperature environments [5]. The detail of this study is shown in the report presented at Offshore Goteborg '81 [5].

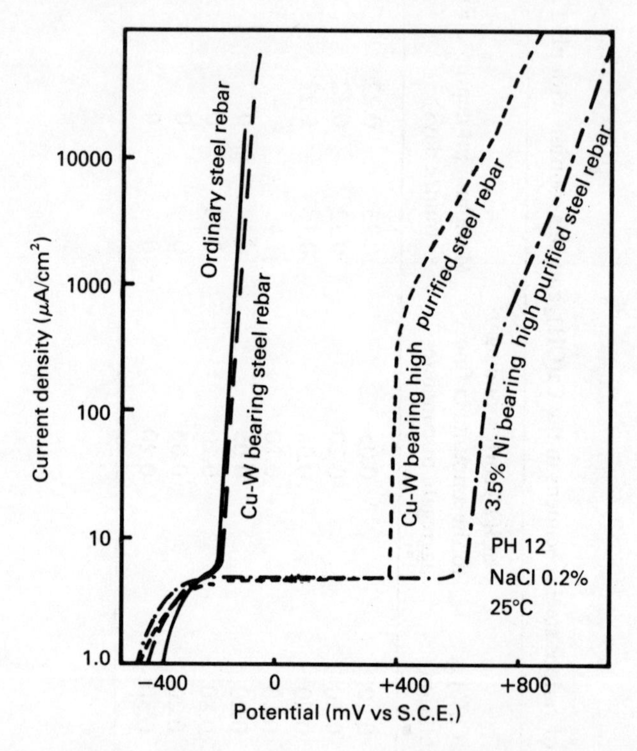

Fig. 3 – Anodic polarization curves of various bar specimens immersed in the aqueous solution of $Ca(OH)_2$ and 0.2% NaCl with pH 12 at room temperature.

The new technique for appraising concrete damage within a short period

In general, the concrete construction tends to be damaged by the occurrence and the growth of the concrete cleavage caused by the expansion force of the corrosion products of the embedded rebars. In order to accelerate such a concrete damage, we prepared the concrete mortar with high NaCl content (NaCl(%) in fine aggregate is 0.8%) which is corresponding to the case of mixing with sea water. In addition we prepared the ordinary steel rebar and the above promising rebars, and polished their surface with emery paper after they had been descaled, degreased, dried and held in a desiccator. After embedding them in the

Table 4 – Mix proportions and the mechanical strength of concrete.

Mix proportions, unit weight (kg/m^3)				NaCl(%) in fine aggregate	Slump (cm)	Compressive strength (kg/mm^2 at 28 days)
Portland cement	Water	Fine aggregate	Coarse aggregate			
288	173	875	955	0.8	18	310–322

Table 5 – Comparison of the concrete cleavage among various concrete blocks embedding various hot rolled rebars.

Embedded rebars	Maximum width (mm) of concrete cleavage		
	After 35 days (10 cycles)	After 50 days (15 cycles)	After 70 days (20 cycles)
Ordinary steel rebar	0.10	0.20	1.30
Cu-W-bearing high purified steel rebar	0	0.08	0.10
3.5% Ni-bearing high purified steel rebar	0	0.04	0.10

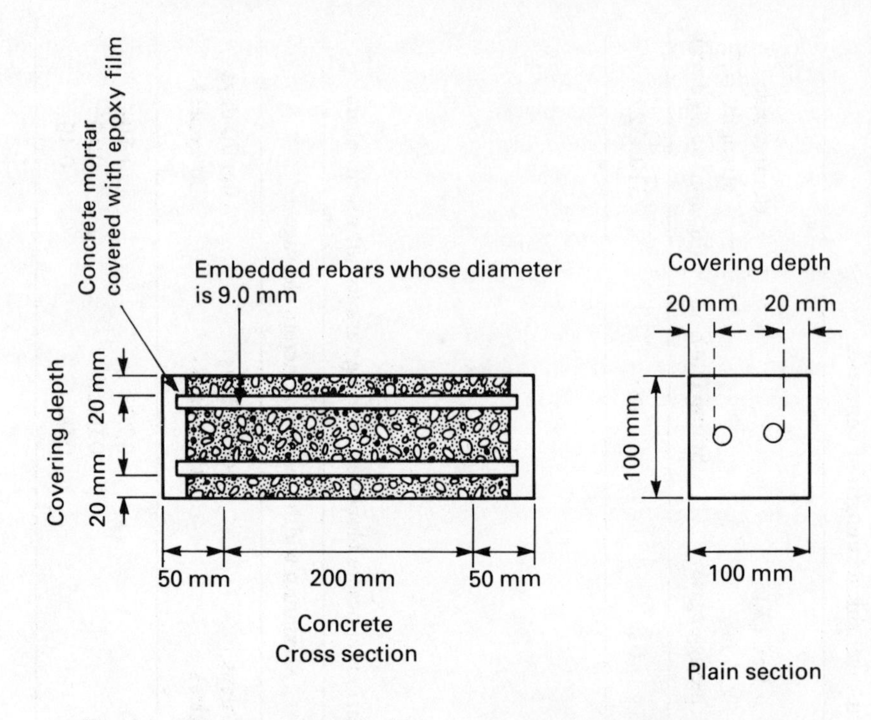

Fig. 4 – The arrangement of rebars embedded in the concrete blocks prepared for the laboratory test accelerating the occurrence and the growth of concrete cleavage.

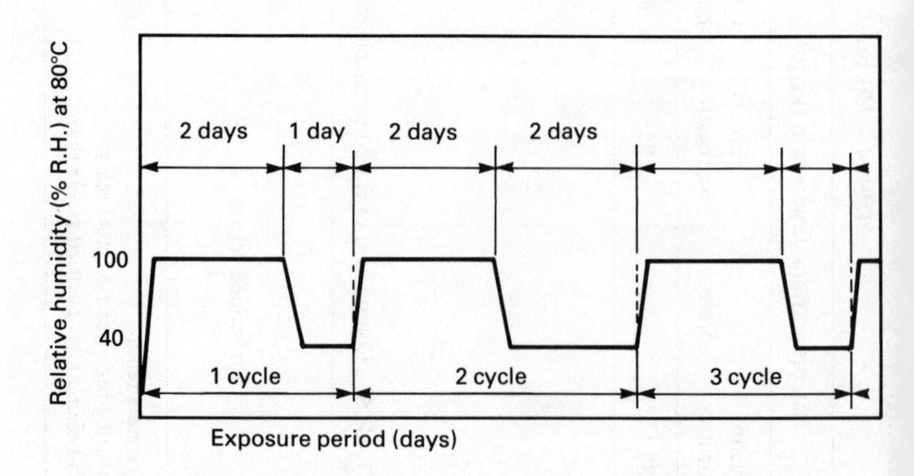

Fig. 5 – Exposure condition of the steel reinforced concrete blocks in the newly developed laboratory test.

concrete mortars, they were cured for 28 days in water. The arrangement of the embedded rebars and mix proportions of the concretes are shown in Fig. 4 and Table 4. The concrete blocks prepared in this way were set in the thermostat at the high temperature, high humidity condition and exposed in the period from 35 days to 70 days. The exposure condition is shown in Fig. 5. The results obtained are shown in Table 5 and Fig. 6. From the results obtained, it was found that after 35 days exposure test, the concrete blocks embedding the ordinary steel rebars caused the occurrence of the concrete cleavage but such concrete cleavage was not detected in the concrete rebars embedding the above promising rebars. This is due to the difference of the amounts of the corrosion products of these rebars embedded in concretes which tend to cause concrete cleavage, as shown in Fig. 6.3.

The concrete block embedding ordinary steel rebars

The concrete block embedding Cu–W bearing high purified steel rebars

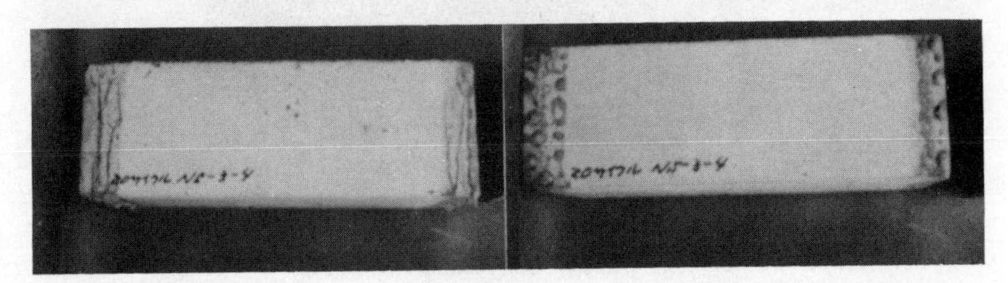

The concrete block embedding 3.5% Ni bearing high purified steel rebars

Fig. 6.1 – The concrete cleavage caused by the expansion force of the corrosion products of the embedded rebars after 70 days' exposure test.

Fig. 6.2 – The occurrence of the cleavage through the embedded rebars of the concrete block embedding ordinary steel rebar after 70 days.

Ordindary steel rebars

Cu–W bearing high purified steel rebars

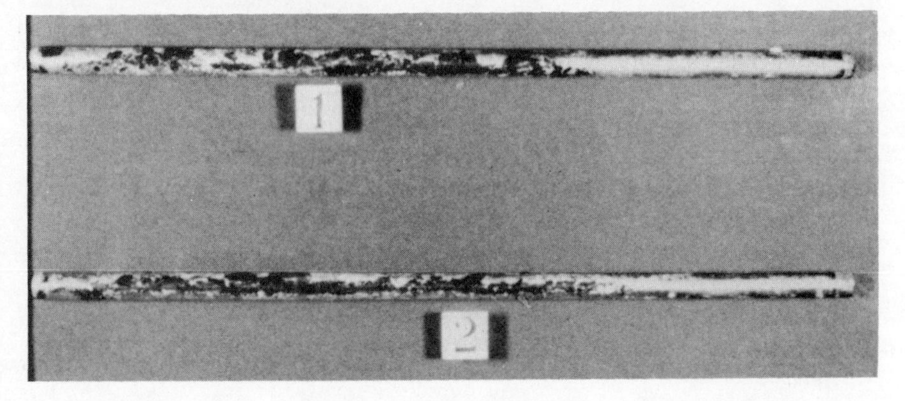

Fig. 6.3 – The corrosion behaviour of various rebars embedded in the concrete blocks with high chloride ion, exposed at high temperature, high humidity for 35 days.

In addition, after 75 days' exposure the concrete rebars embedding the ordinary steel rebars caused the remarkable growth of the concrete cleavage and its maximum width reached as much as 1.3 mm but those embedding the above promising rebars showed the tendency to delay the growth of such concrete cleavage and their maximum width reached as little as 0.1 mm. Fig. 6 represents very clearly the difference of such concrete cleavage growth among these three concrete blocks embedding ordinary-, Cu-W-bearing high purified- and 3.5% Ni-bearing high purified steel rebars. In order to confirm that the occurrence and the growth of the concrete cleavage are caused by the expansion force of the corrosion products, we examined the weight loss of the rebars caused by the corrosion attack due to chloride ion. The amounts of corrosion products tend to increase in proportion to the weight loss of the rebars. After crushing the concretes, we set off the embedded rebars from the concretes, dipped into acidic solution with inhibitor and eliminated corrosion products. The weight loss was determined by weighing the difference between the original rebars and the above corroded rebars. Table 6 shows the results obtained by the 50 days' exposure test of the concrete blocks. From Table 6, it is clear that the weight loss of the above promising rebars is much less than that of the ordinary steel rebars. This means that they tend to create the corrosion products much less than the ordinary steel rebars, where steel-reinforced concretes are exposed to high chloride ion attack. This is in good agreement with the experimental results that the concrete blocks embedding the above promising rebars tend to delay the occurrence of the concrete cleavage. The high resistance of the Cu-W-bearing- and 3.5% Ni-bearing high purified steel rebars to concrete deterioration might be due to the following mechanism.

Table 6 – Weight loss of the rebars embedded in concrete blocks with high chloride ion exposure at high temperature, high humidity conditions for 50 days.

Embedded rebars	Original weight (g) of the rebars	Weight (g) of the corroded rebars, whose rusts were eliminated	Weight loss (g) through exposure test
Ordinary steel rebar	131.3	122.5	8.8
Cu-W-bearing high purified steel rebar	136.0	134.0	2.0
3.5% Ni-bearing high purified steel rebar	134.44	134.42	0.02

(1) *The effect of low Si, low P and ultra low S in steel rebars* [6]
The main effect of low Si and low P is to delay rust formation because the decrease of Si and P tends to decrease the content of Si^{4+} and P^{5+} in the FeO layer covering the rebar surface in such a high pH region as concrete and to stabilize its p-type semiconductor characteristic which supplies positive holes trapping electrons of Cl^- ion and therefore prevents the direct attack of Cl^- ion on the steel rebars. The main effect of ultra low S is also to delay rust formation due to the extreme decrease of the amounts of α-MnS which is the origin of the rust initiation and tends to form γ-Mn_2O_3 accelerating the rust-formation process.

(2) *The effect of nickel*
Ni0 is the same type of semiconductor as the FeO p-type semiconductor, and tends to stabilize its effect by increasing positive holes which tend to rapidly trap the electrons of Cl^- ion and to delay its direct attack on the steel rebar [6]. In this way, the bearing of Ni tends to delay rust formation.

According to the report [7], the bearing of Ni in steels as much as 2.0% tends to form the soft rust layers and to decrease the adherent hard rusts which cause extreme pH decrease that rust layer and accelerate the corrosion rate as rust formation proceeds in ocean environments. Therefore, it is supposed that 3.5% Ni-bearing is much more favourable for decreasing the corrosion rate and the amounts of the corrosion products of the steel. In addition, the steel rebars are embedded in the concrete and therefore Ni tends to accumulate in the rust layer and to increase its effect more and more. In this way, the bearing of Ni as much as 3.5% is very effective not only for delaying rust formation but also for decreasing the corrosion products whose expansion force tends to cause concrete cleavage.

(3) *The effect of W*
As already mentioned, the bearing of W tends to create WO_4^{2-} ion inhibitor from the steel rebar itself which lessens the rust formation proceeding [2]. Accordingly, once the rust layers are formed and their amounts are increasing, the amounts of WO_4^{2-} ion inhibitor tend to increase in the rust layers and to accumulate in the layers with lapse of time. The increase of WO_4^{2-} ion inhibitor is very effective for decreasing the corrosion products of the embedded rebars. In this way, the bearing of W is effective for the decrease of the corrosion products whose expansion force tends to cause the concrete cleavage.

3. SUMMARY

(1) We proved the Cu-W-bearing- and 3.5% Ni-bearing high purified steel rebars were promising for concrete construction exposed to chloride ion attack, by using the laboratory corrosion test to appraise the corrosion behaviour of the embedded rebars. However, we could not confirm whether these promising

rebars were effective for preventing the concrete deterioration caused by the corrosion of the embedded rebars. Therefore, we developed the new laboratory accelerated testing method to create concrete cleavage within a short period, from 35 to 70 days, by preparing the steel reinforced concrete blocks with high chloride ion corresponding to the case of mixing with sea water and exposing them in the high temperature, high humidity condition.

(2) By using this technique, we compared the concrete deterioration behaviour among the concrete blocks embedding the ordinary steel rebar and the promising Cu-W-bearing- and 3.5% Ni-bearing high purified steel rebars. And it was confirmed that the Cu-W-bearing- and 3.5% Ni-bearing high purified steel rebars show the tendency to delay the occurrence and the growth of concrete cleavage caused by the expansion force of the corrosion products of the embedded rebars much more than the ordinary steel rebar. Therefore, these steel rebars are promising for the concrete construction exposed to chloride ion attack.

ACKNOWLEDGEMENTS

The authors would like to express their thanks to Dr Hideya Okada, Director of Nippon Steel Corporation.

REFERENCES

[1] Kōichi Kishitani, Study of the corrosion behaviour of the steel bars embedded in concrete, No. 1, May 1968; No. 3, June 1969, published by Architectural Institute of Japan.

[2] Haruo Shimada, Hideya Okada and Seiya Nishi, OTC No. 3194, May 1978, at Houston, USA.

[3] Howard F. Finley, *Corrosion* **17** (3) (1961) pp. 104t–108t.

[4] Haruo Shimada, Hideya Okada and Seiya Nishi, *Proceedings of the 5th International Congress on Marine Corrosion and Fouling 1980, at Barcelona, Spain*, pp. 238–253.

[5] Haruo Shimada, Yoshiaki Sakakibara and Hiroshi Hirotani, *Proceedings of the First International Conference on Marine Technology, at Goteborg, Sweden*, Paper No. 42, pp. 1–41.

[6] Haruo Shimada and Yoshiaka Sakakibara, Paper No. ICTTE-82-114, May 1982, at Pittsburgh, USA.

[7] Akihiro Tamada and Masayuki Tanimura, *Corrosion Engineering*, **21** (11) (1972), pp. 513–522.

Index

A

AC impedance measurements, 180, 268
aggregates, Middle Eastern, 102
  salt-contaminated, 143
aggressive environments, 93, 143
air-entrainment and carbonation, 115
alkali content of cements and galvanised reinforcement corrosion, 343
analysis, destructive, of
  deterioration, 270
  of structural condition, 193
anodic reaction, 120

B

Bahrain, survey of structures, 103
Berlin Congress Hall collapse, 79
blending agents, influence on alkalinity, 147
bridge decks, 358
  cathodic protection, 287

C

calcium chloride damage, 193
  hydroxide and passivity, 134
carbonation, 226
  atmospheric, 172
  'continuous' and 'intermittent' corrosion differences, 164
  depth, 228
  depth measurement, 178
  of concrete, 339
  of repair mortars, 245
  of mortar, accelerated, 159
  in Middle Eastern conditions, 101
carbon dioxide penetration, 214, 335, 410
cathodic oxygen, 119
  protection criteria, 291
  protection of bridge decks, 287
  reaction, 123
cement type, influence on electrochemical behaviour, 393
chloride and carbonation, joint effect, 102
  diffusivity, 148
  measurement, 228
  penetration, 39, 61, 75, 121, 206, 212, 335, 410
  penetration, accelerated, 258
chlorides and durability, 143
  effect, 227
computer, use in data analysis, 201
concrete disintegration, 94
'Concrete in the Oceans' research programme, 39, 59, 119
concrete, repair, 263
core sampling, 178
corrosion, active, low potential, 152
  assessment, 51
  cell, theoretical model, 62
  electrochemical, 20
  general, 13
  localised, 14
  monitoring, 255
  monitoring system design, 256
  onset, 39
  pitting, 152
  rate of development, 39
  resistance, 62
  uniform, 14
covermeter readings, 211
crack healing, 134

D

damaged structures, investigation and repair, 223
deicing salts, damage by, 193, 287
densified silica-cement coating as corrosion protectant, 333
desert environment, rebars for, 419
design of corrosion monitoring system, 256
deterioration, causes, 224

E

electrical resistivity measurements, 178
electrochemical behaviour of steel, influence of cement type, 393
 corrosion, 120
 measurements, 124, 178, 396
 noise measurement, 181
 potential measurements, 201
 reactions producing rust, 21
Elstree test site, 65
epoxy mortars in concrete repairs, 274

F

fly-ash cements and electrochemical behaviour of steels, 394
fracture surface examination, 82, 388
Friedel's salt formation, 144

G

galvanic coupling, 131
galvanised reinforcements, 343, 357, 407

H

hollow leg effect, 131, 209
highly-resistive concrete surface layers, 207
hydrogen absorption, 384
 induced stress corrosion cracking, 79

I

immersion effects, 63
inspection techniques, 76

L

linear polarization resistance measurements, 159, 179
Loch Linnhe deep-water immersion zone facility, 47, 64

M

magnesium sulphate, effect on pH of concrete, 96
marine environment, 19
 environment, rebars for, 419
 growth, 64
 structures, durability, 91
Middle East, carbonation of concrete in, 101
 conditions, 143
monitoring of corrosion, 175, 255
mortar, accelerated carbonation, 159
 repair-systems, 235

N

nickel in rebars, 433
non-destructive testing, 197
North Sea, 61, 193

O

offshore structures, 39
oxygen, cathodic, 119
 diffusion, 25, 75

P

passivation, chemical, 239
passivity, 134, 151
performance, prediction, 211
permeability of concrete, 75, 102
 of repairs, 239
 test, 228
pH of concrete, 95
polarization fields, 207
pore solution analysis, 146
Portland cement in concrete repairs, 274
 splash zone facility, 45, 65
potential of steel in concrete, 301
potentiodynamic sweep measurements, 179
potentiostatic polarization curves, 398
pozzolanic materials and electrochemical behaviour of steel, 393
precorrosion, endurance after, 383

R

rebars, epoxy, 357
 for offshore, seaside and desert environments, 419
 galvanised, 343, 357, 407
redundancy in structures, 17
renovation of United Arab Emirates sewage plant, 247
repair, cement-based, 230
 execution, 230
 materials selection, 217
 mortars, evaluation, 236
 of Billhay Lane footbridge, 249
 of concrete, 263
 of damaged structures, 223
 of St Joseph's Primary School, Dundee, Scotland, 250
 preparation for, 229
 procedure selection, 217
 resin-based, 232
 systems, mortar, 235
resistivity of concrete, 75
 measurements, 210
roofs, flat, 12

S

safety, 11
salt-contaminated aggregates, 143
penetration, 97
Schmidt hammer test, 199
seawater, 19, 92, 407
serviceability of structures, 11
silica-cement coating, densified, 333
Southall test site, 65
spalling, 62
splash zone, 64, 209
strength and durability, balance, 12
structural condition analysis, 193
sulphates and durability, 143

T

tests, mechanical and ultrasonic, 177
tidal effects, 132
Tongue Sands fort survey, 75
Tukey's test, application to Schmidt hammer results, 199
Tungsten oxide inhibitor, 419

U

ultrasonic pulse velocity measurements, 177, 211

V

visual inspection, 177, 223

W

wires, prestressed, corrosion behaviour, 379

Z

zinc galvanised rebars and seawater corrosion, 407